Thermal and Flow Measurements

Thermal and Flow Measurements spans the basics of measurement methods to the advanced diagnostic applications in experiments and testing. It provides a foundational understanding of how to measure flow rates, temperature, pressure, and other key parameters in fluid and thermal processes.

Offering in-depth technical treatments of underlying physics, ranging from optics and electromagnetics to quantum mechanics, this book demonstrates how to design and apply measurements in mechanical, aerospace, and chemical engineering processes. It features new sections on image-based particle velocimetry, high-speed velocimetry, satellite-based methods for scaled measurements, and AI in image analysis, advanced laser diagnostics, and remote sensing. It also includes end-of-chapter problems and exercises.

This book is intended for senior undergraduate and graduate engineering students studying experimental methods and measurement.

Instructors will be able to utilize a Solutions Manual and Figure Slides for their course.

Thermal and Flow Measurements

Second Edition

T.-W. Lee

CRC Press
Taylor & Francis Group
Boca Raton London New York

CRC Press is an imprint of the
Taylor & Francis Group, an **informa** business

SUPPORT MATERIAL

Designed cover image: Courtesy of Prof. Gokul Pathikonda, Arizona State University

Second edition published 2026
by CRC Press
2385 NW Executive Center Drive, Suite 320, Boca Raton FL 33431

and by CRC Press
4 Park Square, Milton Park, Abingdon, Oxon, OX14 4RN

CRC Press is an imprint of Taylor & Francis Group, LLC

First edition published by CRC Press 2008

Library of Congress Cataloging-in-Publication Data
Names: Lee, T.-W. (Tae-Woo) author
Title: Thermal and flow measurements/T.-W. Lee.
Description: Second edition. | Boca Raton, FL : CRC Press, 2026. | First
edition published by CRC Press 2008. | Includes bibliographical references and index.
Identifiers: LCCN 2025035238 (print) | LCCN 2025035239 (ebook) |
ISBN 9781041008224 hardback | ISBN 9781041008231 paperback |
ISBN 9781003611837 ebook
Subjects: LCSH: Fluid dynamics | Thermodynamics
Classification: LCC TA357.L439 2026 (print) | LCC TA357 (ebook)
LC record available at https://lccn.loc.gov/2025035238
LC ebook record available at https://lccn.loc.gov/2025035239

ISBN: 978-1-041-00822-4 (hbk)
ISBN: 978-1-041-00823-1 (pbk)
ISBN: 978-1-003-61183-7 (ebk)

DOI: 10.1201/9781003611837

Typeset in Times
by codeMantra

Access the Support Material: Routledge.com/9781041008224

Contents

Preface to the Second Edition

Experimental methods in thermal and fluid mechanics are regaining their importance, as the cost of computational fluid mechanics becomes more expensive and their limits are found for complex processes. In addition, computational and theoretical results almost always need to be checked with experimental data. Thus, this revision will provide guidance and basic knowledge for contemporary researchers and engineers for experimental work.

Many advances have been made in thermal-fluid diagnostics due to the integration of technology into lasers, imaging devices, and image/data analysis, particularly in certain applications such as high-speed reacting and multiphase flows. AI methods are now readily available for image/data analysis. These topics are included as additional chapter sections to keep the reader in step with these advances.

The original version of this book was an ambitious task, covering a wide range of topics. Some revisions and corrections are deemed necessary at this point for the main text and for the problem sets/solutions.

About the Author

Dr. T.-W. Lee is an Associate Professor of Aerospace and Mechanical Engineering in the School for Engineering of Matter, Transport and Energy at Arizona State University, USA. His areas of expertise include spray atomization, turbulence, and environmental monitoring, with innovative methods to solve important thermal-fluid problems. Dr. Lee has published more than 100 peer-reviewed publications and authored three books. His research has been funded by the National Science Foundation, Department of Energy, Federal Aviation Administration, and engine and power generation companies. The Salt River Project (SRP) has funded more than 50 research projects since 2001, ranging from renewable energy and fuel cells to steam and hydro-power. He has worked with a range of diagnostic methods from micro-sensors, laser Doppler velocimetry, particle image velocimetry, laser-induced fluorescence/incandescence, FT-IR, lidars, and radars. His recent work is on AI/machine-learning methods for high-density image data for various thermal, fluid, and environmental applications.

1 Introduction

1.1 THERMAL AND FLOW MEASUREMENTS

We begin with some brief thoughts on the topic of this book: thermal and flow measurements.

Experimentation drives engineering and scientific knowledge. It is human nature to try to make sense of the complex events and occurrences around us and to either harness or avoid the powerful forces at work. In science and engineering, this is manifested through numerous theories being devised, with only a few of them surviving the rigors of experimental observation over time. Experimentation is also fundamentally important in the development of theories; it is experimental observation that allows the formulation of theories. We use, for example, Newton's second law to describe a wide range of apparently disparate phenomena, from fluid acceleration by pressure gradient to the motion of celestial objects. Newtonian dynamics was developed to synthesize and generalize the observed motions of celestial objects and, through demonstrations and observations, it was proven to apply universally to the motion of all objects in the nonrelativistic and nonquantum-mechanical realm, including the "falling of the apple."

Thermal and flow processes are ubiquitous. When we speak of thermo-fluid sciences, they include thermodynamics, heat transfer, and fluid mechanics. In instances where the thermal energy is sufficiently high to cause chemical transformations, we must add chemical reactions to the list of subjects describing thermal and flow processes. Our world is driven by energy processes, some of them driven by solar energy and others anthropogenically generated. Energy conversion may be achieved through burning of fossil fuels (petroleum and coal) or through nuclear, solar, wind, or other processes; they all include some aspects of thermal and flow physics. Even in an area such as nuclear power generation, the two-phase flow in the cooling system is an important thermal-fluid process. These phenomena are not limited to the man-made world, as atmospheric dynamics, such as weather systems, are driven by the nonuniform heating of the earth's surface and the atmosphere. Many industrial processes are designed to formulate new chemicals or products through thermal and flow processes. Therefore, there is a great need in many sectors of our society and industry to characterize these processes through experimental and observational means.

Complex phenomena can be characterized through experimentation. In spite of the impressive progress in computational and theoretical methods, for real thermal-fluid processes occurring in complex geometry, the detailed characteristics can only be obtained through experimentation. For turbulent flows in complex geometry, there have been some great advances in turbulence modeling. However, even these models need to be validated through experimental data when the geometry or flow parameters are altered. Adding heat transfer or chemical reactions to these turbulent flows makes the process even more complex, usually requiring physical testing to

DOI: 10.1201/9781003611837-1

ascertain certain features of the computational models. For two-phase flows such as fuel sprays, the challenges for computational and analytical methods become even greater. Experimental data, if obtained through careful setup and diagnostics, provide a solid basis for understanding and analyzing complex thermal and flow processes.

In spite of the recent advances in computational modeling and simulations, complex thermal-fluid processes require validation with experimental data. In addition, computational fluid dynamic simulations have become quite expensive and time-consuming to perform and then visualize, as they produce a very large amount of data.

Measurement methods require cross-disciplinary knowledge. As we shall see in this book, one goes through quite a range of engineering and scientific subjects when dealing with experimental and diagnostic methods. To start, the experimental apparatus needs to be designed so that the desired thermal-fluid processes can be reproduced while allowing for variation in the control parameters, such as flow speed, temperature, and pressure. Depending on the diagnostic method to be used, access for physical or optical probes must be designed into the experimental apparatus. For structural integrity and operational safety, one needs to be knowledgeable about structural mechanics and other basic engineering approaches to design. The final experimental setup needs to be functional in those regards but should also maintain a level of professionalism and aesthetic elegance.

Even in relatively simple measurement devices, such as temperature and pressure probes, where the structural response of the sensor elements is converted to electronic signals, there are components whose design and use will be significantly enhanced if one has an understanding of the physical and engineering processes embedded in these devices. Many modern diagnostic methods, including flow visualization, involve optical and spectroscopic devices. Needless to say, one needs to be fluent in at least basic optics and molecular structure to be an effective user of these methods. Image-based data acquisition is a powerful tool nowadays, providing intuitive visual data. Being able to manipulate and analyze these data requires one to be well-versed in image processing techniques, and indeed, many of the imaging methods can be processed to yield quantitative data. Finally, micro- and nanoscale devices are making some inroads into sensor development. Fabrication of these devices typically involves steps used in microelectronics. These are just some examples where interdisciplinary knowledge is needed in understanding and using diagnostic methods for thermal and flow processes.

1.2 SOME EXAMPLES OF THERMAL AND FLOW MEASUREMENTS

Let us look at some examples of thermal and flow measurements. Figure 1.1 shows a flow visualization image of the flow around a 1/48th-scale model of an F-18 aircraft. The experiment is performed in a water tunnel to approximate the Reynolds number of the actual high-speed flows. The water has a density about 1000 times that of the air at sea level and a viscosity about 55 times that of air, so one can achieve nearly the same Reynolds number ($Re = \frac{\rho UL}{\mu}$) at about 1/20th of the flow velocity, U. The color streaks are from the dyes injected at the nose of the aircraft and also at the leading edge of the wings. The aircraft is at a high angle of attack, so the flow becomes

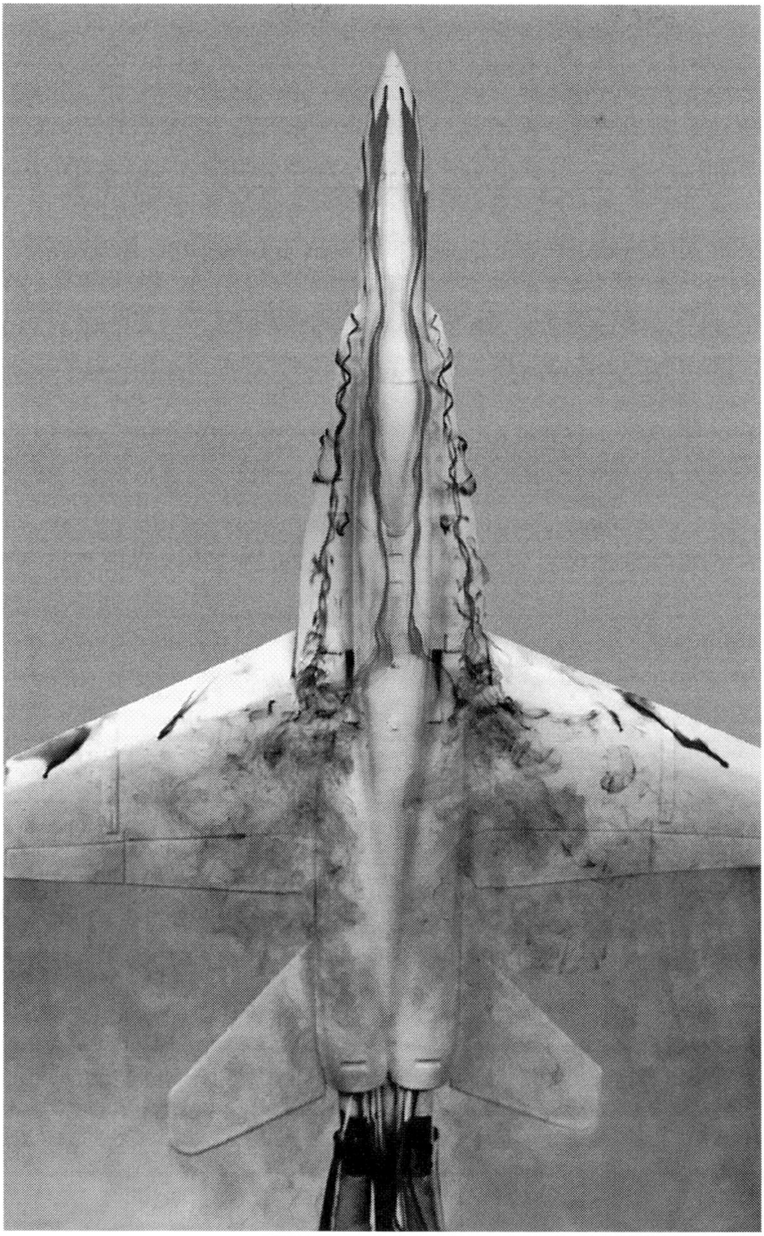

FIGURE 1.1 An example of flow visualization. This image shows a plastic 1/48th-scale model of an F-18 aircraft inside a water tunnel, more formally known as the NASA Dryden flow visualization facility. Water is pumped through the tunnel in the direction of normal airflow over the aircraft. Colored dyes are pumped through tubes with needle valves. The dyes flow back along the airframe and over the airfoils, highlighting their aerodynamic characteristics. The aircraft can also be moved through its pitch axis to observe airflow disruptions while simulating flight at high angles of attack. (Photo courtesy of NASA Dryden.)

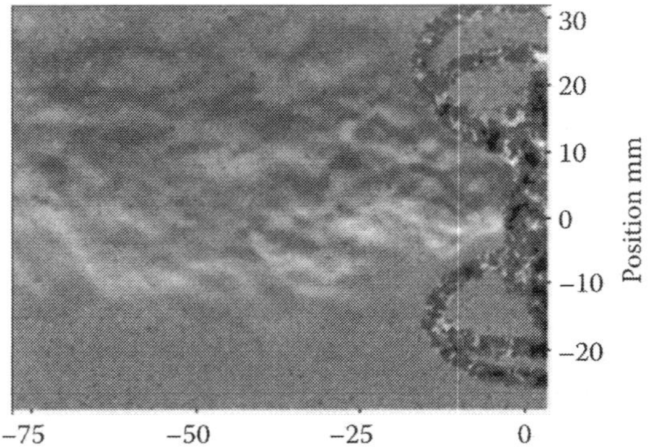

FIGURE 1.2 A digital Schlieren image of a helium jet in air. The rings on the right are the tubes that feed helium to the injector head.

detached from the top of the wing and creates a wake-like appearance. The vortical structure generated by the leading edge of the wing-body contour is also clearly shown, following a nearly straight flow in the forward part of the aircraft. As is evident in this example, visual and intuitive information about the flows can be obtained by flow visualization. As will be demonstrated in Chapter 3, much flow visualization data can be digitally processed to provide quantitative information as well. Thus, flow visualization or more broadly image-based diagnostic methods furnish a powerful tool for thermal and fluid engineers. Further examples of flow visualization can be seen in Chapter 3.

Figure 1.2 is a so-called Schlieren image of a helium jet in air that shows the density gradient arising from the mixing of low-density helium and high-density air. Schlieren is a classical technique; however, the image in Figure 1.2 was obtained through so-called *background-oriented* Schlieren imaging, which is a digital method of visualizing the density gradients in the flow without the complex optical setup associated with classical Schlieren imaging.

Figure 1.3 shows infrared images of (a) a circuit board and (b) the gas-turbine engine exhaust port. Temperature, pressure, and flow rates are some of the most basic parameters in thermal and fluid processes, and, depending on the accuracy and temporal and spatial resolution, many different methods are available for measuring these parameters. Infrared imaging is a technique used to measure the temperature field of a remote object without requiring direct contact because it is based on the radiation emitted from the object's surface. From your heat transfer class, you may recall that a surface at temperature (T_s) emits a total radiation of

$$E = \varepsilon\sigma T_s^4 \tag{1.1}$$

Therefore, the higher the surface temperature, the greater the emitted radiation will be; in fact, much greater, as dictated by the fourth-power dependence on temperature. A radiation sensor is needed to relate the emitted radiation to the surface temperature. However, two elements of these measurements need to be considered. First, at

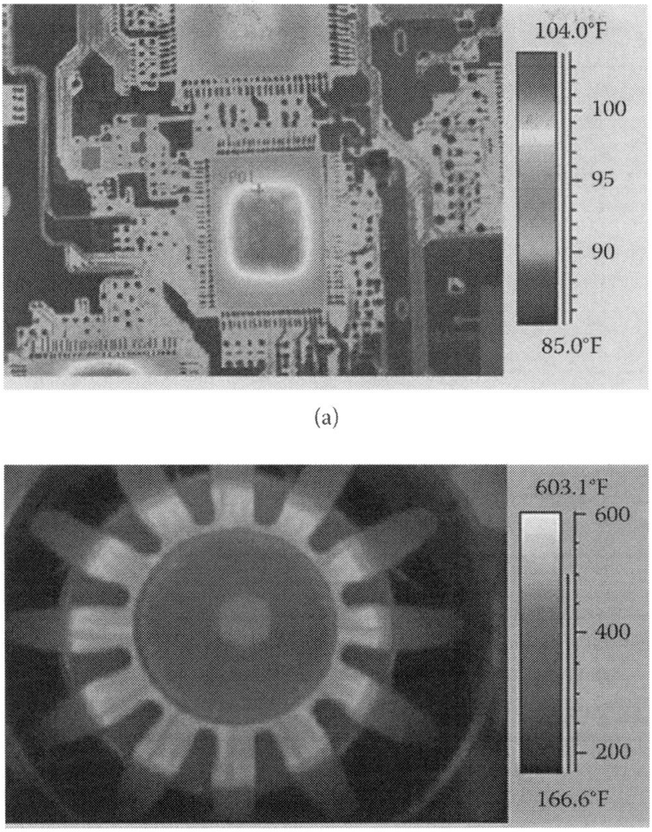

(a)

(b)

FIGURE 1.3 Infrared images of (a) a circuit board and (2) a gas-turbine engine viewed from the exhaust port. Both images are color-coded to an approximate temperature scale. (Images courtesy of SPI Corp., www.imaging1.com.)

low temperatures below about 800 K (527°C), most of the emitted radiation is in the infrared. Second, the emitted radiation varies widely depending on the emissivity, ε, of the surface, as given by Equation 1.1. Thus, the radiation sensor requires infrared sensitivity, and the emissivity of the surface needs to be known or calibrated. In Figure 1.3, the infrared intensity detected by the imaging system is color-coded to an approximate temperature scale.

In pure fluid mechanical processes, the flow velocity and pressure field furnish all the necessary information to characterize the flow. In thermo-fluid processes, these same variables are also fundamentally important, as they dictate the thermal and convection processes. For these reasons, determination of the flow velocity field is an important element in thermal-fluid measurements. Figure 1.4 shows an example of a set of images used for particle image velocimetry. The image of the seeded particles (appearing as bright dots) is transposed by the flow velocity vector at each point, and

FIGURE 1.4 Particle image velocimetry.

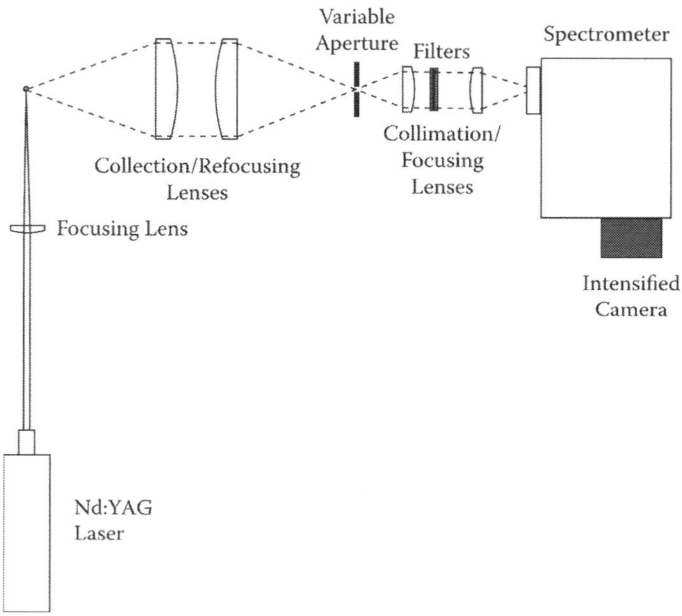

FIGURE 1.5 Experimental setup for spontaneous Raman scattering and laser-induced breakdown spectroscopy.

a sequence of two images therefore contains information on the flow velocity field. The displacement, Δs, of the particle image during a time interval, Δt, between the two images, gives the velocity of the particle at that point through $\nu = \Delta s/\Delta t$. The flow pattern, however, is not readily discernible from visual inspection of the sequence of images in Figure 1.4, and only through digital image analysis (cross-correlation) can the flow velocity vectors be effectively extracted from these particle images. Other widely used velocimetry techniques include laser Doppler velocimetry and hot-wire anemometry, details of which are discussed in Chapter 4.

Similar to velocity measurements, precise temporally and spatially resolved measurements of scalar properties such as temperature and chemical composition require elaborate diagnostic setups, often involving optical components. Figure 1.5 shows a cascade of optical components used for both *spontaneous Raman scattering* and *laser-induced breakdown spectroscopy*. Many commonly used laser optical

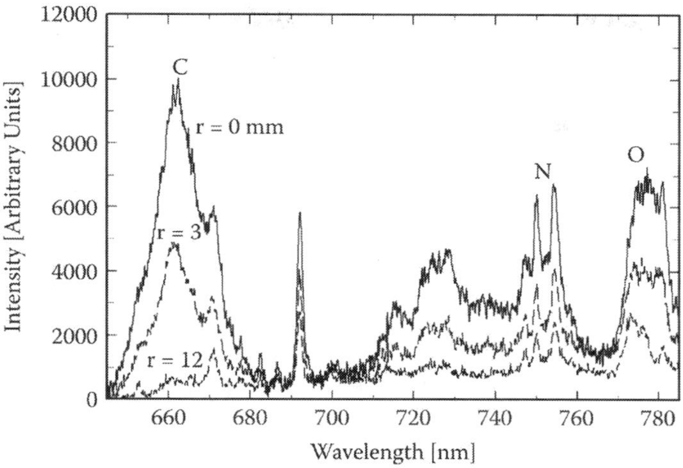

FIGURE 1.6 Typical laser-induced breakdown spectrum.

diagnostic techniques share similar optical setups, except that different laser output and detection wavelength combinations are used. In the setup shown in Figure 1.5, the laser light source is an Nd:YAG laser emitting a green beam (532 nm wavelength) at an energy of ca. 500 mJ per pulse. Spontaneous Raman scattering refers to the light scattered from molecules that carry specific wavelength signatures (with respect to the incident light wavelength) corresponding to their molecular structure, so that the identity and amount of the molecules can be determined by deciphering the resulting Raman scattering spectra. Laser-induced breakdown spectroscopy is also a spectroscopic technique in which the light emanating from the focal point of the laser bears spectral signatures of the constituent atomic species. The series of lenses, filters, and field stops (or apertures), ending with a spectrometer, is designed to collect a very low-level signal, spectrally decompose it, and amplify the signal for further analysis. Figure 1.6 shows an example of a laser-induced breakdown spectrum obtained from such a setup. The spectrum was taken from a methane–air flame, and atomic peaks of nitrogen, carbon, and atoms are clearly visible in the spectrum.

Modern applications of thermal and flow measurement methods are not limited to classical laboratory-scale experiments. With the advent of micro- and nanoscale devices that operate in fluid environments, characterization of the interaction of these devices with fluids is an important issue. In addition, concerns about environmental factors such as meteorology, pollutant dispersion, and climate change necessitate constant monitoring of large-scale atmospheric thermal and flow processes. Many of the optical diagnostic methods to be discussed in Chapter 5 can be scaled up through telescopic optical systems for monitoring large-scale processes. One prominent example is lidar (light detection and ranging), an acronym generically used for optical systems designed for long-range observations of atmospheric phenomena. A lidar system to detect aerosol distribution in the atmosphere is shown in Figure 1.7. It consists of a high-power, high-repetition-rate laser, transmitting and receiving optics with a mirror rotation system to scan the space above the lidar, and a high-speed

FIGURE 1.7 A photograph of an aerosol lidar.

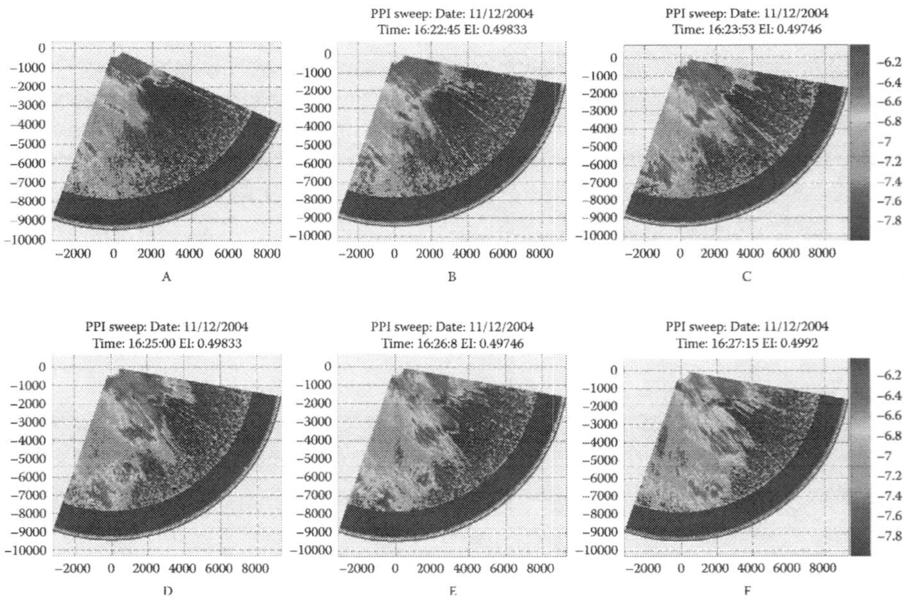

FIGURE 1.8 Typical lidar images showing the advance of the aerosol cloud in the atmosphere.

data acquisition system to collect and store the data. Figure 1.8 shows typical digitally processed images of aerosol distribution in a time sequence. The convection of the aerosol cloud by wind is visible. In lidar, an angular volume is swept to acquire spatial distribution, whereas the distance information is obtained from the time interval at which the *backscattered* signal reaches the detector. That is, by timing the return of the lidar signal with respect to the outgoing laser pulse, the distance of the returned signal can be obtained.

1.3 CHARACTERISTICS OF MEASUREMENT SYSTEMS

When we choose or design a measurement system, the following performance characteristics need to be matched to the measurement needs:

Accuracy. What are the average error and the range of error with respect to the "true" values associated with the instrument? The requirements for accuracy vary by application. If one is only interested in global monitoring of the flow, then simple, low-cost instrumentation may suffice, whereas, for researching detailed thermal and flow characteristics, highly sophisticated diagnostic methods with intensive data processing may be needed. Regardless of the degree of sophistication, no instrument gives the exact, true value, and there are many sources of error (instrument, calibration, apparatus, data acquisition, human, etc.). Therefore, in engineering and scientific measurements, it is standard practice to estimate and report measurement uncertainties, which specify the accuracy with which the measurements were made. More on measurement uncertainties will be discussed later in this chapter. The term *signal-to-noise* is commonly used in association with accuracy and, as the phrase implies, means the level of signal versus noise. Obviously, a low signal-to-noise ratio can impede acquisition of accurate data.

Resolution. This is a concept slightly different from accuracy. Resolution means "what is the smallest increment in the measured quantity?" One simple way of describing the resolution is the number of usable, significant digits of the measured quantity. An instrument with high resolution may provide data down to many decimal points. However, the concept of resolution also applies to time and space. That is, what are the smallest time and spatial scales resolved by the instrument? The temporal scale is, in many instances, referred to as the response time and can be considered an important system characteristic in itself.

Sensitivity. Again, this is a concept different from accuracy and resolution. Sensitivity means that the instrument has sufficient gain or response to provide a readable output in response to small changes in the measured parameter.

Repeatability. No measurement is valid unless it can be duplicated by the instrument and also by other operators using different instruments. For this reason, measurement methods need to be capable of providing repeatable, unbiased data.

Linearity. Although not a requirement in a measurement system, a linear response to signals makes data conversion easier. In many sensors, a linear response is highly desirable.

Signal output interface. Regardless of the instrument, if the signal is not easily converted into the kind that can be efficiently interfaced with data-processing units (typically a personal computer), then the burden of that work falls on the operator. Associated with this topic is the data conversion to the target quantity, requiring calibration or validation. In many instances, fully dedicated software is utilized, as in data acquisition and processing, particle image velocimetry, and two- and three-dimensional image analyses.

Cost, size, and user-friendliness. Even in the largest laboratories, the cost issue always comes up. Therefore, one must strike a balance between performance and cost or find ingenious ways of using low-cost instrumentation for high-performance applications to the extent that it does not compromise data quality. The dimensions of the instrument are also a consideration for measurements performed in tight spaces or those requiring low weight. No instrument is of any use unless someone can deploy it correctly and effectively; complicated measurement systems are not only time-consuming to learn to use but also prone to operator errors.

We will discuss time response and uncertainty analysis in Sections 1.4 and 1.5, respectively. To illustrate some of the basic performance characteristics in the foregoing list, we begin by looking at Figure 1.9, which shows an instrument's response to an input signal, such as a voltage output from a thermocouple in response to a temperature change. Calibration data are usually provided by the thermocouple manufacturer, and for most thermocouples, the voltage–temperature relationships are readily available in textbooks and other literature. For the J-type thermocouple shown in Figure 1.9, the regression through the voltage–temperature calibration data is expressed as a polynomial function.

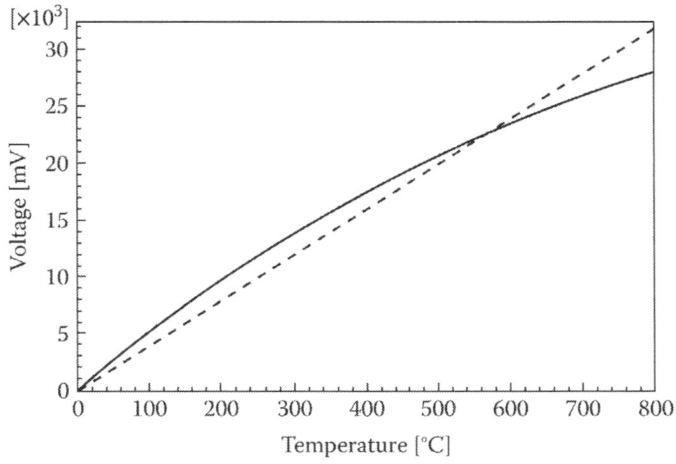

FIGURE 1.9 A voltage response of a J-type thermocouple with respect to temperature change.

$$V = 50.381187815 \times T + 0.003047583693 \times T^2 - 8.568106572 \times 10^{-5} \times T^3$$
$$+ 1.3228195295 \times 10^{-7} \times T^4 - 1.7052958337 \times 10^{-10} \times T^5$$
$$+ 2.0948090697 \times 10^{-13} \times T^6 - 1.2538395336 \times 10^{-16} \times T^7$$
$$+ 1.5631725697 \times 10^{-20} \times T^8$$

V = voltage output in mV
T = temperature in °C

In the temperature data shown in Figure 1.9, the response deviates from linearity, and the slope is relatively steep, meaning that the sensitivity is high. The slope is usually used as a measure of sensitivity. Because the preceding voltage–temperature relationship deviates from a linear relationship (there are several other terms in addition to the linear term), the response is said to be nonlinear, although the departure from linearity is not severe in this case. The *linearity error* is the deviation of the output from a linear response.

$$\text{linearity error} = y_o(x) - y_L(x) \tag{1.2a}$$

$y_o(x)$ = actual output
$y_L(x)$ = linear output

Another way to quantify the deviation from linearity is to use the maximum expected linearity error ($\%e_{L,\max}$), given by

$$\%e_{L,\max} = \frac{e_{L,\max}}{\text{FSO}} \times 100 \tag{1.2b}$$

$e_{L,\max}$ = maximum deviation from the linear output, $\max[y_o(x) - y_L(x)]$
FSO = full-scale output range = $\max[y_o(x)] - \min[y_o(x)]$

Similarly, output repeatability can be defined in terms of the standard deviation of the signal output for a given input.

$$\%e_R = \frac{\sigma_R}{\text{FSO}} \times 100 \tag{1.2c}$$

σ_R = standard deviation of $y_o(x_i)$ obtained during repeated calibration at input x_i
FSO = full-scale output range

Most modern measurement systems use digital data acquisition and processing using PC-based hardware and software. This means that the measured quantity (the voltage in Figure 1.9) cannot be measured continuously in time or amplitude, but only in discrete steps. This limitation is what is referred to as the *resolution*. Depending on the measurement needs, hardware with sufficient resolution needs to be provided. The existing standard hardware digitizes the signal in binary steps, i.e., in products of 2 (2^0, 2^1, 2^2, 2^3, 2^4, …). The resolution is, for this reason, typically given in terms of

2^n and is called n-bit resolution. Thus, a data acquisition unit with a 2^8 resolution has 256 different voltage levels that can be resolved. With a measurement range of 0–5 V with a 2^8 resolution, the smallest change in voltage that the system can measure is $5/2^8$ or 0.0196 V (= 19.6 mV). On the other hand, if the range is set at 0–1 V, then the resolution improves to 3.9 mV.

1.4 TIME RESPONSE OF MEASUREMENT SYSTEMS

Like other engineering systems, measurement devices have intrinsic temporal responses depending on the constituent component characteristics. An ideal instrument would have an instantaneous response to the changes in the parameter being measured; however, real instruments fall short of this ideal, and one needs to arrange instrumentation so that the time constant of the measurement is smaller than the time scale of the flow to be characterized. A simplified analysis of instrument time response can be made using differential equations.

The simplest possible measurement system can be described by an algebraic equation (or a zeroth-order differential equation). If y_o represents the instrument output and x_i the input, then the system response is

$$a_0 y_o = b_0 x_i \tag{1.3a}$$

Or simply,

$$y_o = \frac{b_0}{a_0} x_i \tag{1.3b}$$

An example of such a simple instrument is a displacement potentiometer, as shown in Figure 1.10. A sliding contact on a resistance strip measures the voltage across the contact point and the ground, giving the position of the contact point in terms of the voltage output, y_o. If we assume that the voltage readout is instantaneous, then the instrument response is also instantaneous and can be described by a direct relationship, as in Equation 1.3.

A first-order system is described by a differential equation.

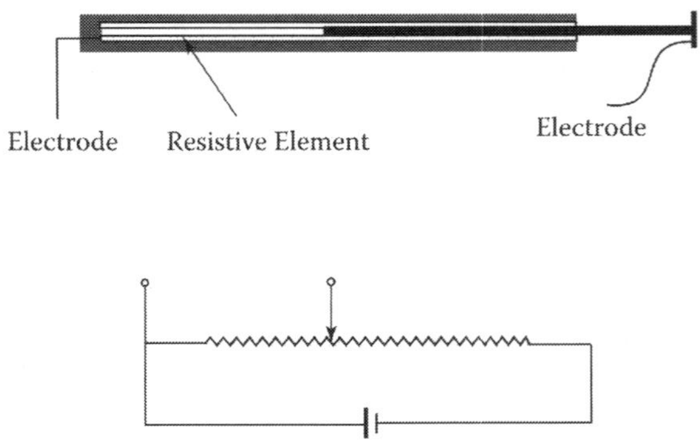

FIGURE 1.10 A simple displacement potentiometer.

$$a_1 \frac{dy_o}{dt} + a_0 y_o = b_0 x_i \tag{1.4a}$$

or

$$\frac{a_1}{a_0}\frac{dy_o}{dt} + y_o = \frac{b_0}{a_0} x_i \tag{1.4b}$$

For this first-order system, we can define two important parameters.

$$\tau = \frac{a_1}{a_0} = \text{time constant} \tag{1.5}$$

$$K = \frac{b_0}{a_0} = \text{sensitivity} \tag{1.6}$$

The relevance of these two terms is evident if we write the solution of the differential equation for Equation 1.4. If at $t=0$, $y_o=x_i=0$, but x_i goes to x_{is} through a step function at $t=0^+$ as shown in Figure 1.11, then the solution is

$$y_o = \frac{b_0}{a_0} x_{is}\left[1-\exp\left(\frac{-t}{a_1/a_0}\right)\right] = Kx_{is}\left(1-e^{-\frac{t}{\tau}}\right) \tag{1.7}$$

Therefore, a large sensitivity K results in a large response y_o, and a small time constant τ means a rapid response of the system to reach Kx_{is}, as shown in Figure 1.11.

An example of a first-order system is the thermocouple response to a temperature change. The thermocouple generates a voltage signal proportional to the temperature at the thermocouple bead. If we consider the time rate of change of the thermocouple bead temperature (T) due to heat transfer with the surrounding fluid, we can write a first-order ordinary differential equation as follows:

$$mc_p \frac{dT}{dt} = hA\left(T_f - T\right) \tag{1.8}$$

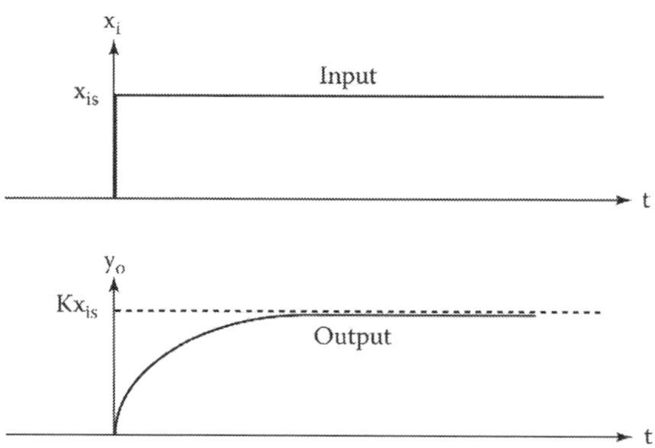

FIGURE 1.11 A first-order system response to a step input at $t=0$.

m = mass of the thermocouple bead
c_p = specific heat of the thermocouple bead material
T = thermocouple temperature
h = heat transfer coefficient
A = surface area of the thermocouple bead
T_f = fluid temperature

The right-hand side is Newton's law of cooling for convection heat transfer between the fluid at temperature T_f and the thermocouple bead. Equation 1.8 describes the transient response of the thermocouple in coming to thermal equilibrium with the surrounding fluid. We can rewrite Equation 1.8 in the same form as Equation 1.4a.

$$mc_p \frac{dT}{dt} + hAT = hAT_f \tag{1.9}$$

Then, we can associate the coefficients with those of Equation 1.4a.

$$\begin{aligned} a_0 &= hA \\ a_1 &= mc_p \\ b_0 &= hA \end{aligned} \tag{1.10}$$

And, we can write the time constant in terms of the above coefficients.

$$\tau = \frac{a_1}{a_0} = \frac{mc_p}{hA} = \frac{\rho V c_p}{hA} = \frac{\rho r c_p}{3h} \tag{1.11}$$

ρ = thermocouple bead density
V = thermocouple bead volume
r = thermocouple bead radius

It can be observed from Equation 1.11 that the thermocouple response decreases linearly with the thermocouple bead radius. Also, a large heat transfer coefficient tends to reduce the time constant. Other parameters such as density and specific heat are determined by the choice of thermocouple material.

We can also consider a slightly more complex system that can be described by a second-order differential equation.

$$a_2 \frac{d^2 y_o}{dt^2} + a_1 \frac{dy_o}{dt} + a_0 y_o = b_0 x_i \tag{1.12}$$

This equation for second-order sensor systems is analogous to the classical spring and damped-mass system, where y_o represents the position of a mass ($m = a_2$) subject to a spring-like force of $a_0 y_0$ and a damping force proportional to dy_o/dt ($b_0 x_i$ would be the external forcing applied to the mass). The classical damped-mass system has an equation in terms of the mass (m), spring constant (k), damping constant (λ), and external force $F(t)$.

$$m\frac{d^2y_o}{dt^2} + \lambda\frac{dy_o}{dt} + ky_o = F(t) \tag{1.13}$$

We can again define some characteristic parameters of second-order systems in terms of the coefficients a_0, a_1, a_2, and b_0.

$$K = \frac{b_0}{a_0} = \text{ static sensetivity}$$

$$\omega_n = \left(\frac{a_0}{a_2}\right)^{1/2} = \text{ undamped natural frequency, rad/sec} \tag{1.14}$$

$$\zeta = \frac{a_1}{2(a_0a_2)^{1/2}} = \text{ damping ratio}$$

The static sensitivity, K, again represents the linear response of the system in the absence of dynamic effects ($y_o = Kx_i$ if $a_1 = a_2 = 0$). ω_n is the angular frequency of the system oscillation if the damping term was zero ($a_1 = 0$), and ζ is the factor that describes the amount of damping in the system. For $\zeta = 0$, the system would oscillate permanently, whereas for high ζ, the system oscillation would be suppressed, and the system response approaches its steady-state value.

For a step input (x_i goes from 0 to x_0 at $t = 0$), a closed-form solution of Equation 1.12 exists and can be written in terms of the parameters in Equation 1.14.

$\zeta > 1$ (overdamped):

$$\frac{y_o}{Kx_0} = 1 - e^{-\zeta\omega_n t}\left[\cos h\left(\omega_n t\sqrt{\zeta^2 - 1}\right) + \frac{\zeta}{\sqrt{\zeta^2 - 1}}\sin h\left(\omega_n t\sqrt{\zeta^2 - 1}\right)\right] \tag{1.15}$$

$\zeta = 1$ (critically damped):

$$\frac{y_o}{Kx_0} = 1 - e^{-\omega_n t}\left(1 + \omega_n t\right) \tag{1.16}$$

$\zeta < 1$ (underdamped):

$$\frac{y_o}{Kx_0} = 1 - e^{-\zeta\omega_n t}\left[\frac{1}{\sqrt{1-\zeta^2}}\sin\left(\omega_n t\sqrt{1-\zeta^2} + \phi\right)\right] \tag{1.17}$$

$$\phi = \sin^{-1}\left(\sqrt{1-\zeta^2}\right)$$

These system responses can be plotted as a function of time for various values of ζ and are shown in Figure 1.12. The dynamic effect is least (i.e., the system response is the optimum) in Figure 1.12 when ζ is close to 0.7, with an overshoot of approximately 5% that eventually disappears as the system response approaches the steady-state value of Kx_0.

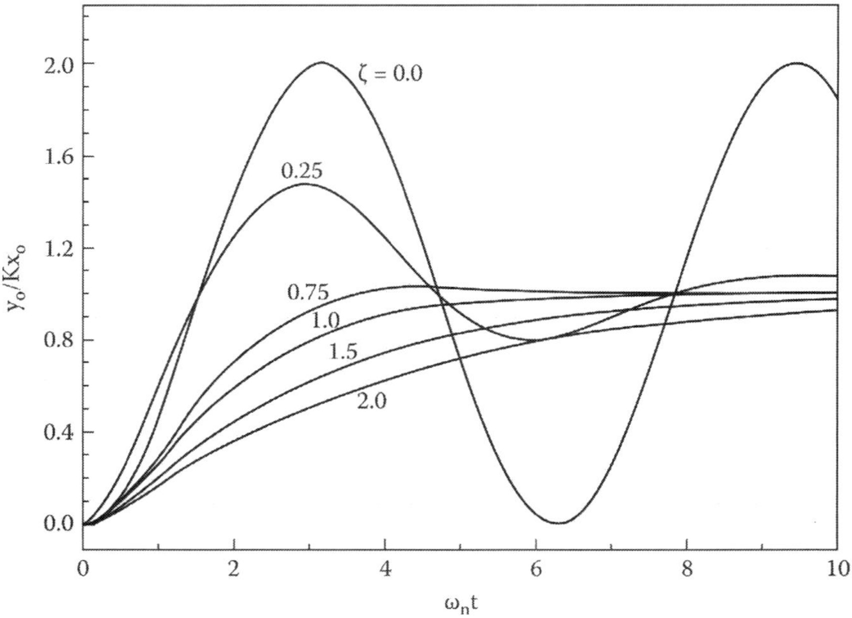

FIGURE 1.12 System response of a second-order measurement system.

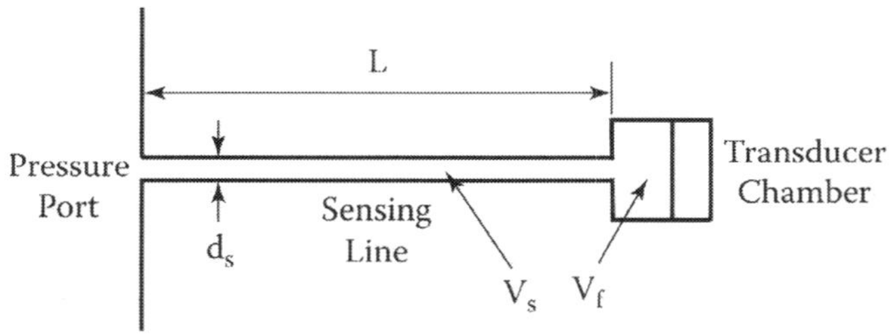

FIGURE 1.13 An example of a second-order response in the pressure-sensing line.

The system response for a sinusoidal input ($x_i = x_0 \sin\omega t$) also has a closed-form solution.

$$\frac{y_o}{Kx_0} = \frac{1}{\left[\left(1-\omega^2/\omega_n^2\right)^2 + \left(2\zeta\omega/\omega_n\right)^2\right]^{1/2}} \sin\left(\omega t + \phi\right)$$

$$\phi = -\tan^{-1}\frac{2\zeta\omega/\omega_n}{1-\omega^2/\omega_n^2} \tag{1.18}$$

An example of a second-order response is the interaction of the pressure-sensing line with the pressure sensor volume, as shown in Figure 1.13. The sensing line typically has a natural frequency, f_n,, due to its internal acoustics, and also damping due to viscous effects along the wall, represented by the damping ratio, ζ (Hougen et al., 1963).

$$f_n = \frac{\omega_n}{2\pi} = \frac{c}{2\pi L\sqrt{0.5 + V_f / V_s}} \tag{1.19}$$

$$\zeta = \frac{32\mu L}{2\rho c d_s^2}\sqrt{0.5 + \frac{V_f}{V_s}} \tag{1.20}$$

c = sonic velocity (speed of sound)
L = length of the sensing line
V_s = volume of the sensing line
V_f = volume of the transducer chamber
ρ = fluid density
d_s = diameter of the sensing line

From Equation 1.19, it is evident that a short length, L, and a large volume, V_s, lead to a high frequency, f_n, meaning a rapid response of the measurement system.

Example 1.1

Find the spring constant of the pressure transducer diaphragm that will respond to pressure fluctuations of 1 kHz with an accuracy of 1% or better. The transducer damping ratio is characterized as 0.1, and the mass of the transducer diaphragm is 0.01 kg.

Solution

We consider this problem to be represented by the damped-mass system described in Equation 1.13 with a sinusoidal pressure fluctuation of frequency $f = 1$ kHz. This gives the external forcing circular frequency of $\omega = 2\pi f = 6280$ rad/s. Using Equation 1.18 and setting the maximum amplitude for the ratio of output to input (y_o/Kx_i) to 1.01 for 1% allowable error, we get

$$\frac{y_o}{Kx_0} = \frac{1}{\left[\left(1 - \omega^2/\omega_n^2\right)^2 + \left(2\zeta\omega/\omega_n\right)^2\right]^{1/2}} = 1.01$$

We can solve for ω/ω_n in

$$\frac{1}{\left[\left(1 - \omega^2/\omega_n^2\right)^2 + \left(2\zeta\omega/\omega_n\right)^2\right]^{1/2}} = 1.01$$

$$\frac{\omega}{\omega_n} = 0.1$$

With ω being 6280 rad/s, we get ω_n of 62,800 rad/s.

In Equation 1.14, the circular frequency in terms of the mass and spring constant is

$$\omega_n = \left(\frac{k}{m}\right)^{1/2}$$

Thus, the required spring constant is

$$k = m\omega_n^2 = 0.01\ \text{kg} \times \left(62{,}800\,\text{rad/s}\right)^2 = 3.944 \times 10^8\ \text{N/m}$$

1.5 TIME-SERIES ANALYSIS AND SIGNAL PROCESSING

Computer-based, solid-state data acquisition systems are widely used in experiments. These data acquisition hardware systems are intrinsically *quantized* both in time and in amplitude. By quantized, we mean that the measurements are not continuous but only capable of being taken in discrete increments. The resolution then depends on how fine those discrete increments are, and such resolution also introduces certain limits on measurement capabilities. For constant or slowly varying signals, temporal resolution is not a serious issue. However, for dynamic signals with high-frequency components, it becomes necessary to consider the system's temporal resolution.

The *sampling rate theorem* (sometimes referred to as the *Nyquist criterion*) states that the data sampling rate in time must be greater than twice the highest frequency component of the signal in order to resolve the waveform of that frequency. This can be visualized in Figure 1.14, where a sinusoidal wave with a frequency of 10 Hz is shown, and a low sampling rate (11 Hz) produces a completely erroneous waveform due to intermittent sampling. Only when the sampling rate is greater than 20 Hz can sufficiently features of the waveform be captured, and a higher sampling frequency will provide a more faithful representation of the true signal.

The resolution limit also applies to the amplitude axis, where the signal amplitude is measured in discrete steps. Therefore, the captured signal is discrete in both time and amplitude, and we can only capture the best approximation of the true

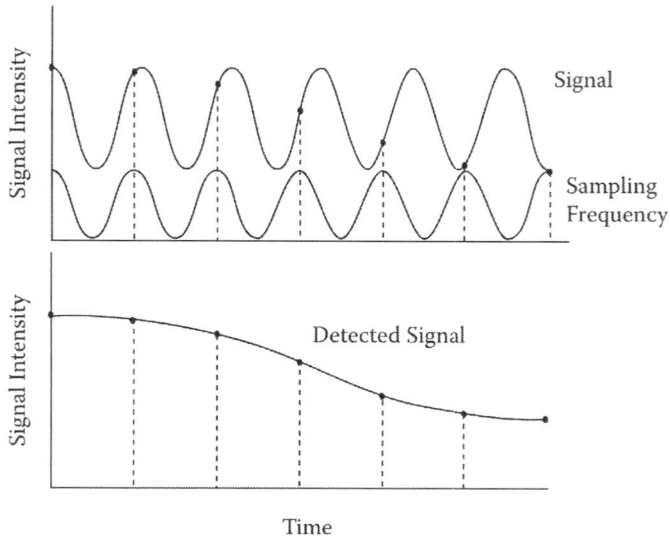

FIGURE 1.14 An illustration of the sampling rate theorem.

signal by increasing the resolution. Due to the standard solid-state architecture that uses binary formats, the amplitude resolution is given in terms of the number of bits used by the data acquisition system in resolving the amplitude, e.g., 8-, 10-, or 12-bit data conversion corresponding to 256, 512, or 1024 levels in the data amplitude. The term *dynamic range* is used to express the amplitude range in terms of dB.

$$\text{dynamic range } = 20 \log_{10}\left(2^N\right) [\text{dB}] \tag{1.21}$$

N = number of bits in the data converter

The convention for dB comes from measurements of acoustical energy, which is proportional to the square of the pressure, and therefore requires a factor of 20 in front of the logarithmic sign. It is frequently more intuitive to simply use the full range as given by 2^N.

In practice, one may encounter machine-generated sinusoidal, square, sawtooth, or other types of regular waveforms, as shown in Figure 1.15. More often, the signal has random or mixed frequency components due to noise and low-frequency modulations, also depicted in Figure 1.15. The latter type of signal (many frequency components with noise) is frequently encountered and represents so-called time-series data. There are several ways to analyze such time-series data.

One approach is simply to take statistical moments of the data, starting with time averages, root-mean squares, and sometimes higher moments, if needed.

$$\text{Time-mean } = \bar{y} = \frac{\int_{t1}^{t2} y(t)\,dt}{\int_{t1}^{t2} dt} = \frac{\int_{t1}^{t2} y(t)\,dt}{t2 - t1} \tag{1.22a}$$

$y(t)$ = signal acquired as a function of time
$t1$ = time at the start of measurement
$t2$ = time at the end of measurement

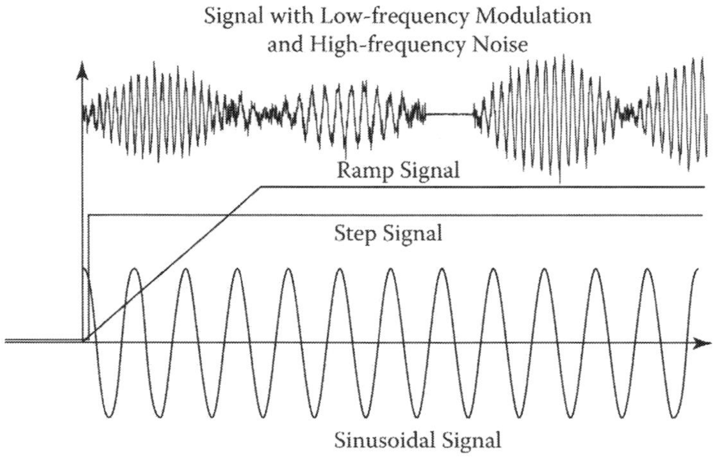

FIGURE 1.15 Some regular-shaped and random waveforms.

Because the data are discrete, the foregoing integral must be converted to a summation.

$$\text{Time-mean } = \bar{y} = \frac{\sum_{i=1}^{N} y_i \Delta t_i}{\sum_{i=1}^{N} \Delta t_i} = \frac{1}{N}\sum_{i=1}^{N} y_i \tag{1.22b}$$

The second part of the foregoing equation holds only if Δt_i is the same for all time intervals, which is usually the most convenient case.

The time mean gives the arithmetic average of the signal amplitudes, and the deviation of the signal about the mean is represented by the root-mean square.

$$\text{Root-mean square } = y_{\text{rms}} = \sqrt{\frac{\int_{t1}^{t2} y^2(t)\,dt}{t2 - t1}} \tag{1.23a}$$

or

$$\text{Root-mean square } = y_{\text{rms}} = \sqrt{\frac{1}{N}\sum_{i=1}^{N} y_i^2} \tag{1.23b}$$

In Equation 1.23b, we again assume that the time interval is the same between all of the data points.

Sometimes, in turbulence or particle measurements, the statistical distribution of the data is represented using skewness and kurtosis to characterize the shape of the data distribution. Skewness represents the degree of distortion from a symmetrical distribution. For a perfectly symmetrical distribution, the skewness is zero. Kurtosis is a measure of the smoothness of the data distribution. A data distribution with a sharp peak at some point has a positive kurtosis, whereas a *flat* distribution has a negative kurtosis.

$$\text{skewness } = g = \frac{N}{(N-1)(N-2)}\sum_{i=1}^{N}\left(\frac{y_i - \bar{y}}{\sigma}\right)^3 \approx \frac{\sum_{i=1}^{N}(y_i - \bar{y})^3}{\sigma^3 N} \tag{1.24a}$$

$$\text{where } \sigma = \sqrt{\frac{\sum_{i=1}^{N}(y_i - \bar{y})^2}{N-1}} = \text{ standard deviation} \tag{1.24b}$$

$$\text{kurtosis } = K = \frac{N(N+1)}{(N-1)(N-2)(N-3)}\sum_{i=1}^{N}\left(\frac{y_i - \bar{y}}{\sigma}\right)^4 - \frac{3(N-1)^2}{(N-2)(N-3)} \tag{1.25}$$

To determine the relationship between two variables, x and y, in the data, the correlation coefficient is used.

$$R_{xy} = \frac{\sum_{i=1}^{N}(x_i - \bar{x})(y_i - \bar{y})}{\left[\sum_{i=1}^{N}(x_i - \bar{x})^2 \sum_{i=1}^{N}(y_i - \bar{y})^2\right]^{1/2}} \tag{1.26}$$

Looking at Equation 1.26, it is evident that if the variable x is exactly followed by y, then the correlation coefficient would be 1. Conversely, if the variable x was completely independent of y, then R_{xy} would be zero. Other variations of the above correlation coefficient are also used to characterize turbulent flows.

In addition to the foregoing statistical parameters, an important approach for time-series data is frequency analysis, which gives the frequency content of the signal. An example of a signal containing many frequency components is turbulent flow, where eddies of different sizes (length scales) produce various time scales of flow fluctuations and, therefore, frequency content in velocity measurements. Another example is the sound spectrum, where many frequency components may contribute to the overall sound level. The most common method to analyze the frequency content is the Fourier analysis, or discrete Fourier analysis for digitized data. The mathematical background for the Fourier analysis can be found in many engineering mathematics books, and here we present only a summary applicable to signal frequency analysis.

A periodic function $y(t)$ with a period T can be expanded or written in terms of sine and cosine functions with increasing frequency.

$$y(t) = A_0 + \sum_{n=1}^{\infty}\left(A_n \cos\frac{2\pi nt}{T} + B_n \sin\frac{2\pi nt}{T}\right) \tag{1.27}$$

The coefficients, A_0, A_n, and B_n, are obtained by multiplying by 1, $\cos\dfrac{2\pi mt}{T}$, and $\sin\dfrac{2\pi mt}{T}$, respectively, and integrating from $-T/2$ to $T/2$.

$$A_0 = \frac{1}{T}\int_{-T/2}^{T/2} y(t)\,dt \tag{1.28a}$$

$$A_n = \frac{1}{T}\int_{-T/2}^{T/2} y(t)\cos\frac{2\pi nt}{T}\,dt \tag{1.28b}$$

$$B_n = \frac{1}{T}\int_{-T/2}^{T/2} y(t)\sin\frac{2\pi nt}{T}\,dt \tag{1.28c}$$

The Fourier series in Equation 1.27 represents a construction of the function $y(t)$ in terms of cosine and sine functions of frequencies of n/T, along with a constant (time-independent) component. The contribution of the time-independent, or so-called dc component, is A_0, and the frequency components are represented by A_n and B_n, depending on whether they resemble the cosine or sine function shapes. If the function is a superposition of sinusoidal waves of different frequencies, then the Fourier series returns an exact replication of the function in terms of the expression in Equation 1.27. In reality, the data do not have such exact representations, and errors will exist for finite-length representations using Equation 1.27. Nonetheless, the frequency components will show up in Fourier representations.

We can make a more compact form of the Fourier series by using Euler's formula for cosine and sine functions.

$$\cos x = \frac{e^{ix} + e^{-ix}}{2}; \quad \sin x = \frac{e^{ix} - e^{-ix}}{2i} \tag{1.29a}$$

This expression can be easily derived from the exponential function of a complex number.

$$e^{ix} = \cos x + i \sin x \tag{1.29b}$$

Using Equation 1.29a in Equation 1.27, we obtain

$$y(t) = A_0 + \sum_{n=1}^{\infty} \left(A_n \frac{e^{i2\pi nt/T} + e^{-i2\pi nt/T}}{2} + B_n \frac{e^{i2\pi nt/T} - e^{-i2\pi nt/T}}{2i} \right) \tag{1.30a}$$

Collecting the terms of the exponentials and noting that $1/i = -i$, we get

$$y(t) = A_0 + \sum_{n=1}^{\infty} \left(\frac{A_n - iB_n}{2} e^{i2\pi nt/T} + \frac{A_n + iB_n}{2} e^{-i2\pi nt/T} \right) \tag{1.30b}$$

Defining new coefficients

$$\begin{aligned} c_0 &= A_0 \\ c_n &= \frac{A_n - iB_n}{2} \\ c_{-n} &= \frac{A_n + iB_n}{2} \end{aligned} \tag{1.30c}$$

We finally arrive at

$$y(t) = \sum_{-\infty}^{\infty} c_n e^{i2\pi nt/T} \tag{1.30d}$$

$$c_n = \frac{1}{T} \int_{-T/2}^{T/2} y(t) e^{-i2\pi nt/T}$$

In signal acquisition, the signal is discrete in time, so we need a discretized version of Equation 1.30d, called the discrete Fourier transform (DFT). The signal $y(t)$ is considered time-series data with a discrete sampling interval Δt and a total number of data points, N.

$$y(t) = \{y(m\Delta t)\} = \{y(t=0), y(\Delta t), y(2\Delta t), y(3\Delta t), \ldots\}, \quad m = 0, 1, 2, 3, \ldots, N-1 \tag{1.31}$$

Then, the magnitude of the frequency component f_n is found from the discrete form of the coefficient c_n in Equation 1.30d.

$$c_n(f_n) = \sum_{m=1}^{N-1} y(m\Delta t)e^{-i2\pi mn/N} \quad n = 1,2,\ldots,N-1 \tag{1.32}$$

$$f_n = n\Delta f; \quad \Delta f = \frac{1}{N\Delta t}$$

The frequency analysis through the discrete Fourier transform provides a display of the *spectral content* of the signal. This has many uses in thermal and fluid measurements, such as turbulence spectra, noise analysis, and flow-induced oscillations. Yet another use of spectral analysis is to identify the sources of interference. For example, with respect to a target signal at a certain frequency, the signal spectra will show the frequencies due to interference and noise. Identification of such frequencies enables systematic elimination or at least reduction of such interference and noise through filtering. Random noise tends to be of high frequency, and a low-pass filter removes these high-frequency components. Any low-frequency modulation of signal that interferes with the accurate signal acquisition can similarly be removed or reduced by employing a high-pass filter. An example of signal *cleaning* using filters is shown in Figure 1.16. Various hardware-based frequency filters are commercially available, and digital filtering algorithms can also be applied once the signal is digitized.

1.6 ERROR ESTIMATES AND UNCERTAINTY ANALYSIS

Similar to the analysis of time-series data, to estimate the errors in a data set we begin with some basic statistical measures of the data. The mean value of the numbers represented in a finite data set is

$$\text{Mean value } = \bar{y} = \frac{1}{N}\sum_{i=1}^{N} y_i \tag{1.33}$$

The standard deviation of the data is a measure of the spread of the data with respect to the mean.

$$\text{Standard deviation } = \sigma = \sqrt{\frac{1}{N}(y_i - \bar{y})^2} \tag{1.34}$$

If one can apply a so-called probability density function (pdf) for the distribution of the variable, then the foregoing statistical parameters can be obtained from the pdf. The pdf is defined as the probability of occurrence of the variable y in an interval between y and $y+dy$.

$$\text{Probability } (y_i \le y \le y_i + dy) = P(y_i \le y \le y_i + dy) \equiv f(y_i)\,dy \tag{1.35}$$

From the pdf, the probability of finding the variable in any interval $[a,b]$ can be evaluated.

$$P(a \le y \le b) = \int_a^b f(y)\,dy \tag{1.36}$$

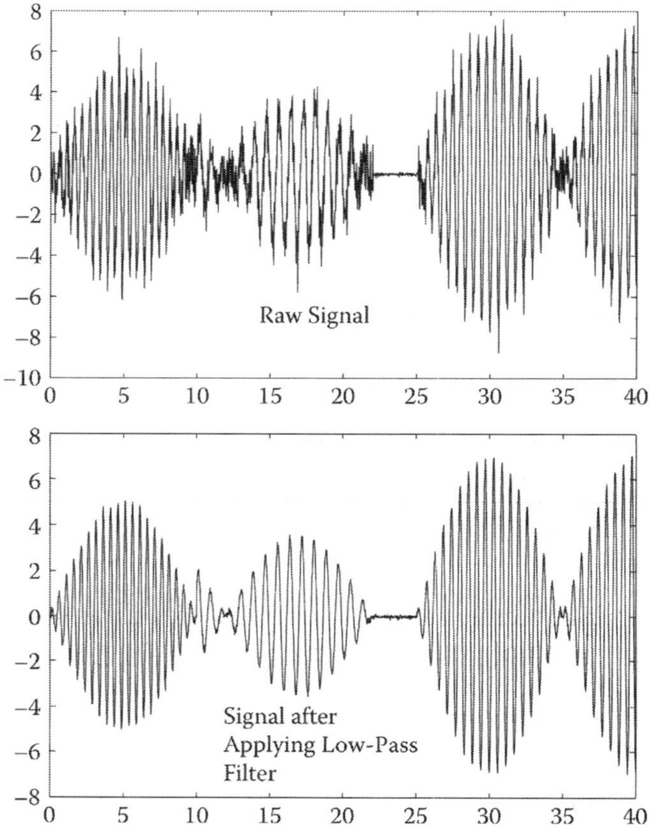

FIGURE 1.16 Low-pass filtering to reduce noise.

It follows that the probability of finding the variable in the interval [−∞, ∞] is equal to one.

$$P(-\infty \le y \le \infty) = \int_{-\infty}^{\infty} f(y)\,dy = 1 \tag{1.37}$$

Moreover, any statistical "moments" of the variables, such as the mean and standard deviation, can be obtained from the pdf as well.

$$\bar{y} = \int_{-\infty}^{\infty} y f(y)\,dy \tag{1.38}$$

$$\sigma = \sqrt{\int_{-\infty}^{\infty} (y-\bar{y})^2 f(y)\,dy} \tag{1.39}$$

The Gaussian distribution is one of the most useful pdfs in engineering and science and describes the dispersion of a variable with respect to its mean due to random effects. The randomness in the data distribution implicitly requires that the data set be large and that both positive and negative deviations from the mean value be equally probable. The Gaussian distribution has the following form of the probability density function:

$$f(y) = \frac{1}{\sigma\sqrt{2\pi}} e^{-\frac{(y-\bar{y})^2}{2\sigma^2}} \tag{1.40}$$

The Gaussian pdf function is centered at the mean, $\bar{y}$, and its "width" depends on the standard deviation, σ. Figure 1.17 shows the shape of the Gaussian pdf for various standard deviations for a mean value of 2.5. The larger the standard deviation (σ), the flatter the curve becomes. As we will see later, the Gaussian pdf is useful for providing error estimates because a random variable in a large data set tends to be distributed according to the Gaussian distribution.

Another useful pdf for engineering applications is the binomial distribution. It is applicable for predicting the likelihood of an event that has only two possible outcomes, e.g., "success" or "failure." The following binomial pdf gives the probability of finding r events (successes or failures) out of n trials when the probability (p) of that event is fixed and known.

$$P(r) = \frac{n!}{r!(n-r)!} p^r (1-p)^{n-r} \tag{1.41}$$

The use of this pdf is best illustrated through an example.

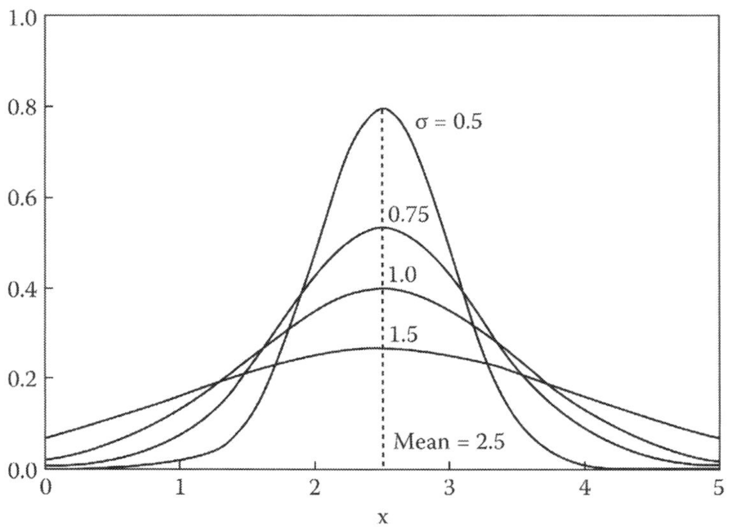

FIGURE 1.17 Gaussian probability density function.

Example 1.2

A hot-wire probe has a history of 10% failure during a 1 h continuous run in a flow tunnel. If a new probe is inserted each time a test is made, what is the probability that 5 will fail out of 20 test runs?

The probability of failure is 10% or 0.1, so we can set $p=0.1$. The number of trials is 20, so $n=20$. We want to find the probability of five failure events, so this number is associated with r. Having set these parameters, we can use the binomial pdf to find the foregoing probability.

$$P(r)=\frac{20!}{5!(20-5)!}(0.1)^5(1-0.1)^{20-5}=0.032$$

So, there is a 3.2% chance that one will need five backup probes to retest the runs during which the probe has failed.

The Poisson distribution is useful in estimating the probability of the number of random events during a specified time or space interval. In contrast to the binomial distribution, which is useful for events that are classified as either a success or a failure, the Poisson distribution allows us to deal with the number of events that may happen over some interval. The probability of the occurrence of that event, likewise, is given in terms of the number of times, λ, of that event during a time or space interval. The statistical average of the number of events of interest during some time or space interval is one way to present this quantity. Then, the probability, $P(x)$, of finding x number of events in an interval is given by

$$P(x)=\frac{e^{-\lambda}\lambda^x}{x!} \tag{1.42}$$

In instances where one desires to know the probability that the event occurs m times or fewer, the foregoing pdf can be summed to find that likelihood.

$$P(x\le m)=\sum_{i=0}^{m}\frac{e^{-p}p^i}{i!} \tag{1.43}$$

Example 1.3

Due to turbulent fluctuations, the temperature at the exit of a combustor varies as a function of time. Through lengthy measurements, the average number of times that the temperature exceeds 1600 K is 10 over a 1 min period. For temperature control, the number of times the temperature exceeds this limit of 1600 K needs to be 15 or below. What is the probability that this will occur over a 1 min period?

The number of occurrences of the event ($T>1600$ K) is 10, $\lambda=10$. The target number, x, is 15.

$$P(15)=\frac{e^{-10}10^{15}}{15!}=0.03472$$

1.6.1 Error Estimates Using Gaussian Distribution

For data involving a random variable, the data will be scattered about the mean, and this scatter, for sufficiently large data sets, will tend to follow the Gaussian distribution. This is the basis for estimating error margins in the data. For example, let us consider an interval from $\bar{y}-2\sigma$ to $\bar{y}+2\sigma$, as shown in Figure 1.18. For this interval, we can calculate the probability that the data will fall in this range, using Equations 1.36 and 1.40, assuming that the data are distributed according to the Gaussian distribution (Equation 1.40).

$$P\left(\bar{y}-2\sigma \le y \le \bar{y}+2\sigma\right)=\int_{\bar{y}-2\sigma}^{\bar{y}+2\sigma}\frac{1}{\sigma\sqrt{2\pi}}e^{-\frac{(y-\bar{y})^2}{2\sigma^2}}dy \tag{1.44}$$

Now, the notation can be simplified and standardized if we define a variable, z.

$$z=\frac{y-\bar{y}}{\sigma} \tag{1.45}$$

Then,

$$dz=\frac{dy}{\sigma} \quad \text{or} \quad dy=\sigma dz \tag{1.46}$$

and the integral in Equation 1.44 can be rewritten as (noting that $y=\bar{y}\pm 2\sigma$ corresponds to $z=\pm 2$)

$$P\left(\bar{y}-2\sigma \le y \le \bar{y}+2\sigma\right)=\int_{\bar{y}-2\sigma}^{\bar{y}+2\sigma}\frac{1}{\sigma\sqrt{2\pi}}e^{-\frac{(y-\bar{y})^2}{2\sigma^2}}dy=\frac{1}{\sqrt{2\pi}}\int_{z=-2}^{z=+2}e^{-\frac{z^2}{2}}dz \tag{1.47}$$

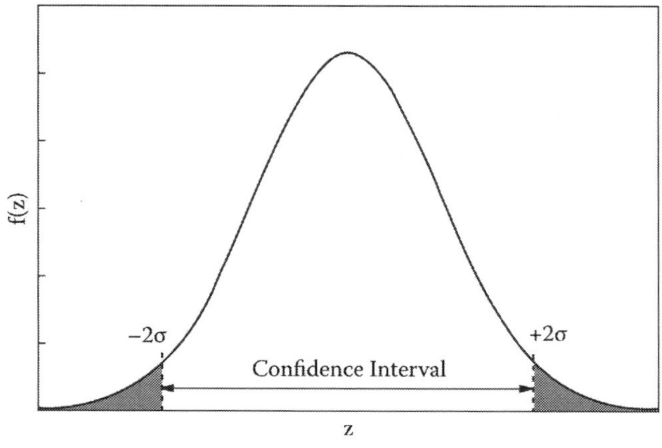

FIGURE 1.18 Error estimate using Gaussian distribution.

The integral on the right-hand side of Equation 1.47 is widely used, and its values (actually, half of its integral values) are tabulated in Table 1.1. To get the integral from $-z$ to z, the listed value is multiplied by 2 using the symmetry of the Gaussian distribution $\left(\frac{1}{\sqrt{2\pi}} \int_{-z}^{z} e^{\frac{z^2}{2}} dz = 2 \times \frac{1}{\sqrt{2\pi}} \int_{0}^{z} e^{\frac{z^2}{2}} dz \right)$.

TABLE 1.1

Integral (or Area) of the Gaussian Distribution from $z = 0$ to z $\left(\frac{1}{\sqrt{2\pi}} \int_{0}^{z} e^{-\frac{z^2}{2}} dz \right)$.

z	0.00	0.01	0.02	0.03	0.04	0.05	0.06	0.07	0.08	0.09
0.0	0.0000	0.0040	0.0080	0.0120	0.0160	0.0199	0.0239	0.0279	0.0319	0.0359
0.1	0.0398	0.0438	0.0478	0.0517	0.0557	0.0596	0.0636	0.0675	0.0714	0.0753
0.2	0.0793	0.0832	0.0871	0.0910	0.0948	0.0987	0.1026	0.1064	0.1103	0.1141
0.3	0.1179	0.1217	0.1255	0.1293	0.1331	0.1368	0.1406	0.1443	0.1480	0.1517
0.4	0.1554	0.1591	0.1628	0.1664	0.1700	0.1736	0.1772	0.1808	0.1844	0.1879
0.5	0.1915	0.1950	0.1985	0.2019	0.2054	0.2088	0.2123	0.2157	0.2190	0.2224
0.6	0.2257	0.2291	0.2324	0.2357	0.2389	0.2422	0.2454	0.2486	0.2517	0.2549
0.7	0.2580	0.2611	0.2642	0.2673	0.2704	0.2734	0.2764	0.2794	0.2823	0.2852
0.8	0.2881	0.2910	0.2939	0.2967	0.2995	0.3023	0.3051	0.3078	0.3106	0.3133
0.9	0.3159	0.3186	0.3212	0.3238	0.3264	0.3289	0.3315	0.3340	0.3365	0.3389
1.0	0.3413	0.3438	0.3461	0.3485	0.3508	0.3531	0.3554	0.3577	0.3599	0.3621
1.1	0.3643	0.3665	0.3686	0.3708	0.3729	0.3749	0.3770	0.3790	0.3810	0.3830
1.2	0.3849	0.3869	0.3888	0.3907	0.3925	0.3944	0.3962	0.3980	0.3997	0.4015
1.3	0.4032	0.4049	0.4066	0.4082	0.4099	0.4115	0.4131	0.4147	0.4162	0.4177
1.4	0.4192	0.4207	0.4222	0.4236	0.4251	0.4265	0.4279	0.4292	0.4306	0.4319
1.5	0.4332	0.4345	0.4357	0.4370	0.4382	0.4394	0.4406	0.4418	0.4429	0.4441
1.6	0.4452	0.4463	0.4474	0.4484	0.4495	0.4505	0.4515	0.4525	0.4535	0.4545
1.7	0.4554	0.4564	0.4573	0.4582	0.4591	0.4599	0.4608	0.4616	0.4625	0.4633
1.8	0.4641	0.4649	0.4656	0.4664	0.4671	0.4678	0.4686	0.4693	0.4699	0.4706
1.9	0.4713	0.4719	0.4726	0.4732	0.4738	0.4744	0.4750	0.4758	0.4761	0.4767
2.0	0.4772	0.4778	0.4783	0.4788	0.4793	0.4799	0.4803	0.4808	0.4812	0.4817
2.1	0.4821	0.4826	0.4830	0.4834	0.4838	0.4842	0.4846	0.4850	0.4854	0.4857
2.2	0.4861	0.4864	0.4868	0.4871	0.4875	0.4878	0.4881	0.4884	0.4887	0.4890
2.3	0.4893	0.4896	0.4898	0.4901	0.4904	0.4906	0.4909	0.4911	0.4913	0.4916
2.4	0.4918	0.4920	0.4922	0.4925	0.4927	0.4929	0.4931	0.4932	0.4934	0.4936
2.5	0.4938	0.4940	0.4941	0.4943	0.4945	0.4946	0.4948	0.4949	0.4951	0.4952
2.6	0.4953	0.4955	0.4956	0.4957	0.4959	0.4960	0.4961	0.4962	0.4963	0.4964
2.7	0.4965	0.4966	0.4967	0.4968	0.4969	0.4970	0.4971	0.4972	0.4973	0.4974
2.8	0.4974	0.4975	0.4976	0.4977	0.4977	0.4978	0.4979	0.4979	0.4980	0.4981
2.9	0.4981	0.4982	0.4982	0.4983	0.4984	0.4984	0.4985	0.4985	0.4986	0.4986
3.0	0.4987	0.4987	0.4987	0.4988	0.4988	0.4988	0.4989	0.4989	0.4989	0.4990

TABLE 1.2
Various Confidence Intervals and Associated Confidence Levels

Confidence Interval	z	Confidence Level [%]
$[\bar{y}-\sigma, \bar{y}+\sigma]$	1	68.26
$[\bar{y}-1.96\sigma, \bar{y}+1.96\sigma]$	1.96	95
$[\bar{y}-2\sigma, \bar{y}+2\sigma]$	2	95.44
$[\bar{y}-2.575\sigma, \bar{y}+2.575\sigma]$	2.575	99
$[\bar{y}-3\sigma, \bar{y}+3\sigma]$	3	99.74
$[\bar{y}-3.5\sigma, \bar{y}+3.5\sigma]$	3.5	99.96

Using Table 1.1, we can see that the probability of finding the data in the interval $y = [\bar{y} - 2\sigma, \bar{y} + 2\sigma]$ or $z = [-2, 2]$ is

$$P(\bar{y} - 2\sigma \le y \le \bar{y} + 2\sigma) = P(-2 \le z \le +2) = 2 \times (0.4772) = 0.9544 \qquad (1.48)$$

Again, half of the integral value is looked up for $z = 2.0$ from Table 1.1 and then multiplied by 2 to get the full integral value from −2 to +2. Thus, there is a 95.44% probability that the data will be found in the range $[\bar{y} - 2\sigma, \bar{y} + 2\sigma]$; we call this range the *confidence interval,* and the associated probability the *confidence level.* Using these terms, we can state that the confidence level is 95.44% for the confidence interval $[\bar{y} - 2\sigma, \bar{y} + 2\sigma]$. The most commonly used confidence interval in stating the error estimate of the data corresponds to a confidence level of 95%. Again, from Table 1.1, we can look up the z value that gives 0.95/2 = 0.475, which is $z = 1.96$. Table 1.2 lists various confidence intervals and associated confidence levels. It can be seen in Table 1.2 that when the data fall within 2.575 standard deviations of the mean, the confidence level is 99%, and at z of 3.5, nearly all of the data (99.96%) are contained within the confidence interval.

Example 1.4

For the following voltage data, determine the data error (confidence interval) associated with the 95% and 99% confidence levels.

$$\text{Data} = y = [2.187, 2.347, 2.201, 2.401, 2.007, 2.281, 2.061, 2.380]\text{V}$$

The mean and the standard deviation can be calculated using calculator functions or a spreadsheet program: $\bar{y} = 2.233$ V, $\sigma = 0.136$. Using the Gaussian distribution function in Table 1.2, the 95% confidence level should have $1.96\sigma = 0.266$, and the 99% confidence level should have $2.575\sigma = 0.350$. So we present the data as follows:

$$y = 2.233 \pm 0.266\,\text{V with 95\% confidence}$$

$$y = 2.233 \pm 0.350\ \text{V with 99\% confidence}$$

or

$$y = 2.233 \text{ V with } \pm 11.9\% \text{ error with } 95\% \text{ confidence}$$

$$y = 2.233 \text{ V with } \pm 15.7\% \text{ error with } 95\% \text{ confidence}$$

However, for a small sample size ($N=8$), the use of the Gaussian distribution may be inappropriate, as it assumes a large sample size (infinite N). Therefore, for a realistic sample size, a modification to the error estimates is needed, and the Student-*t* distribution is often used. The use of the Student-*t* distribution involves a multiplicative factor $t_{N,P}$ to account for finite sample size, where N is the number of samples and P is the required confidence level. The factor, $t_{N,P}$, for $N=1$–60 and for $P=50$, 90, 95, and 99% confidence levels is listed in Table 1.3. The data above are then presented in the form

$$y = \bar{y} \pm t_{N,P}\sigma \text{ with } P\% \text{ confidence}$$

TABLE 1.3
The Factor $t_{N,P}$ Based on the Student-*t* Distribution for Error Estimates of Finite Data y Sets ($y = \bar{y} \pm t_N\sigma$ with P% Confidence)

N = Number of Samples	P = 50% Confidence Level	90%	95%	99%
1	1.000	60,314	12.706	63.657
2	0.816	2.920	4.303	9.925
3	0.765	2.353	3.182	5.841
4	0.741	2.132	2.770	4.604
5	0.727	2.015	2.571	4.032
6	0.718	1.943	2.447	3.707
7	0.711	1.895	2.365	3.499
8	0.706	1.860	2.306	3.355
9	0.703	1.833	2.262	3.250
10	0.700	1.812	2.228	3.169
11	0.697	1.796	2.201	3.106
12	0.695	1.782	2.179	3.055
13	0.694	1.771	2.160	3.012
14	0.692	1.761	2.145	2.977
15	0.691	1.753	2.131	2.947
16	0.690	1.746	2.120	2.921
17	0.689	1.74	2.110	2.898
18	0.688	1.734	2.101	2.878
19	0.688	1.729	2.093	2.861
20	0.687	1.725	2.086	2.845
21	0.686	1.721	2.080	2.831
30	0.683	1.697	2.042	2.750
40	0.681	1.684	2.021	2.704
50	0.680	1.679	2.010	2.679
60	0.679	1.671	2.000	2.660
∞	0.674	1.645	1.960	2.575

That is, for the data set in the example, we get, after using Table 1.3 for $N=8$ and $P=95$ and 99%,

$$y = 2.233 \pm (2.306)(0.136) = 2.233 \pm 0.314\,\text{V with 95\% confidence}$$

$$y = 2.233 \pm (0.355)(0.136) = 2.233 \pm 0.456\,\text{V with 99\% confidence}$$

We can see that, due to the small data size, the error estimates are significantly larger than those obtained by assuming an infinite data set. We can also see that for $N=\infty$, the factor $t_{N,P}$ is identical to that obtained from the Gaussian distribution function in Table 1.2.

1.6.2 Data Regression

Regression refers to fitting a linear, polynomial, exponential, or other function to the data to display the data trend. If the relationship between data x and y is expected to be linear (e.g., a zeroth-order system as in Equation 1.3), then a linear regression is first tried to fit the data to a line. Similarly, for a first-order system as in Equation 1.7, an exponential fit may be attempted. Regression is typically implemented using the method of least squares. If there are N number of data (x_i, y_i), then we seek a linear function that has the minimum degree of deviation from the data set.

$$Y = Ax + B \tag{1.49}$$

Let us denote the deviation from the data and the foregoing function as e_i at x_i.

$$e_i = Y_i - y_i, \text{ at } x_i \tag{1.50}$$

The sum of the squares of e_i is

$$E \equiv \sum_{i=1}^{N} e_i^2 = \sum_{i=1}^{N} (Y_i - y_i)^2 = \sum_{i=1}^{N} (Ax_i + B - y_i)^2 \tag{1.51}$$

The least squares will occur when E is at a minimum, or $\partial E/\partial A = \partial E/\partial B = 0$

$$\frac{\partial E}{\partial A} = \sum_{i=1}^{N} 2(Ax_i + B - y_i)x_i = 0 \tag{1.52a}$$

$$\frac{\partial E}{\partial B} = \sum_{i=1}^{N} 2(Ax_i + B - y_i) = 0 \tag{1.52b}$$

The foregoing two equations can be solved for A and B.

$$A = \frac{N\sum_{i=1}^{N} x_i y_i - \sum_{i=1}^{N} x_i \sum_{i=1}^{N} y_i}{N\sum_{i=1}^{N} x_i^2 - \left(\sum_{i=1}^{N} x_i\right)^2} \tag{1.53a}$$

$$B = \frac{\sum_{i=1}^{N} x_i^2 \sum_{i=1}^{N} y_i - \sum_{i=1}^{N} x_i \sum_{i=1}^{N} x_i y_i}{N\sum_{i=1}^{N} x_i^2 - \left(\sum_{i=1}^{N} x_i\right)^2} \tag{1.53b}$$

A slight variation of the preceding least-squares linear fit is a linear function that must go through the origin ($x=0$, $y=0$). A similar expression can be derived for this instance.

$$Y = Ax \tag{1.54a}$$

$$A = \frac{\sum_{i=1}^{N} x_i y_i}{\sum_{i=1}^{N} x_i^2} \tag{1.54b}$$

Polynomial, exponential, and other regression coefficients can also be found, and these regression functions are embedded in many spreadsheet and data analysis software. Useful functions for regression analysis are also available on MATLAB.

1.6.3 Uncertainty Analysis

In engineering measurements, we are always dealing with some level of error, and this leads to uncertainties in the final results. The best we can do is quantify the limits of these errors and provide an estimate of the uncertainty in the results. Measurement errors can be classified into two categories: *systematic* errors and *random* errors.

Systematic errors (sometimes called bias errors) are due to system defects such as deviation from linearity, hysteresis, time lag, or calibration errors. For example, if a linear calibration is provided for a pressure transducer voltage as a function of the applied pressure, there will be small errors associated with the fact that the actual response deviates from a perfectly linear function of the pressure. Thermocouples are a good example of potential error due to time lag because they require a finite amount of time to equilibrate to the surrounding temperature, depending on the thermocouple diameter and fluid conditions. Digitization errors arise because only discrete levels of the signals can be read, and any smaller variations are discarded by the system. Sensitivity errors occur when a finite change in the input variable causes only a small increment in the signal that cannot be discerned by the system. Random errors refer to minute fluctuations in the system that cause random scatter in the data. There could be many sources of random errors, not all of which are easily quantified; therefore, one way to estimate random errors is to take many measurements and use the error estimates outlined earlier in this section. Some examples of sources of random errors are minor oscillations away from the set flow conditions, power supply fluctuations in the electrical instrumentation, random electron motion in detectors, and something as exotic as cosmic ray particles impinging on sensitive detectors.

Both systematic and random uncertainties can occur through the following phases of the experiment: (1) calibration, (2) diagnostic, (3) data acquisition, and (4) data reduction. *Calibration* uncertainty is associated with the fact that some errors can occur even during the calibration procedure, and any measurements based on this calibration will carry this uncertainty. The *diagnostic* method itself may have uncertainties. For example, a concentration measurement could be devised based on laser-induced fluorescence of some tracer dye molecules. However, the fluorescence signal may depend not only on the concentration of the tracer molecule but also on pressure and temperature. *Data acquisition* uncertainties cover a wide range of factors such as apparatus operation conditions, sensor operation, spatial and temporal resolution, signal transmission, signal digitization, and signal conditioning. *Data reduction* uncertainties refer to those arising during postprocessing of data, such as data regression, interpolation, differentiation, and other mathematical operations involving raw data. This is where *elemental uncertainty*, that is, uncertainty associated with a single variable, can propagate into post-processed parameters. This effect is called *propagation of uncertainties*, and we will describe how to treat such an effect later in this section.

As noted previously, elemental uncertainty is the total uncertainty associated with a measurement of a single variable. Both systematic and random uncertainties can contribute to the total uncertainty, and to estimate it, we take the square root of the sum of the squares.

$$U_x = \sqrt{(B_x)^2 + (P_x)^2} \tag{1.55}$$

U_x = total uncertainty for variable x
B_x = systematic uncertainty for variable x
P_x = random uncertainty for variable x

This step of combining the systematic and random uncertainties is analogous to calculating the vector length from x and y component lengths. As discussed earlier, there can, in turn, be many contributing factors for both systematic and random uncertainties, and we use the same approach of taking the square root of the squares of the individual uncertainties.

$$B_x = \sqrt{\sum_{i=1}^{M} B_i^2} \tag{1.56}$$

B_i = individual systematic uncertainty (i = 1, 2, …, M)

$$P_x = \sqrt{\sum_{i=1}^{M} P_i^2} \tag{1.57}$$

P_i = individual random uncertainty (i = 1, 2, …, N)

Example 1.5

A pressure transducer has a range (or full scale) of 0–3000 kPa. The following specifications give the manufacturer's uncertainty in terms of the % FS (full scale). Also, calibration measurements at 1200 kPa showed an uncertainty of ±20 kPa with 95%

confidence. The analog-to-digital (A/D) conversion has a resolution that can give a potential error with a standard deviation of ±5 kPa. Find the total uncertainty in the measurements using this transducer.

Sensitivity uncertainty	±0.50% FS
Linearity	±0.25% FS
Response	Up to ±15 kPa

The form of Equations 1.55–1.57 suggests that the classification of the uncertainties into systematic and random is informative but does not affect the total uncertainty. However, in dealing with uncertainties, it is standard procedure to use the same uncertainty intervals. In this example, the random measurement error of ±20 kPa with 95% confidence is "mixed" with an A/D conversion uncertainty of one standard deviation of ±5 kPa, which is *not* at the 95% confidence level. To convert one standard deviation to a 95% confidence interval, we use Table 1.2 to apply 1.96*s* to obtain the 95% confidence interval.

$$P_{A/D} = 1.96\left(5\ \text{kPa}\right) = 9.8\ \text{kPa}$$

So, using Equation 1.57,

$$P = \sqrt{\left(20\ \text{kPa}\right)^2 + \left(9.8\ \text{kPa}\right)^2} = 22.27\ \text{kPa}$$

The two systematic errors are given in terms of FS.

$$B_{\text{sensitivity}} = 0.5\%\left(3000\,\text{kPa}\right)/100 = 15\ \text{kPa}$$

$$B_{\text{linearity}} = 0.25\%\left(3000\,\text{kPa}\right)/100 = 7.5\ \text{kPa}$$

Using Equation 1.56,

$$B = \sqrt{\left(15\ \text{kPa}\right)^2 + \left(7.5\ \text{kPa}\right)^2 + \left(15\ \text{kPa}\right)^2} = 22.5\ \text{kPa}$$

And the total uncertainty is calculated using Equation 1.55.

$$U = \sqrt{\left(22.27\ \text{kPa}\right)^2 + \left(22.5\,\text{kPa}\right)^2} = 31.65\ \text{kPa}$$

Thus, this pressure transducer measurement is associated with an uncertainty of 31.65 kPa or approximately 1% FS.

The foregoing example and Equations 1.55–1.57 deal with uncertainty associated with measuring *one* variable (the pressure in the previous example). In many engineering measurements, we need to measure several variables to obtain the final quantity of interest, and one question regarding uncertainty estimation is: "How do we account for uncertainties for each variable?" We can start by writing a general equation for the final variable, y, as a function of many measured variables, x_i.

$$y = f\left(x_1, x_2, \ldots, x_n\right) \tag{1.58}$$

We can see the result of the uncertainty, Δx_i, associated with each variable x_i is by using the chain rule to calculate Δy from Equation 1.58.

$$\Delta y = \frac{\partial f}{\partial x_1}\Delta x_1 + \frac{\partial f}{\partial x_2}\Delta x_2 + \cdots + \frac{\partial f}{\partial x_n}\Delta x_n = \sum_{i=1}^{n}\frac{\partial f}{\partial x_i}\Delta x_i \tag{1.59}$$

We can use this equality to calculate the final uncertainty, U_y, in terms of the elemental uncertainties, U_i.

$$\left(U_y\right)_{\max} = \Delta y = \sum_{i=1}^{n}\left|\frac{\partial f}{\partial x_i}\Delta x_i\right| = \sum_{i=1}^{n}\left|\frac{\partial f}{\partial x_i}U_i\right| \tag{1.60}$$

Two comments about the foregoing "maximum" uncertainty are that (1) we take the absolute values because the negative uncertainty does not reduce the overall uncertainty, and (2) this equation gives us the maximum uncertainty because we are adding all of the possible uncertainties assuming that they will all occur at the same time.

An alternate way of calculating the final uncertainty is actually similar to summing systematic and random uncertainties by taking the square root of the sum of the squares. Here, we can again take the square root of the sum of the squares of the elemental uncertainties, U_i.

$$U_y = \left(\sum_{i=1}^{n}\left[\frac{\partial f}{\partial x_i}U_i\right]^2\right)^{1/2} \tag{1.61}$$

Example 1.6

The following data and uncertainties are given for a mass flow rate measurement

density $= \rho = 1.0 \pm 0.05$ kg/m^3 with 95% confidence
velocity $= V = 5.0 \pm 0.25$ m/s with 95% confidence
flow area $= A = 0.5 \pm 0.001$ m^2 with 95% confidence
Estimate the uncertainty in the mass flow rate.

$$m = \rho VA$$

$$\frac{\partial m}{\partial \rho} = VA = (5.0)(0.5) = 2.5$$

$$\frac{\partial m}{\partial V} = \rho A = (1.0)(0.5) = 0.5$$

$$\frac{\partial m}{\partial A} = \rho V = (1.0)(5.0) = 5.0$$

Using Equation 1.60,

$$\left(U_y\right)_{\max} = \left|\frac{\partial m}{\partial \rho}U_\rho\right| + \left|\frac{\partial m}{\partial V}U_V\right| + \left|\frac{\partial m}{\partial A}U_A\right| = (2.5)(0.05) + (0.5)(0.25)$$

$$+ (5.0)(0.001) = 0.23 \text{ kg/s}$$

Using Equation 1.61, we get

$$U_y = \sqrt{(2.5\times 0.05)^2 + (0.5\times 0.25)^2 + (5.0\times 0.001)^2} = 0.176 \text{ kg/s}$$

1.7 DIMENSIONAL ANALYSIS AND SIMILITUDE

Two issues that often arise in experimentation are (1) the large number of variables and (2) the scale of the experiment. For example, in considering the drag force on an object, we can consider the following variables: fluid density (ρ), fluid velocity (V), viscosity (μ), object surface area (S), and object size (D). Evaluating the effects of each of these variables is not only time-consuming but also clouds the issues. Secondly, in many instances, the experimental scale cannot match that of the real objects in terms of geometrical dimensions, as well as the magnitudes of the velocity, pressure, etc. Dimensional analysis is very useful for reducing the number of variables in testing and data analysis, whereas similitude provides a systematic method of scaling the experiment while preserving the physics of the phenomenon.

We can use the boundary-layer equation to illustrate the origin of dimensional analysis.

$$\rho u \frac{\partial u}{\partial x} + \rho \text{v} \frac{\partial u}{\partial y} = \mu \frac{\partial^2 u}{\partial y^2} \tag{1.62}$$

We can non-dimensionalize the variables by using the characteristic length (L) and the free-stream velocity (U_∞):

$$x^* = x/L \tag{1.63a}$$

$$y^* = y/L \tag{1.63b}$$

$$u^* = u^*/U_\infty \tag{1.63c}$$

$$v^* = v^*/U_\infty \tag{1.63d}$$

Substituting Equations 1.63 into Equation 1.62 results in the following form of the boundary-layer equation:

$$\rho \frac{U_\infty^2}{L} u^* \frac{\partial u^*}{\partial x^*} + \rho \frac{U_\infty^2}{L} \text{v}^* \frac{\partial u^*}{\partial y^*} = \mu \frac{U_\infty}{L^2} \frac{\partial^2 u*}{\partial y^{*2}} \tag{1.64}$$

This equation becomes nondimensional by dividing by $\rho \dfrac{U_\infty^2}{L}$.

$$u^* \frac{\partial u^*}{\partial x^*} + \text{v}^* \frac{\partial u^*}{\partial y^*} = \frac{\mu}{\rho U_\infty L} \frac{\partial^2 u^*}{\partial y^{*2}} \tag{1.65}$$

The set of parameters on the right-hand side is the inverse of the Reynolds number, and it represents the ratio of convective (fluid momentum) to the viscous effects.

Equation 1.65 is a dimensionless equation; therefore, the Reynolds number is also dimensionless.

$$\mathrm{Re} = \frac{\rho U_\infty L}{\mu} \tag{1.66}$$

Examination of Equation 1.65 from this perspective indicates that the larger the Reynolds number, the less important the viscous term becomes.

Dimensionless parameters can be obtained without going through the normalization of governing equations (as was the case in the previous example of the boundary-layer flow). Using the aerodynamic drag force on an object, for example, we can simply form an arbitrary function in terms of the relevant variables: fluid density (ρ), fluid velocity (V), viscosity (μ), object surface area (S), and object size (D).

$$\text{Drag force } = F_D = c\rho^{\alpha 1} U^{\alpha 2} S^{\alpha 3} \mu^{\alpha 4} D^{\alpha 5} \tag{1.67}$$

Here, c and α's are constant and arbitrary exponents to be determined through dimensional analysis. We use the basis of dimensional analysis; the mathematical relationships between thermal-flow parameters are dimensionally consistent. That is, we have the dimension of force on the left-hand side of Equation 1.67, and so must we on the right-hand side. Denoting the units of mass M, length L, and time T, we can list the following dimensions for the variables in Equation 1.67.

$$F_D : ML/T^2$$

$$\rho : M/L^3$$

$$U : L/T$$

$$S : \ L^2$$

$$\mu : M/LT$$

$$D : \ L$$

Substituting the foregoing in Equation 1.67, we get an equation for the units.

$$\frac{ML}{T^2} = \left(\frac{M}{L^3}\right)^{\alpha 1} \left(\frac{L}{T}\right)^{\alpha 2} \left(L^2\right)^{\alpha 3} \left(\frac{M}{LT}\right)^{\alpha 4} (L)^{\alpha 5} \tag{1.68}$$

For the dimensions on both sides to match, we must have the same exponents for M, L, and T.

$$M: \quad 1 = \alpha 1 + \alpha 4$$

$$L: \quad 1 = -3\alpha 1 + \alpha 2 + 2\alpha 3 - \alpha 4 + \alpha 5$$

$$T: \quad -2 = -\alpha 2 - \alpha 3$$

We have three equations, so we can only solve for three variables, say $\alpha 1$, $\alpha 2$, and $\alpha 3$, in terms of $\alpha 4$ and $\alpha 5$.

$$\alpha 1 = 1 - \alpha 4$$

$$\alpha 2 = 2 - \alpha 4$$

$$\alpha 3 = 1 - \alpha 4/2 - \alpha 5/2$$

Substituting this result back in Equation 1.67 and collecting the terms with the same exponents, we arrive at

$$F_D = \rho U^2 Sc \left(\frac{\mu}{\rho U D} \right)^{\alpha 4} \left(\frac{D^2}{S} \right)^{\frac{\alpha 4 + \alpha 5}{2}} \tag{1.69}$$

Again, the Reynolds number appears out of the dimensional analysis, and the leading terms on the right-hand side are the dynamic pressure and the surface area. In fact, Equation 1.69 can be rearranged to define the drag coefficient, C_D.

$$C_D = \frac{F_D}{\frac{1}{2}\rho U^2 S} = f\left(\mathrm{Re}, D^2/S\right) \tag{1.70}$$

Equation 1.70 is quite meaningful in that we need to vary only the Reynolds number and the size parameter, D, to characterize the drag force. The drag coefficient in Equation 1.70, which is the drag force normalized by $\frac{1}{2}\rho U^2 S$, is then a universally applicable parameter under dynamically similar conditions, and we have learned all this just through a brief dimensional analysis without knowledge of the fluid mechanical processes or any experimental data.

Table 1.4 lists some of the dimensionless parameters encountered in thermal and flow processes.

In conducting a scaled-down experiment, we must maintain dynamic similarity between the model and the actual flow. *Similitude* refers to the matching of the dynamic aspects. For example, Equation 1.70 shows that the drag coefficient provides the general drag characteristics. If we construct a model at a different scale to investigate the drag properties, then we need to be able to observe the same drag coefficient. This means that

$$\left(\frac{F_D}{\frac{1}{2}\rho U^2 S} \right)_{\text{actual}} = \left(\frac{F_D}{\frac{1}{2}\rho U^2 S} \right)_{\text{model}} \tag{1.71}$$

If the main drag mechanism is the viscous shear stress at the surface, then the drag force may be written as

$$F_D = \mu \frac{\partial u}{\partial y} S \tag{1.72}$$

Substituting this in Equation 1.72 and writing the representative units for each term, we get

TABLE 1.4
Frequently Encountered Dimensionless Parameters in Thermal and Flow Measurements

Name	Symbol	Definition	Variables	Meaning	Applications
Biot number	Bi	$\frac{hL}{k}$	h = heat transfer coefficient L = wall thickness k = thermal conductivity	Ratio of the convection to conduction heat transfer	Heat transfer
Drag coefficient	C_D	$\frac{F_D}{\frac{1}{2}\rho U^2 S}$	F_D = drag force ρ = density U = velocity S = surface area	The drag force normalized by the dynamic pressure and surface area	Aerodynamics, fluid mechanics
Force coefficient	C_F	$\frac{F}{\frac{1}{2}\rho U^2 S}$	F = force	Force (lift, drag, etc.) normalized by the dynamic pressure and surface area	Aerodynamics, fluid mechanics
Friction factor	F	$\frac{\tau_w}{\frac{1}{2}\rho U_\infty^2}$	τ_w = wall shear stress	Wall shear stress normalized by the dynamic pressure	Fluid mechanics
Froude number	Fr	$\frac{U^2}{gL}$	g = gravitational acceleration L = characteristics length scale	Ratio of the fluid momentum to the gravity force	Fluid mechanics, heat transfer
Fourier number	Fo	$\frac{\alpha t}{L^2}$	t = time a = thermal diffusivity = $k/\rho c_p$	Dimensionless time. Ratio of the heat conduction rate to the rate of heat storage	Transient heat transfer
Grashoff number	Gr	$\frac{g\beta(T_s - T_\infty)L^3}{v^2}$	b = thermal expansion coefficient	Ratio of the buoyancy to viscous forces	Heat transfer
Length-to-diameter ratio	L/D	L/D	L = length D = diameter	Ratio of the length to diameter	Fluid mechanics
Lewis number	Le	$\frac{\alpha}{D_{AB}}$	D_{AB} = mass diffusivity of species A in B	Ratio of the thermal to mass diffusivity	Heat/mass transfer. Combustion

(Continued)

TABLE 1.4 (*Continued*)
Frequently Encountered Dimensionless Parameters in Thermal and Flow Measurements

Name	Symbol	Definition	Variables	Meaning	Applications
Mach number	M	U/a	a = speed of sound	Ratio of the fluid velocity to the speed of sound	Compressible flows, transonic, supersonic, hypersonic flows
Nusselt number	Nu	$\frac{hL}{k}$	h = heat transfer coefficient L = length scale R = thermal conductivity	Dimensionless temperature gradient at the surface. Characteristics of convection heat transfer	Heat transfer
Peclet number	Pe	$\frac{UL}{\alpha}$	Pe = RePr	Ratio of the fluid momentum to thermal diffusivity	Heat transfer
Prandtl number	Pr	$\frac{\nu}{\alpha} = \frac{c_p\mu}{k}$	c_p = specific heat μ = viscosity	Ratio of the momentum to thermal diffusivity	Heat transfer
Pressure coefficient	C_p	$\frac{p - p_\infty}{\frac{1}{2}\rho U_\infty^2}$	p = pressure	Nondimensional pressure	Aerodynamics
Rayleigh number	Ra	GrPr	Gr = Grashof number Pr = Prandtl number	Ratio of the buoyancy to thermal diffusion effects.	Heat transfer
Relative roughness	ε/Δ	ε/Δ	ε = mean surface roughness	Nondimensional surface roughness	Fluid mechanics
Reynolds number	Re	$\frac{\rho UL}{\mu}$		Ratio of the momentum to viscous force	Fluid mechanics, aerodynamics
Schmidt number	Sc	$\frac{\nu}{D_{AB}}$	D_{AB} = binary diffusivity	Ratio of the momentum to mass diffusivities	Mass transfer, fluid mechanics
Weber number	We	$\frac{\rho U^2 L}{\sigma}$	σ = surface tension	Ratio of the fluid momentum to surface tension force	Fluid mechanics, spray atomization

$$\left(\frac{\mu \dfrac{\partial y}{\partial y} S}{\dfrac{1}{2} \rho U^2 S} \right)_{\text{actual}} = \left(\frac{\mu \dfrac{\partial x}{\partial y} S}{\dfrac{1}{2} \rho U^2 S} \right)_{\text{model}} \rightarrow \left(\frac{\mu \dfrac{U}{L} L^2}{\rho U^2 L^2} \right)_{\text{actual}} = \left(\frac{\mu_L^U L^2}{\rho U^2 L^2} \right)_{\text{model}} \tag{1.73}$$

Or simply,

$$\left(\frac{\rho UL}{\mu} \right)_{\text{actual}} = \left(\frac{\rho UL}{\mu} \right)_{\text{model}} \rightarrow (\text{Re})_{\text{actual}} = (\text{Re})_{\text{model}} \tag{1.74}$$

Thus, for frictional drag, achieving dynamic similitude simply means that we must match the Reynolds numbers for the experiment and the actual conditions. If we scale the geometric dimension, L, down, we must increase either the velocity or the density (or decrease viscosity) to match the Reynolds number. Sometimes this is done by using water with a much higher density and somewhat higher viscosity. Similarly, if the pressure drag or buoyancy effects are important, then we need to match the pressure coefficients, Froude number, or other nondimensional parameters (shown in Table 1.4).

1.8 BASIC SCIENCE AND ENGINEERING IN THERMAL AND FLOW MEASUREMENTS

As noted at the outset of this chapter, being an experimentalist requires an understanding of a variety of topics in science and engineering. From apparatus design and fabrication to instrument use, optical diagnostics, signal processing, and data interpretation, one comes across a wide range of topics, as we will explore in this book. This makes thermal and flow measurements a quite challenging and interesting topic in engineering. The following is an abbreviated list of topics encountered in thermal and flow measurements, and are therefore included in this book.

- Solid-state electronic devices (some examples given in Appendix A)
- Basic fluid mechanics, thermodynamics, and heat transfer
- Optics (Appendix B)
- Electromagnetics and electromagnetic wave propagation (Appendix C)
- Quantum mechanics and atomic/molecular structure (Appendix D)
- Data acquisition and signal processing
- Image processing and analysis

Many modern instrumentations convert the signal into an electrical output, and therefore, being able to condition and acquire such electrical signals typically involves the use of solid-state devices and circuitry. Because electrical circuitry is a required course in most engineering curricula, we will briefly describe solid-state devices and circuitry in Appendix A and also as they occur in various chapters of this book. One needs a good knowledge of fluid mechanics, thermodynamics, and heat transfer to be able to design and run thermal-fluid measurements and interpret data, and it is understood that the readers of this book possess this set of knowledge.

Flow visualization methods, by definition and as discussed in Chapter 3, require optical devices to acquire data in image formats. Also, optical diagnostic methods (Chapters 4 and 5) available for thermal and flow measurements are based on the application of optical devices. We will cover the basic fundamentals of optics in Appendix B.

Many of the devices used in thermal and flow measurements, such as lasers, light sources, filters, and optical components, interact with the light or electromagnetic waves, which cover a wavelength range much wider than that of visible light. Electromagnetic wave propagation and interaction with the medium will be discussed in Appendix C, starting from the basic equations of electromagnetics. For detailed measurements using spectroscopic methods (Chapter 5), it is very helpful to understand the atomic and molecular structure and associated energy levels. These aspects are mathematically describable using quantum mechanics, which will be presented in Appendix D. Some basic elements of signal processing were discussed in this chapter. Concepts of image processing and analysis can be found in Chapter 3.

PROBLEMS

1.1. Consider the thermocouple voltage response given by the equation below, where V is the voltage output in mV and T is the temperature in °C.

$$V = 50.381187815 \times T + 0.003047583693 \times T^2 - 8.568106572 \times 10^{-5} \times T^3$$

$$+ 1.3228195295 \times 10^{-7} \times T^4 - 1.7052958337 \times 10^{-10} \times T^5$$

$$+ 2.0948090697 \times 10^{-13} \times T^6$$

$$- 1.2538395336 \times 10^{-16} \times T^7 + 1.5631725697 \times 10^{-20} \times T^8$$

For T from 100°C to 750°C, a linearized voltage response is used where $V = 52 \times T$. Determine the maximum expected linearity error ($\%e_{L,\max}$), using Equation 1.2b.

1.2. Thermocouple material has a density of 7870 kg/m^3, specific heat of 447 J/kgK, and a bead diameter of 1 mm. If this thermocouple bead is placed in a fluid with a temperature of 500°C, with a heat transfer coefficient (h) of 5 W/m^{2}K, determine the time constant for the thermocouple bead to reach steady-state temperature. Also, plot the temperature response as a function of time using Equation 1.8.

$$mc_p \frac{dT}{dt} = hA\left(T_f - T\right) \tag{1.8}$$

m = mass of the thermocouple bead
c_p = specific heat of the thermocouple bead material
T = thermocouple temperature
h = heat transfer coefficient
A = surface area of the thermocouple bead
T_f = fluid temperature

1.3. Find the spring constant of a pressure transducer diaphragm that will respond to pressure fluctuations of 5 kHz with an accuracy of 1% or better.

The transducer damping ratio is 0.1, and the mass of the transducer diaphragm is 0.005 kg.

1.4. It is desired to acquire voltage data in the range of $V=0$–5 V, using a 12-bit data acquisition system. What is the resolution of the system, in terms of mV?

1.5. We have the following time-series data for pressure in a closed vessel. Find the time mean, root-mean square (rms), skewedness, and kurtosis of the data.

Time (s)	Pressure (kPa)
0.1	101.67
0.2	105.86
0.3	106.54
0.4	107.25
0.5	104.22
0.6	95.88
0.7	96.45
0.8	99.57
0.9	102.45
1.0	103.21
1.1	100.50
1.2	99.87
1.3	94.25
1.4	96.54
1.5	99.25
1.6	102.63
1.7	101.79
1.8	104.37
1.9	105.25
2.0	102.77
2.1	104.75
2.2	100.85
2.3	97.45
2.4	99.75

1.6. A signal train containing both the frequency and noise components can be synthesized using the MATLAB functions "sin" for the sine wave and "randu" for the random number generator. An example is

$$s(i) = \mathrm{A1} * \sin\left(2 * \mathrm{pi} * \mathrm{f1} * \mathrm{t}\right) * \mathrm{A2} * \sin\left(2 * \mathrm{pi} * \mathrm{f2} * \mathrm{t}\right) + \mathrm{N}^{*}\mathrm{randn};$$

s(*i*) is the signal array with a time index i, and Al, A2 are the amplitudes of the sinusoidal wave with frequencies f1 and f2, respectively. N represents the amplitude of the noise content.

a. Using Al=2.0, A2=5.0, f1=0.05, f2=2.0, with N=0.75, write a MATLAB program to display a simulated signal.
b. Write a MATLAB® program to perform and display the fast-Fourier Transform (fft) of the simulated signal.

1.7. A thermocouple temperature probe has a history of 5% failure during a 1-h continuous run in a high-temperature flow channel. If a new probe is inserted each time a test is made, what is the probability that one will fail out of ten test runs of 1 h duration each?

1.8. A pressure transducer is used to monitor pressure oscillations in a combustor. Test data indicate that the average number of instances that the pressure exceeds the combustor design limit of 1 MPa is 10, over a 10-min period. During operation, the number of times that the pressure exceeds this limit of 1 MPa needs to be 15 or below. Using the Poisson probability distribution, determine the probability that the above (15 occurrences or fewer) will occur over a 10-min period.

1.9. For a data set following the Gaussian distribution, what is the probability that the data falls in the range $\pm\sigma$ with respect to the mean, where σ is the standard deviation of the data?

1.10. For a 97.5% confidence level, what is the range of data in terms of the mean and standard deviation of the data?

1.11. For the data set in Problem 1.5, determine the data error (confidence interval) associated with 95% and 99% confidence levels using the Gaussian distribution.

1.12. For the data set in Problem 1.5, determine the data error (confidence interval) associated with 95% and 99% confidence levels using the Student-*t* distribution.

1.13. For the following temperature data as a function of time, find the least-square fit coefficients *A* and *B* for a linear function, $T = At + B$.

t, Time (ms)	*T*, Temperature (K)
0.1	101.67
0.2	112.86
0.3	125.54
0.4	127.25
0.5	144.22
0.6	155.88
0.7	160.45
0.8	175.57
0.9	182.45
1.0	197.21
1.1	205.50
1.2	209.87
1.3	219.25
1.4	226.54
1.5	236.25
1.6	245.63
1.7	247.79
1.8	264.37
1.9	275.25
2.0	280.77

1.14. A pressure transducer has a range (or full scale) of 0–2500 kPa. The specifications below provide the manufacturer's uncertainty in terms of the % FS (full scale). Also, calibration measurements at 1000 kPa show an uncertainty of ±50 kPa with 95% confidence. The analog-to-digital (A/D) conversion has a resolution that can give a potential error with a standard deviation of ±15 kPa. Find the total uncertainty in the measurements, using this transducer.

Sensitivity Uncertainty	±1.50% FS
Linearity	±2.25% FS
Response	Up to ±25 kPa

1.15. The following data and uncertainties are given for temperature and pressure measurements.

$$\text{Temperature } = T = 350 \pm 24 \text{ K with 95\% confidence}$$

$$\text{Pressure } = p = 155.0 \pm 5 \text{ kPa with 95\% confidence}$$

Estimate the uncertainty in the product of pressure and temperature, pT.

1.16. For fluid dynamics measurements, it is desired to match the Reynolds number of the flow between the actual flow and the test flow. The actual Reynolds number is 100,000 in air at $T = 300$ K and $p = 1$ atm, for a characteristic flow speed of 75 m/s and object dimension of 15 m. If we were to reproduce the actual conditions using water with a 1/20th scaled model, what should be the water flow speed?

BIBLIOGRAPHY

Doebelin, E.O., *Measurement Systems: Application and Design*, 5th ed., McGraw Hill Higher Education, New York, 2004.

Hougen, J., Martin, O., and Walsh, R., Dynamics of pneumatic transmission lines, *Control Engineering*, 10, p. 114, 1963.

Incropera and Dewitt, D.P., *Fundamentals of Heat and Mass Transfer*, 5th ed., John Wiley, New York, 2002.

2 Basic Measurements of Pressure, Temperature, and Flow Rates

2.1 PRESSURE MEASUREMENTS

Pressure is exerted upon a surface due to the random molecular motion and the momentum exchange when the molecules *bounce* on the surface. In the limit of zero molecular motion of the molecules at an absolute temperature of 0 K or a perfect vacuum, there would be zero pressure. Some of the commonly used units of pressure are Pa (= N/m^2), psi (= lbf/in.^{2}), bar (1 bar = 100 kPa), and atm (1 atm = 101.325 kPa). A distinction is made in pressure measurements whether the pressure is measured relative to the atmospheric pressure or zero pressure (vacuum). The pressure measured with respect to the atmospheric pressure is called the gauge pressure, and therefore, the gauge pressure (p_g) is equal to the absolute pressure (p_{abs}) minus the atmospheric pressure (p_o).

$$p_g = p_{abs} - p_o \tag{2.1}$$

For some applications, the pressure is measured in the units of the columnar height of a fluid with a corresponding pressure at the bottom of that fluid column. This has its origin in pressure measurements using fluid columns (manometers). This concept will be used later to describe the operation of manometers. From hydrostatics, the weight of the fluid of depth h is going to exert a pressure

$$p_{abs}(h) = p_o(h_o) + \rho g h \tag{2.2}$$

p_o = reference pressure at h_o
r = fluid density
g = gravitational acceleration = 9.81 m/s^2

The weight of the fluid, ρghA, exerts a force that becomes ρgh when divided by the area, A. From Equation 2.2, the fluid depth, h, is

$$h = \frac{p_{abs} - p_o}{\rho g} \tag{2.3}$$

Using mercury and water density of 13,595.1 kg/m^3 (at 0°C) and 998.207 kg/m^3 (at 20°C), respectively, Equation 2.3 gives the following conversion factors:

DOI: 10.1201/9781003611837-2

1 atm abs	= 760 mm Hg abs
	= 10.3508 m H_2O abs
	= 407.513 in H_2O abs
	= 29.92 in Hg abs

If the absolute pressure is less than the atmospheric pressure, then the term *vacuum* pressure is used, which is typically given as negative gauge pressure for moderate vacuum levels. For *high* vacuum, meaning very low pressure, absolute pressure is expressed in units of torr (1 torr = 1 mm Hg) or μm Hg.

When the pressure is measured in a fluid flow, static and stagnation pressures, in general, will differ due to the momentum of the mean flow (see Figure 2.1). In practical terms, the pressure measurement will give a different reading depending on the orientation of the pressure sensor with respect to the flow. From the conservation of momentum,

$$p_t = p + \frac{1}{2}\rho V^2 \tag{2.4}$$

p_t = stagnation or the total pressure
p = static pressure
$\frac{1}{2}\rho V^2$ = dynamic pressure (V = fluid velocity)

As shown in Figure 2.1, static pressure is measured in a flow with the pressure "tap" opening parallel to the direction of the flow to minimize flow stagnation effects.

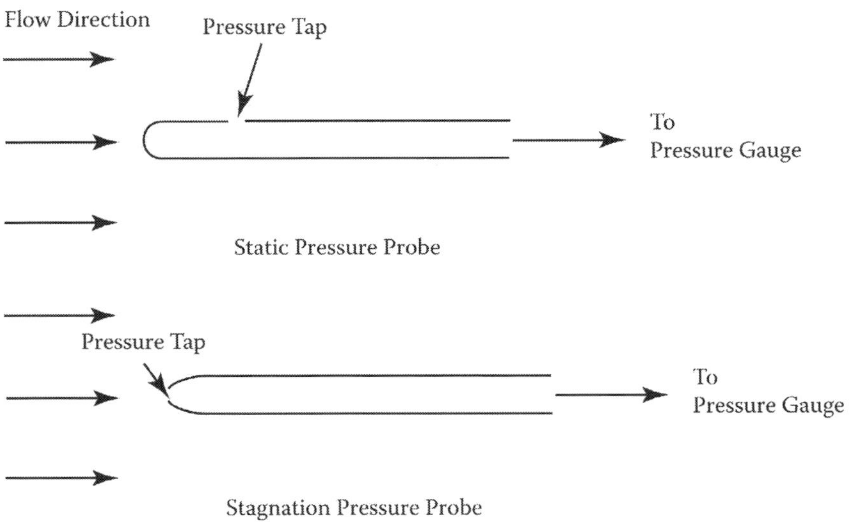

FIGURE 2.1 Pressure taps to measure the static and stagnation pressure components in a flow.

The difference between static and stagnation temperature is due to the fluid momentum, as shown in Equation 2.4, and therefore, the pressure tap for stagnation pressure measurements faces the incoming flow. The difference between static and stagnation pressure can be used as a basis for flow speed measurements (to be discussed later in this chapter).

2.1.1 Manometers

Manometers are simple devices that can yield reasonably accurate differential pressure measurements. Well-type and U-tube manometers, two of the widely used types, are shown in Figure 2.2. For the well-type manometer in Figure 2.2a, the force balance at the bottom of the liquid column is

$$p_1A - p_2A - \rho ghA = 0 \tag{2.5}$$

ρ = density of the manometer liquid
A = cross-sectional area of the liquid column
h = the liquid-column heights
g = gravitational acceleration = 9.81 m/s^2

For steady-state pressure measurements, the liquid column is stationary, so the net force is equal to zero, as in the foregoing equation. If point 2 in Figure 2.2 is vented to the ambient so that p_2 is the local atmospheric pressure (p_o), then the gauge pressure at point 1 is given by the weight of the liquid column divided by the cross-sectional area:

$$p_{g1} = p_1 - p_2 = p_1 - p_o = \rho gh \tag{2.6}$$

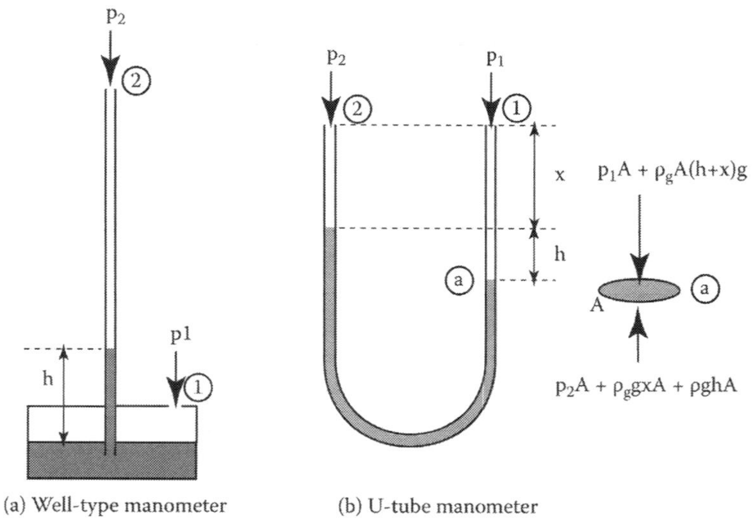

FIGURE 2.2 Basic manometers. (a) well-type and (b) U-tube.

Knowing the density and gravitational acceleration, the gauge pressure can be obtained by measuring the height of the liquid column. If the absolute pressure is to be determined from a manometer, then an independent measurement of the local atmospheric pressure is needed, or point 2 must be sealed to a vacuum. It can be seen that, to gain a large range of pressure measurable with a manometer, a fluid with a high density should be used. To gain sensitivity, a large change in the fluid column displacement with respect to a small pressure change, a less dense fluid or inclined manometers can be used. For the inclined manometer, the height term h in Equation 2.6 is replaced by $h/\sin\theta$, where θ is the angle of inclination with respect to the horizontal plane. The reading of the fluid column on the inclined plane will be magnified with decreasing angle θ.

A U-tube manometer (Figure 2.2b) also measures the differential pressure between 1 and 2. We need to consider the force balance at point "a" in Figure 2.2b. The pressure at 1, along with the weight ($\rho_g gxA$) of the fluid above the liquid surface at "a," exerts a downward force. The pressure force at 2 is transmitted by the manometer fluid and exerts an upward force, along with the weights of the fluid ($\rho_g gxA$) and the manometer liquid (ρghA). The force balance at "a" is then

$$-p_1A-\rho_g(h+x)gA+p_2A+\rho ghA+\rho_g gxA=0 \tag{2.7}$$

ρ = density of the manometer liquid
ρ_g = density of the fluid whose pressure is to be measured
A = cross-sectional area of the liquid columns
h = height difference of the manometer liquid

The terms containing x cancel, and, after dividing by A, we get the differential pressure, p_1-p_2.

$$p_1-p_2=(\rho-\rho_g)gh \tag{2.8}$$

If the density of the fluid whose pressure is to be measured is small in comparison to the manometer liquid density ($\rho_g << \rho$), then Equation 2.8 reduces to Equation 2.6.

Example 2.1

For a well-type manometer using water inclined at 60°, the liquid-column position increases by 5 cm on the inclined plane. What is the change in pressure?

Water has a density of 996.5 kg/m^3 at 27°C. We use Equation 2.6 except that in place of h, we must use $\Delta y\sin\theta$ if Δy is the change in the liquid-column position on the inclined plane.

$$p_1-p_2=\rho g\Delta y\sin\theta=(996.5\,\text{kg/m}^3)(9.81\,\text{m/s}^2)(0.05\,\text{m})\sin 60^\circ=42.33\,\text{Pa}$$

If p_2 corresponded to the ambient pressure, then the gauge pressure reading is +42.33 Pa.

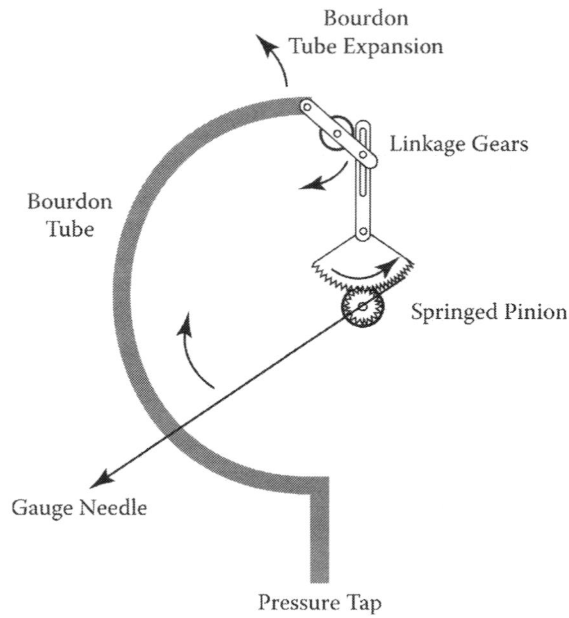

FIGURE 2.3 Dial pressure gauge using the Bourdon tube.

2.1.2 Pressure Transducers Based on Elastic Strain

There are a large number of pressure sensors based on the elastic deformation or strain of materials when subjected to a pressure force. The classical "dial" pressure gauges are based on the mechanical deformation of a curved metal tube with an elliptical cross-section, called the Bourdon tube. When pressure is applied to the Bourdon tube, it expands in a highly repeatable manner. The deformation due to the Bourdon tube expansion is connected through mechanical linkages to a rotating indicator, as shown in Figure 2.3. High-precision pressure gauges of this type are designed for linearity and repeatability.

Most modern pressure measurements are, however, made through the elastic deformation of diaphragms and corresponding strain measurements on the diaphragm surface. An example of a diaphragm-based pressure transducer with strain gauges is shown in Figure 2.4. A flat or corrugated plate is used with clamped support around the circumference. Corrugated plate designs allow for a larger range of linear response and also less stress. Some of the desirable properties of strain-based pressure sensors are

1. There should be a sufficient amount of strain in the pressure range to be measured so that the strain sensor can pick up the effect of pressure with a reasonably high level of sensitivity.
2. The diaphragm strain should be linear with respect to pressure increase.

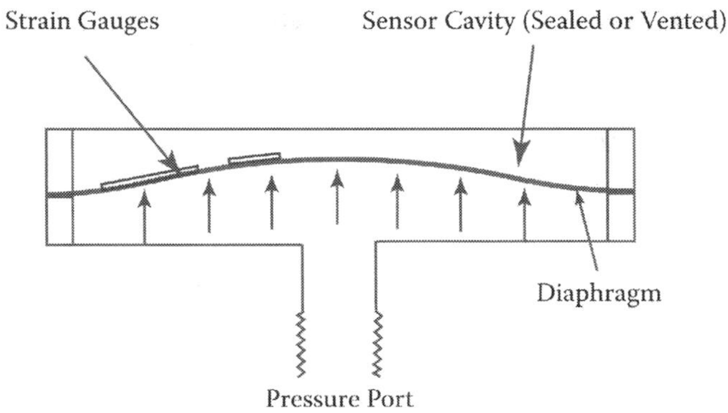

FIGURE 2.4 Pressure transducer using a strain gauge.

3. The diaphragm should have a reasonably high natural frequency so that the response to pressure changes is fast for dynamic measurements.
4. The diaphragm material should be compatible with the fluid.

For thin metal diaphragms, the pressure–strain relationship is linear so long as the vertical displacement is less than about one-third of the diaphragm thickness. The following relationship then gives the vertical displacement under pressure load and is thus used to relate the strain and pressure.

$$y_{\max} = \frac{2(p_1 - p_2)(1-\nu^2)r^4}{16E_m t^3} \tag{2.9}$$

y_{max} = maximum vertical displacement, i.e., the displacement at the plate center
$p_1 - p_2$ = pressure difference across the plate
ν = Poisson's ratio for the plate material
r = plate radius
E_m = Young's modulus of the plate material
t = plate thickness

The deformation of the diaphragm is sensed by the strain gauges fixed to the diaphragm, as shown in Figure 2.4. Strain-gauge elements become elongated or compressed depending on the diaphragm curvature, and such geometrical changes lead to a resistance change that can be electrically detected. The principle of strain gauges is discussed in Section 2.3.

Example 2.2

A pressure-transducer diaphragm is made of a metal plate with a Young's modulus of 207 GPa and a Poisson's ratio of 0.27. It is designed to measure a maximum absolute pressure of 10 MPa using a diaphragm thickness of 0.5 mm. Determine the radius of the diaphragm so that the vertical deflection does not exceed 25% of the thickness.

Equation 2.9 furnishes the necessary relationship among the foregoing diaphragm parameters. We set $y_{max} = 0.25t$ in Equation 2.9 and solve for r.

$$r = \left[\frac{0.25t^4(16E_m)}{2(p_1 - p_2)(1 - v^2)}\right]^{1/4} = \left[\frac{0.25(0.5\text{mm})^4(16)(207\times10^3\text{MPa})}{2(10\text{MPa} - 0)(1 - 0.27^2)}\right]^{1/4} = 7.27\text{mm}$$

One of the desired properties of a pressure-transducer diaphragm is the temporal response for measurements of rapidly changing pressure. One way to characterize the temporal response is by using the natural frequency of a circular clamped diaphragm, which is given by

$$\omega_n = 64.15\sqrt{\frac{E_m t^2}{12\rho(1 - v^2)r^4}} \tag{2.10}$$

ρ = density of the plate material

A high natural frequency means that the plate is able to respond rapidly to pressure changes for dynamic measurements.

Strain-gauge pressure transducers can cover pressure ranges as low as 1000 Pa and up to 1400 MPa. Typical accuracy ranges from 0.1% to 0.25% of full scale. Diaphragm hysteresis can add about a 0.25% drift (with respect to full scale). Because the diaphragm material properties vary with temperature, corrections are needed if the operating temperature deviates from the design temperature.

An alternative method of extracting the pressure effect on an elastic diaphragm is to use a capacitance sensor. A capacitance-based pressure transducer is shown in Figure 2.5. The capacitance developed between two parallel plates is given by

$$C = c\frac{\varepsilon A}{t}[\text{F, Farad}] \tag{2.11}$$

A = overlapping area of the plates
t = distance between the plates
c = proportionality constant ($c = 0.225$ for A in [in.^{2}] and t in [in.], or $c = 0.0885$ for A in [cm^2] and t in [cm])
ε = dielectric constant for the material between the plates ($\varepsilon = 1$ for air)

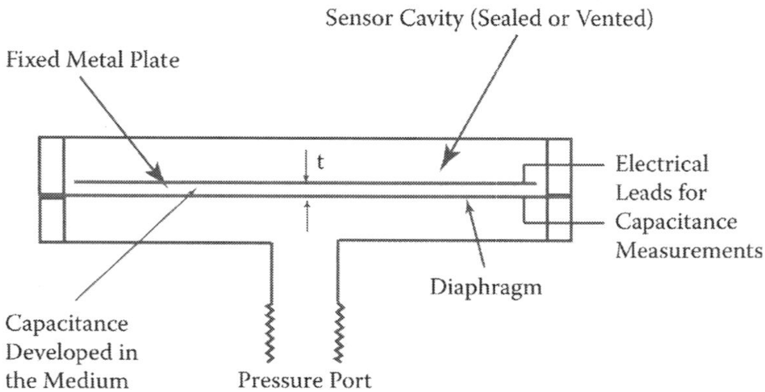

FIGURE 2.5 A schematic of a capacitance-based pressure transducer.

The pressure acting upon the diaphragm alters the effective distance t and causes a change in capacitance, which is converted to a voltage signal using an op-amp circuit (see Appendix A). Capacitance-based transducers have a good dynamic response, along with high accuracy down to 0.1%–0.01% of full scale. Typical measured pressures range from high vacuum to approximately 100 MPa.

2.1.3 Piezoelectric Transducers

Piezoelectric crystals, such as quartz, generate voltage when a force is applied, causing a deformation in the crystalline structure. Conversely, they undergo oscillatory expansion and compression when an alternating voltage is applied. When an electric field (voltage) is applied, the atoms with electric charge distributions (called the *dipole moment*) in the crystal align themselves in the direction of the electric field, leading to shape changes. Conversely, when a force is applied to the crystal, the dipole moment is forced to change, leading to a charge distribution, i.e., an electric field. This effect is used in rapid-response pressure transducers, as shown in Figure 2.6. The charge built by the deformation is

$$q = K_c pA \tag{2.12}$$

K_c = crystal charge sensitivity [Coulombs/N]

The voltage across an electrode attached to the piezoelectric crystal is

$$E = \frac{q}{C} = \frac{K_c pt}{c\varepsilon} = K_p pt \tag{2.13}$$

$$K_p = \frac{K_c}{c\varepsilon} = \text{ voltage sensitivity } [\text{Vm/N}]$$

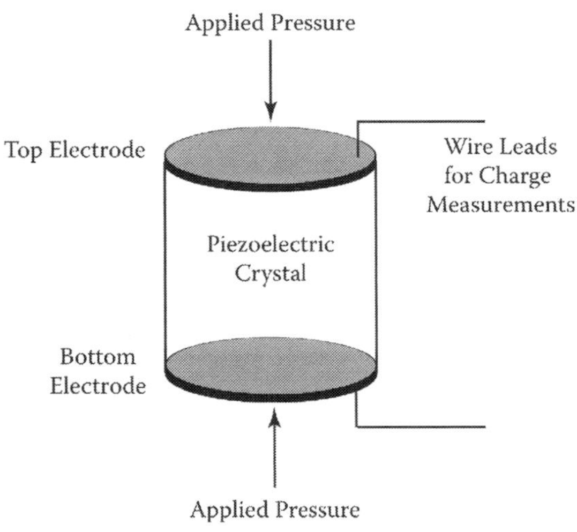

FIGURE 2.6 A piezoelectric pressure transducer.

Equation 2.11 is used to calculate the capacitance, C, of the crystal electrode element. A typical voltage sensitivity of 0.055 Vm/N is achieved using quartz for piezoelectric pressure transducers. Equation 2.13 shows that the thicker the crystal material, the higher the overall sensitivity will be in terms of the voltage output per pressure applied. Because piezoelectric pressure transducers do not rely on diaphragm deflection to measure pressure, they are used in dynamic pressure measurements where rapid pressure fluctuations occur.

2.1.4 Some Implementation Issues

We saw in Equation 2.4 $\left(p_t = p + \frac{1}{2}\rho V^2\right)$ that pressure measurements can yield different results depending on the pressure tap orientation with respect to the flow. For flow around a blunt object, such as a pressure probe, the flow comes to a zero-speed condition at the stagnation point (point 1 in Figure 2.7), and the pressure at that point will approach the total pressure, p_t, in Equation 2.4. At point 2, the flow speed is higher for the flow to go around the blunt object, resulting in a lower static pressure at that point. Free-stream static pressure is recovered only when the streamlines stabilize near point 3 in Figure 2.7.

Thus, to measure the static pressure while minimizing flow acceleration or deceleration, a Prandtl tube can be used (Figure 2.8). As shown in Figure 2.8, the eight pressure taps are located some distance (8–16 probe diameters) downstream of the probe leading edge to minimize the effect of flow acceleration around the leading edge. Conversely, to minimize the effect of flow disturbance due to the support stem, some distance (16 probe diameters) is allowed between the pressure taps and the support stem. For internal flows through a flow channel or surface pressure measurements (e.g., on an airfoil surface), wall pressure taps are used for static pressure measurements, as shown in Figure 2.9. For accurate static pressure measurements, the wall tap hole diameter should be small in comparison to the connecting tube diameter (10% of the tube diameter or less).

The total pressure is measured by orienting the pressure tap normal to the flow direction, as shown in Figure 2.10a. A Pitot tube, shown in Figure 2.10b, can be used in combination with a Prandtl tube to simultaneously measure the total

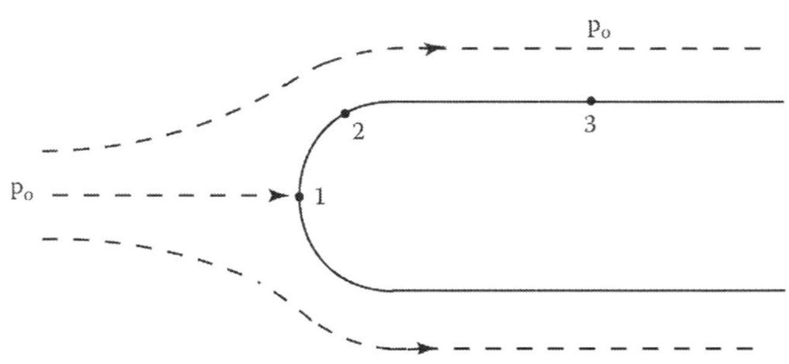

FIGURE 2.7 Flow around a blunt object.

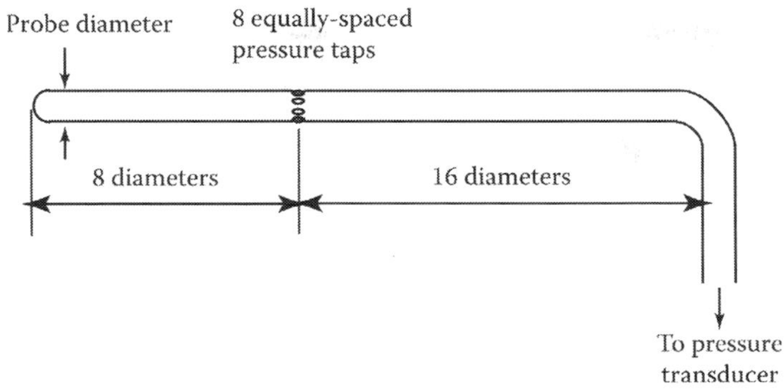

FIGURE 2.8 Prandtl probe for measurements of static pressure in the flow.

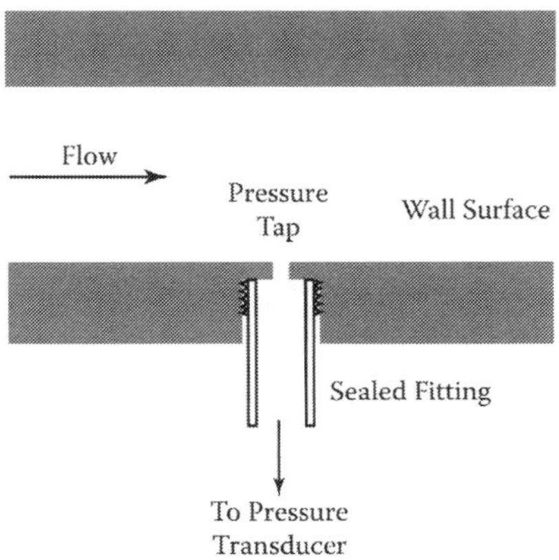

FIGURE 2.9 A probe for static pressure measurements at the wall.

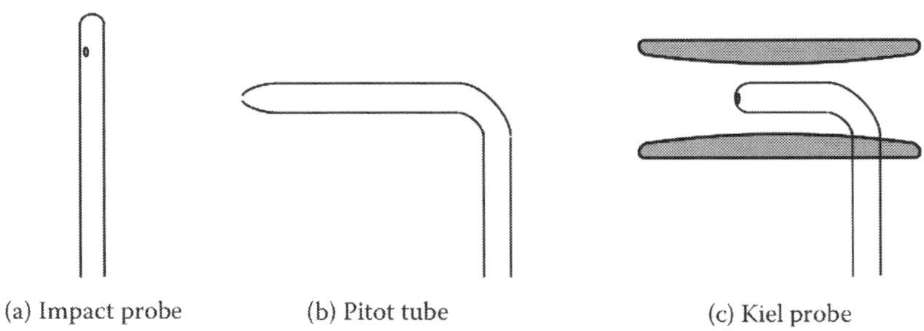

FIGURE 2.10 Total pressure probes.

and static pressure, the difference of which is the dynamic pressure, $\frac{1}{2}\rho V^2$, which can be used to estimate the flow speed. The Kiel probe (Figure 2.10c) uses a shroud to guide the flow to the pressure tap in the normal direction and has the least sensitivity with respect to probe orientation (misalignment errors are negligible within 40° of the flow angle).

2.2 TEMPERATURE SENSORS

There are two major temperature scales: Celsius and Fahrenheit. The Celsius scale is the international standard, based on the freezing point (0°C at the standard pressure of 1 atm = 101.325 kPa) and boiling point (100°C at the standard pressure) of water. The absolute temperature scales, Kelvin (K) and Rankine (°R), are defined with respect to the zero degree corresponding to the absolute zero of internal energy (no motion of any kind). The conversions between these temperature scales can be accomplished using

$$[°C] = \frac{[°F] - 32}{1.8} \tag{2.14a}$$

$$[°F] = 1.8[°C] + 32 \tag{2.14b}$$

$$[K] = [°C] + 273.15 \tag{2.14c}$$

$$[°R] = [°F] + 459.67 \tag{2.14d}$$

Table 2.1 lists some of the major benchmark temperatures.

The methods for measuring temperature can be grouped into the following five categories:

1. Thermal expansion (e.g., liquid-in-glass thermometers, bimetal thermometers)
2. Electrical property change as a function of temperature (thermocouples, thermistors)
3. Thermal radiation (pyrometers)
4. Change in chemical structure (thermochromatic liquid crystals)
5. Pressure change as a function of temperature

TABLE 2.1
Some Benchmark Temperatures

Comment	Kelvin [K]	Celsius [°C]	Fahrenheit [°F]	Rankine [°R]
Absolute zero	0	−273.15	−459.67	0
Boiling point of water at 100 kPa	373.15	100	212	671.67
Freezing point of water at 100 kPa	273.15	0	32	491.67
Average human body temperature	309.95	36.8	98.2	557.9
Surface of the Sun	5800	5526	9980	10,440
Liquid–vapor equilibrium of hydrogen at 1 atm (boiling point)	20.3	−252.87	−423.17	36.54
Melting point of tin at 1 atm	505.1	231.9	449.5	909
Melting point of gold at 1 atm	1337.3	1064.2	1947.1	2407.1
Melting point of copper at 1 atm	1357.7	1084.6	1983.9	2443.9
Melting point of titanium at 1 atm	1941	1668	3034	3494

2.2.1 Temperature Measurements Based on Thermal Expansion of Materials

Most materials have a positive volumetric expansion as a function of temperature; i.e., volume increases with temperature. Mercury, once commonly used in liquid-in-glass thermometers, has a relatively high volumetric expansion coefficient, β, of 18×10^{-5} [/°C], defined as

$$\beta = \frac{\Delta V}{V_o \Delta T} \tag{2.15}$$

ΔV = change in volume
V_o = initial volume
ΔT = temperature change

Such volumetric expansions of mercury or alcohol (which also has a large β) are used to measure temperature by placing the liquid in a small bulb space inside a glass tube connected to a capillary column. The small diameter of the capillary column ensures that a small change in temperature, and therefore in liquid volume, results in a readable change in the liquid-column height. The scales on the capillary column are etched after calibration at known temperature marks, such as the freezing or boiling points of water. Mercury is visible by itself when placed in a capillary column, whereas alcohol is only readily visible when mixed with a small amount of dye (typically red). Mercury undergoes a phase change at −38.87°C or −37.97°F, which obviously limits its use near or below that temperature. Alcohol has a substantially lower freezing temperature of −117.3°C for ethyl alcohol.

Example 2.3

An initial volume of 200 cm³ of mercury, when subjected to an 80°C temperature increase, will undergo a 2.88 cm³ increase in volume, resulting in a final volume of 202.88 cm³.

$$\Delta V = \beta V_o \Delta T = \left(18 \times 10^{-5} /^{\circ}\mathrm{C}\right)\left(200\,\mathrm{cm}^3\right)\left(80^{\circ}\mathrm{C}\right) = 2.88\,\mathrm{cm}^3$$

Bimetallic thermometers use the difference in thermal expansion of two bonded metal strips, as shown in Figure 2.11. A large difference in the thermal expansion coefficient results in high sensitivity. For initially parallel bimetallic strips, the curvature developed depends on the difference in the linear thermal expansion coefficients (β'), the thickness of the bonded strips (t), and, of course, the temperature change (ΔT). The curvature generated can be mechanically linked to provide rotation of a needle, similar to pressure gauges using Bourdon tube expansion (Figure 2.3), or electrically connected to serve as temperature threshold triggers.

$$r \propto \frac{t}{\left(\beta'_A - \beta'_B\right)\Delta T} \tag{2.16}$$

ΔT = temperature change with respect to the initial temperature when $r = \infty$.

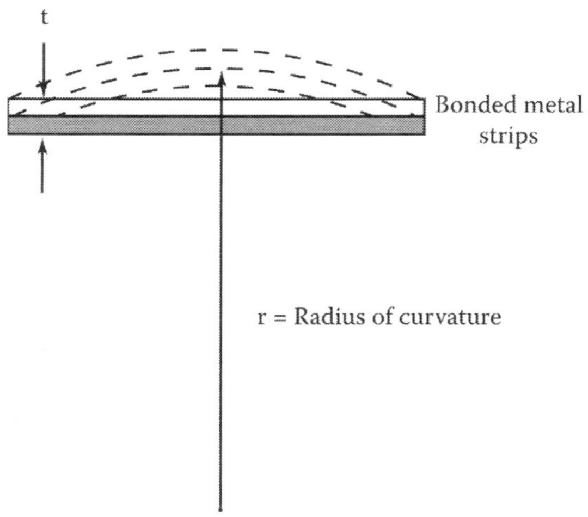

FIGURE 2.11 Bimetallic thermometer.

2.2.2 Thermocouples

In many engineering and scientific applications, thermocouples are used for temperature measurements. They are based on a thermoelectric effect called the Seebeck effect, as shown in Figure 2.12. When two different metal wires are joined together and placed at different temperatures, the difference in the rate of diffusion of electrons that carry thermal energy from one end to the other in the metal wires results in a net electric potential at the open end of the thermocouple circuit. Metals possess different rates of charge carrier diffusion, as observed by the different electrical and thermal conductivities of different metals. The electrons, the primary carriers of thermal energy (and electricity) in metals, are transported at different rates through the metal media in A and B. This net difference in electron transport results in an imbalance of electron concentrations at the gap shown in the top wire segment in the figure.

The electrical potential or voltage thus created is called the electromotive force (emf). The emf developed is typically in the order of a few microvolts per degree of temperature difference. The two "junctions," 1 and 2, of the metal wires need to be at different temperatures to cause a flow of electrons from one junction to the other, analogous to heat conduction requiring a temperature gradient. At least two different metal wires are needed to form a closed loop. The voltage developed in the fundamental thermocouple circuit shown in Figure 2.12 depends on the temperature difference, $\Delta T = T_2 - T_1$, and the difference in the so-called Seebeck coefficients of the two metals, $S_{AB} = S_A - S_B$.

$$\Delta V = (S_A - S_B)(T_2 - T_1) = S_{AB}\Delta T \tag{2.17}$$

So the Seebeck coefficient for the thermocouple, S_{AB}, may be defined as

$$S_{AB} = \lim_{\Delta T \to 0} \frac{\Delta V}{\Delta T} \tag{2.18}$$

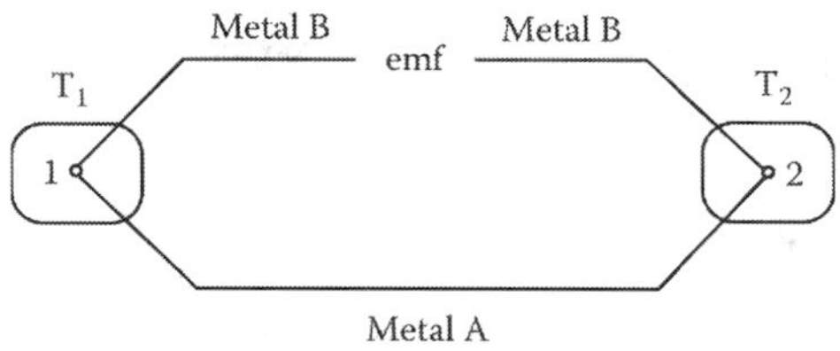

FIGURE 2.12 A basic thermocouple circuit showing the use of the Seebeck effect.

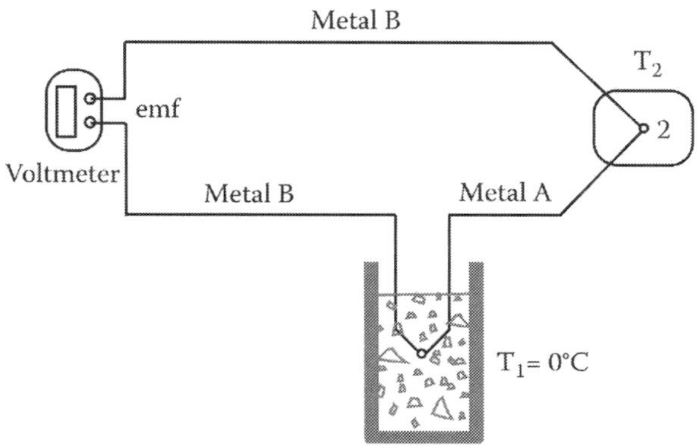

FIGURE 2.13 Thermocouple circuit with a classical ice-bath reference junction.

From an implementation standpoint, we would like to measure T_2 with a known temperature T_1, called a reference junction temperature. The classical approach using an ice bath to immerse the reference junction is shown in Figure 2.13, so that $T_1 = 0.0 \pm 0.01$°C. However, nowadays, thermocouples are connected to a built-in electrical compensation for the ambient temperature, which serves as the reference junction. The electrical compensation involves a temperature-dependent resistor (called a thermistor) or semiconductor temperature sensors (discussed in Section 2.2.3). The thermistors undergo a negative resistance change with increasing temperature so that they alter the developed emf voltage as a function of the ambient temperature (as if the thermocouple circuit had a 0°C reference junction). For example, if the junction temperature increases, then the overall temperature difference decreases and the emf will also decrease according to Equation 2.17. Because the thermistor resistance decreases with increased temperature at the junction, the total voltage of the thermocouple circuit will be compensated to what it would be at the original junction temperature. The electrical compensation introduces an error typically in the ±0.1°C to 0.5°C range. The effective resistance at a semiconductor junction also varies as a function of temperature with a negative slope, and the temperature sensors based on semiconductors can also be used as a compensation element in a thermocouple circuit. With electrical compensation, thermocouple-based temperature detectors are compact, digital, and easy to use, and there are many vendors that offer thermocouple packages and materials for diverse applications (try searching for "thermocouple" on Google).

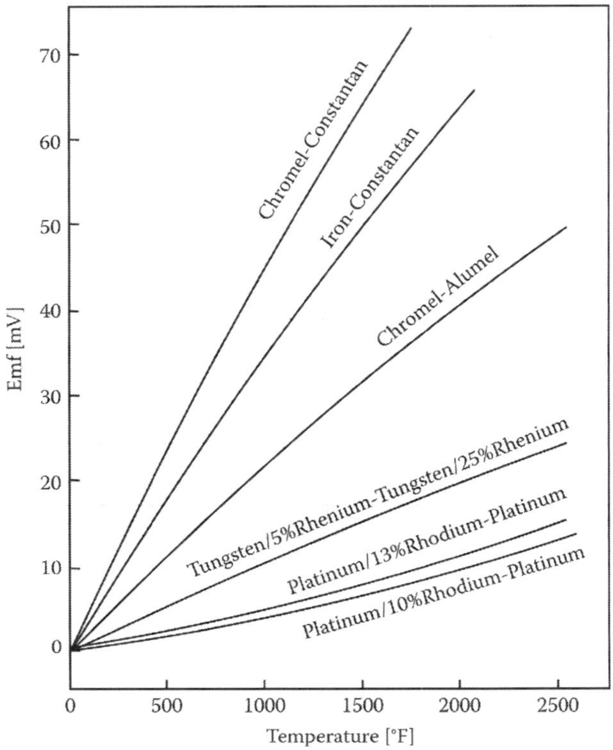

FIGURE 2.14 Thermocouple voltage for some combinations of materials (for a reference junction temperature of 0°C).

There is a variety of thermocouple material combinations with different temperature ranges and sensitivities. Figure 2.14 illustrates the temperature sensitivity of some thermocouple material combinations. It can be seen that there is a wide variety of materials available that give a range of sensitivities (slope of the curves in Figure 2.14). For example, the chromel–constantan pair gives the highest sensitivity and is useful below 1000°C. The iron–constantan combination also gives high sensitivity, but due to the iron content, it should be used only in a non-oxidizing environment. Platinum/10%Rhodium thermocouples have low sensitivity but can withstand high temperatures (<1768°C). Chromel–alumel thermocouples are also durable in high-temperature surroundings below 1372°C. It can also be seen in Figure 2.14 that the typical output of thermocouples is in the order of 10 mV, and compact electronic amplification circuits are usually embedded in the thermocouple readout packages for stable display. Although Figure 2.14 is useful in qualitatively determining the thermocouple sensitivity, the actual conversion of the emf voltage to temperature requires more exact data. The emf response data in Figure 2.14 are fit to polynomial functions, and their coefficients are given. Table 2.2 tabulates these correlations in terms of the polynomial coefficients for various thermocouple combinations.

TABLE 2.2

Thermocouple Types and Their Voltage-Temperature Responses (with a Reference Junction Temperature, Actual or Compensated, of 0°C)

Type	Material and Application	Range [°C]	$E=f[T(°C)]$
B	Pt, 30% vs. Pt, 6% Rh(−)	0 to 1820	$E=(-2.4674601620*10^{-1}*C+5.9102111169*10^{-3}*C^2-1.4307123430*10^{-6}*C^3+2.1509149750*10^{-9}*C^4-31757800720*10-12*C5+2.4010367459*10-15*C^6-9.0928148159*10^{-19}*C^7+1.32990505137*10^{-22}*C^8)*10^{-3}$
E	Chromel(+) vs. Constantan(−)	0 to 1000	$E=(5.8695857799*10*C+4.3311094562*10^{-2}*C^2+5.7220358202*10^{-5}*C^3-5.4020668085*10^{-7}*C^4+1.54259221111*10^{-9}*C^5-2.4850089136*10^{-7}*C^6+2.3389721459*10^{-15}*C^7-1.1946296815*10^{-18}*C^8+2.5561127497*10^{-22}*C^9)*10^{-3}$
J	Fe(+) vs. Constantan(−)	−210 to 760	$E=(+5.037253027*10*C+3.0425491284*10^{-2}*C^2-8.5669750464*10^{-5}*C^3+1.3348825735*10^{-7}*C^4 1.7022405966*10^{-10}*C^5+1.9416091001*10^{-13}*C^6-9.6391844859*10^{-17}*C^7)*10^{-3}$
K	Chromel(+) vs. Alumel(−)	0 to 1372	$E=(1.8533063273*10+3.8918344612*10*C+1.6645154356*10^{-2}*C^2-7.8702374448*10^{-5}*C^3 2.2835785557*10^{-7}*C^4-3.5700231258*10^{-10}*C^5+0.9932909136*10^{-13}*C^6-1.2849848798*10^{-16}*C^7+2.2239974336*10^{-20}*C^8+125\exp[-5((C-127)/65)^2])*10^{-3}$
R	Pt(−) vs. Pt, 13%Rh(+)	630 to 1064	$E=(-2.6418007025*10^2+8.0468680747*10^0*C+2.9892293723*10^{-3}*C^2-2.6876058617*10^{-7})*10^{-3}$
		1064 to 1665	$E=(1.4901702702*10^3+2.8639867552*10^0*C+8.0823631189*10^{-3}*C^2-1.9338477638*10^{-6}*C^3)*10^{-3}$
S	Pt, 10% Rh(+) vs. Pt(−)	630 to 1064	$E=(-2.9824481615*10^2+8.2375528221*10^0*C+1.6453909942*10^{-3}*C^2)*10^{-3}$
		1064 to 1664	$E=(1.2766292175*10^3+3.4970908041*10^0*C+6.3824648666*10^{-3}*C^2-1.5722424599*10^{-6}*C^3)*10^{-3}$
T	Cu(+) vs. constantan(−)	0 to 400	$E=(3.8740773840*10*C+3.3190198092*10^{-2}*C^2+2.0714183645*10^{-4}*C^3-2.1945834823*10^{-6}*C^4+1.1031900550*10^{-8}*C^5-3.0927581898*10^{-11}*C^8+4.5653337165*10^{-14}*C^7-2.7616878040*10^{-17}*C^8)*10^{-3}$

Note: Pt = platinum; Rh = rhodium; Cu = copper; constantan = 55% copper + 45% nickel; chromel = 90% nickel + 10% chromium; alumel = 94% nickel + 3% manganese + 2% aluminum + 1% silicon.

For a thermocouple used in a high-temperature gas environment, the heat radiated from the thermocouple surface may result in some loss of energy from the thermocouple, causing an underestimation of the temperature. The radiation error is usually corrected using a simple energy balance between the heat transferred to the thermocouple bead through convection and the heat lost to the surroundings through radiation. Under steady state, the convection and radiation heat transfer rates to the thermocouple probe will be equal, so that the probe is at some constant temperature, T_{TC}. These rates are, respectively,

$$\begin{aligned}\text{Rate of convection heat transfer} &= hA\left(T_{\infty} - T_{TC}\right)\\ \text{Rate of radiation heat transfer} &= \varepsilon\sigma A\left(T_w^4 - T_{TC}^4\right)\end{aligned} \tag{2.19}$$

h = heat transfer coefficient
A = surface area of the thermocouple probe
T_{∞} = fluid temperature, to be measured by the thermocouple
T_{TC} = thermocouple temperature
ε = emissivity of the thermocouple surface
σ = Stefan–Boltzmann constant = 5.669×10^{-8} W/m^{2}K^4
T_w = temperature of the surrounding enclosure

At steady state, the foregoing two rates will equal, so that both T_{∞} and T_{TC} can be determined through trial-and-error solution of the energy balance equation

$$h\left(T_{TC} - T_{\infty}\right) = \varepsilon\sigma\left(T_w^4 - T_{TC}^4\right) \tag{2.20}$$

Then, T_{∞} is the corrected temperature, and ($T_{\infty} - T_{TC}$) represents the thermocouple error due to radiation loss. In some applications, a radiation shield in the form of a small shroud is used to cover the thermocouple probe to minimize the radiation error. This acts to send the radiation from the thermocouple surface back to the probe so that near-equilibrium in radiation heat transfer rates is achieved.

There can also be conduction loss from the thermocouple probe through the support of the thermocouple when the support is exposed to a medium with a substantially different temperature. However, conduction losses can be easily minimized by using probe geometry to extend the probe away from such conducting media.

The stagnation temperature effect can also cause measurement errors. Analogous to a pressure probe giving stagnation pressure readings, if the probe is oriented normal to the flow, the temperature sensors are subject to the stagnation effect, although the effect is not as significant as in pressure measurements. For isentropic stagnation, the temperature–pressure relationship is

$$\left(\frac{T_t}{T_{\infty}}\right) = \left(\frac{p_t}{p_{\infty}}\right)^{\frac{\gamma-1}{\gamma}} \tag{2.21}$$

For air, the exponent above the pressure ratio is approximately 0.2857, so the temperature change is much less than the pressure change as one goes from the static (∞) to the stagnation (t) condition. Nonetheless, for high-speed gas flows (Mach number greater than about 0.1), a correction for the stagnation effect should be made. The stagnation temperature is defined as the static temperature and the kinetic energy of the flow converted to the temperature unit:

$$T_t = T_\infty + \frac{U^2}{2c_p} \tag{2.22}$$

To correct for the stagnation effect, we introduce a so-called *recovery factor*, *r*, which is the fraction of the kinetic energy transferred to the probe. Using the recovery factor, the corrected temperature is written as

$$T_{TC} = T_\infty + \frac{rU^2}{2c_p} = T_t + \frac{(1-r)U^2}{2c_p} \tag{2.23}$$

The value of the recovery factor, r, ranges from 0.68 for thermocouple wires oriented normal to the flow, to 0.86 oriented in parallel. The use of the foregoing equation, however, does require the flow velocity to be known or estimated. We deal with flow rate and velocity measurements in subsequent sections and in Chapter 4.

2.2.3 Resistance-Based Temperature Sensors

The electrical resistance of a conducting wire of length l and cross-sectional area A is, in general, given by

$$R = \frac{\rho l}{A} \quad [\Omega, \text{Ohm}] \tag{2.24}$$

ρ = resistivity of material [Ωm]

Resistance-temperature detectors, or RTDs, are temperature sensors based on the temperature dependence of resistivity, ρ, a fundamental property of the RTD material (mostly metals). There is also a class of semiconductor materials that exhibit a significant *decrease* in resistivity as temperature increases. Such temperature sensors using materials with reversed resistance-temperature behavior are called *thermistors* and are useful in many applications, including thermocouple junction temperature correction, as noted earlier.

For RTDs of a fixed length and cross-sectional area, the overall resistance, R, changes as a function of temperature.

$$R(T) = R_o\left[1 + \alpha(T - T_o) + \beta(T - T_o)^2\right] \tag{2.25}$$

R_o = reference resistance at T_o (typically 0°C)
α, β = temperature coefficients of resistivity [°C^{-1}]

Usually, the coefficient, β, for the quadratic term in Equation 2.25 is small for most RTD materials (e.g., nickel, platinum, or copper), so that Equation 2.25 can be linearized for convenience with only a small loss of accuracy.

$$R(T) = R_o\left[1 + \alpha(T - T_o)\right] \tag{2.26}$$

For platinum, the error associated with the linearized response equation (Equation 2.26) is less than 0.3% for 0°C–200°C and 1.2% for 200°C–800°C. Table 2.3 shows some of the common materials and their properties for RTDs. Figure 2.15 shows a typical manner in which the metal wire material in Table 2.3 is used in an RTD sensor.

The resistance change in RTDs can be measured using the classical Wheatstone bridge with a current meter, as shown in Figure 2.16. In the Wheatstone bridge

(Figure 2.16), the variable resistance, R_3, is adjusted until the current (I_{AB}), and therefore the voltage across the bridge (V_{BD}), is zero. The fact that V_{BD} is zero means that the voltage drop across the elements must be the same for both sides of the bridge: $V_{AB}=V_{AD}$ and $V_{BC}=V_{DC}$ (if $V_{AB}\neq V_{AD}$, then $V_B\neq V_D$, which would violate V_{BD} being equal to zero). Using Ohm's law ($V=IR$), this condition translates to

$$\begin{aligned} I_{AB}R_1 &= I_{AD}R_3 \\ I_{BC}R_2 &= I_{DC}R_{RTD} \end{aligned} \tag{2.27}$$

Also, the fact that $V_{BD}=0$ means that $I_{BD}=0$, leading to $I_{AB}=I_{BC}$ and $I_{AD}=I_{DC}$. Using this fact and dividing the bottom equation in Equation 2.27 by the top equation, we obtain a relationship between the known resistances and the RTD resistance.

$$\frac{R_{RTD}}{R_3} = \frac{R_2}{R_1} \tag{2.28}$$

For RTDs, accurate measurements of R_{RTD} are required, and the resistance of the wire elements between the resistors in the Wheatstone bridge may introduce errors in using Equation 2.28. The wire resistances (r_1, r_2, and r_3 in Figure 2.16) will alter Equation 2.28 to

$$\frac{R_2}{R_1} = \frac{R_{RTD} + r_3}{R_3 + r_1} \tag{2.29}$$

To compensate for the wire resistances, some variations in the bridge circuitry are used in RTDs. In these circuits, R_1 is set equal to R_2, in which case

$$R_{RTD} = R_3 + r_1 - r_3 \tag{2.30}$$

By adding a resistor or wire length so that $r_1=r_3$, the wire resistance effect is minimized.

As shown in Table 2.3, the metals used in RTDs undergo an *increase* in resistance as the temperature increases. There is a class of ceramic semiconductor materials that change resistance in an inverse manner in response to temperature change, i.e., the resistance *decreases* with increasing temperature. The temperature sensors based on these materials are called thermistors. Typical materials for thermistors are oxides of magnesium, nickel, and cobalt, with resistivities ranging from 100 to 450,000 Ωcm.

TABLE 2.3
Characteristics of Common Materials Used in RTDs

Sensor Material	Case Material	Temperature Range [°C]	Reference Resistance, R_o[Ω]	Temperature Coefficient, α[Ω/Ω°C]	Error Limits [°C]	Response Time [s]
Platinum	Pyrex glass	−190~540	25 at 0°C	0.0039	±0.01	10~30
Platinum	Stainless steel	−200~125	25 at 0°C	0.0039	±1	10~30
Platinum film	Ceramic coating	−50~600	1000 at 0°C	0.0039	±0.25	1
Rhodium-iron	Alumina and glass	−272~200	27 at 0°C	0.0037	±0.04	
Copper	Brass	−75~120	10 at 25°C	0.0038	±0.5	20~60
Nickel	Brass	−0~120	100 at 20°C	0.0067	±0.3	20~60

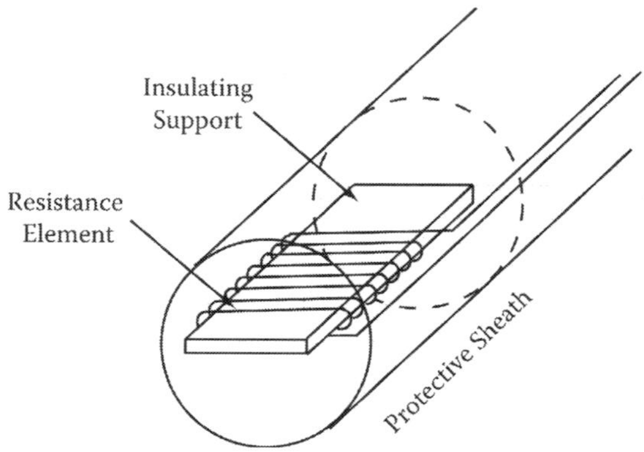

FIGURE 2.15 A typical geometry for a resistance-temperature detector.

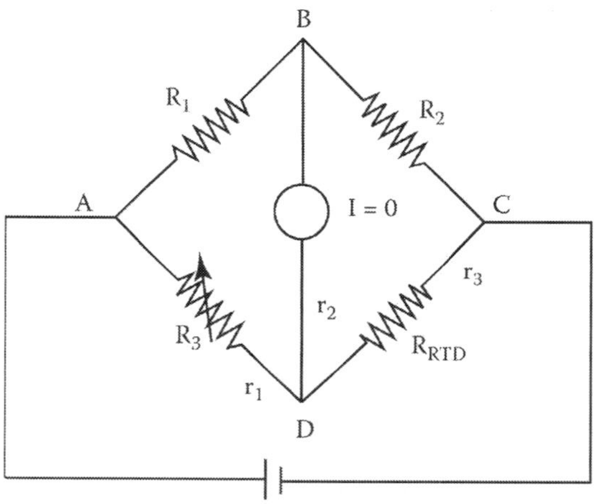

FIGURE 2.16 Bridge circuits used for measuring the resistance change in resistance-temperature detectors.

The resistance–temperature relationship of thermistors is typically given in an exponential form:

$$R = R_o \exp\left[\gamma\left(\frac{1}{T} - \frac{1}{T_o}\right)\right] \tag{2.31}$$

R = resistance at temperature T [Ω]
T = absolute temperature [K]
R_o = reference resistance at a reference temperature T_o
T_o = reference temperature [K]
γ = constant for a given material [K]

The constant, <ι>γ</ι>, ranges in value from 3500 to 4600 [K], and when placed in Equation 2.31, results in a large variation of resistance for a modest change in temperature. Thus, thermistors have high sensitivity in comparison to RTDs. The negative slope of thermistors is useful for compensating for the junction temperature in thermocouple circuitry. Also, thermistors typically have a fast response time in comparison to RTDs and can be used in harsh environments, as thermistors can be coated or encapsulated in protective material such as glass. The nonlinear dependence of the temperature–resistance relationship in Equation 2.30 precludes the direct, linear conversion of resistance to temperature except for a limited range of temperatures. To compensate for the nonlinear response, different thermistor materials are used in series. Otherwise, complete calibration data are needed to convert the resistance change to temperature. Figure 2.17 shows some of the methods in which thermistor materials are implemented into sensors. Because thermistor elements can be produced using microfabrication techniques (see Chapter 8), these devices can be miniaturized to submillimeter scales.

Semiconductor junctions, consisting of silicon or gallium–aluminum–arsenide with different dopant concentrations, also exhibit changes in effective resistance as a function of temperature. This means that, for a fixed current, the voltage across a semiconductor junction will vary as a function of temperature, qualitatively shown in Figure 2.18. Similar to thermistors, the resistance–temperature curve (or voltage–temperature curve in Figure 2.18) has a negative slope. Once the calibration curve for the voltage–temperature relationship is obtained, the voltage output is quite repeatable, with a temperature measurement accuracy down to 50 mK for temperatures between 1.4 and 300 K. Like other semiconductor devices, these sensors cannot be used at temperatures exceeding 100°C–150°C. Based on this approach, integrated circuits (ICs) including these junctions are fabricated to amplify the current output with respect to a supplied voltage (4–30 V). These IC devices are inexpensive and widely used for process monitoring, control, and electronic compensation in thermocouple circuits.

2.2.4 Pyrometer Measurements of Temperature

All materials above the absolute temperature of 0 K emit electromagnetic radiation, sometimes called *thermal radiation*. The latter phrase originates from the fact that the thermal energy of the material results in the emission of photons at the surface.

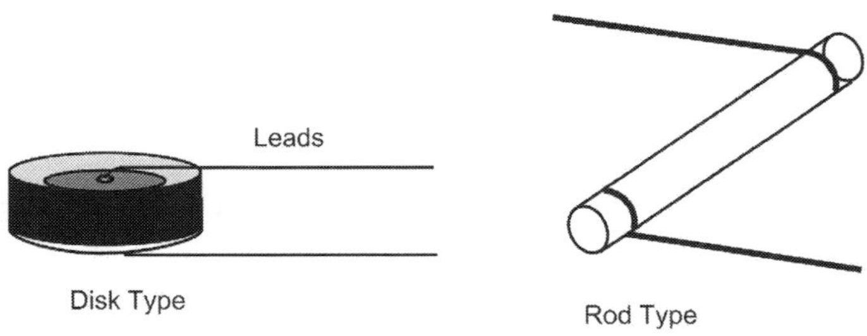

FIGURE 2.17 Temperature sensors using thermistors.

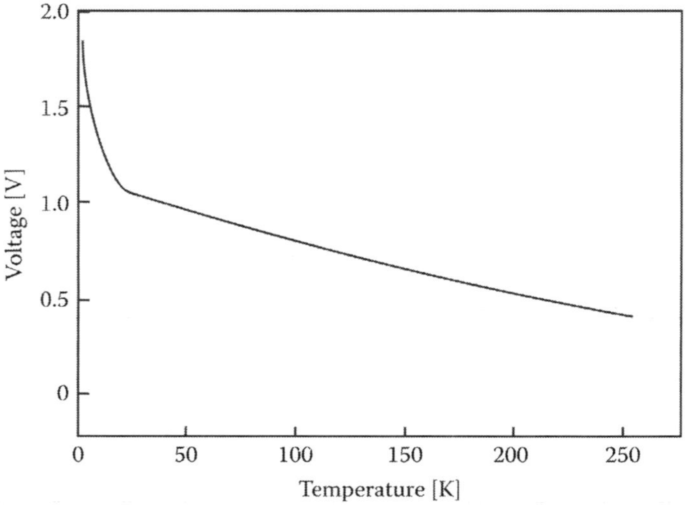

FIGURE 2.18 Temperature dependence of semiconductor junction voltage.

For solid surfaces, the lattice vibration in the molecular structure represents excess (or deficit) energy with respect to the surroundings. Oscillations in the electric field, as observed in lattice vibrations or oscillations in the electronic orbitals in atoms or molecules, result in the emission of photons or electromagnetic radiation (analogous to the antenna effect). By the Second Law of Thermodynamics, the energy differential between the surface and the surroundings needs to come to equilibrium through an exchange of energy between them, and photons or electromagnetic radiation serve as the transmitter of this energy exchange. For an ideal surface called a blackbody, this radiative emission is isotropic, depends only on temperature, and follows the so-called Planck's law.

$$I_{b\lambda} = \frac{2\pi hc^2}{\lambda^5} \frac{1}{\exp\left(\dfrac{hc}{k\lambda T}\right) - 1} \left[\frac{W}{m^2 str \lambda m}\right] \tag{2.32}$$

$I_{b\lambda}$ = blackbody spectral intensity = amount of emissive power [W] per unit surface area [m^2], per unit solid angle [sr], per unit wavelength [λm]
h = Planck's constant = 6.626068×10^{-34} m^2 kg/s
c = speed of light = 299,792.458 km/s
λ = wavelength of the radiation in [λm]
T = temperature in [K]

Figure 2.19 shows the blackbody spectral intensity at various temperatures calculated using Planck's law (Equation 2.32). It can be observed that not only does the amount of radiation vastly change with increasing temperature, but also the radiative spectrum shifts toward shorter wavelengths. For real surfaces, the actual surface emission deviates from blackbody radiation, and the ratio of the actual surface emission to blackbody radiation is defined as the *emissivity*. The emissivity varies as a

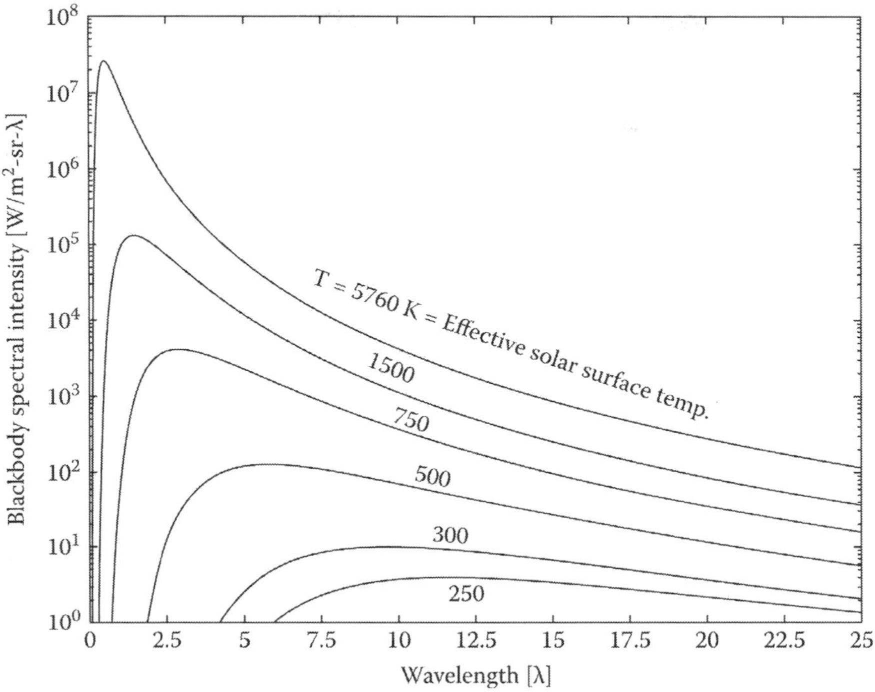

FIGURE 2.19 The blackbody radiation at various temperatures from Planck's law.

function of the surface material, wavelength, angle, and temperature. Integration of Equation 2.32 over the entire wavelength range and across the hemispherical volume over the emitting surface gives the total blackbody emissive power, e_b (see Appendix C on Electromagnetics and Electromagnetic Radiation).

$$e_b = \sigma T^4 \left[W/m^2\right] \tag{2.33}$$

σ = Stefan–Boltzmann constant = 5.6704×10^{-8} W/(m^{2}K^4)

This equation is known as Stefan–Boltzmann's law.

If we differentiate Planck's law (Equation 2.32) with respect to λ and set it equal to zero, it will give us the wavelength at which the maximum radiation in the blackbody spectral intensity occurs. The result is Wien's displacement law.

$$(\lambda)_{\max I_\lambda} T = 2897.8\ \mu\text{mK} \tag{2.34}$$

Example 2.4

For the temperatures shown in Figure 2.19, we can calculate the wavelength at which the peaks in the blackbody radiation occur using Equation 2.34.

We can see that solar radiation has its peak at 0.5 μm (green), so it is no coincidence that humans and most mammals have the most efficient light conversion near that wavelength.

Pyrometry refers to measurements of the emitted radiation. Using the foregoing radiation principles, pyrometry can be used for temperature measurements. However, because emissivity differs from one material to another, pyrometric measurements of temperature require calibration or other methods for correcting these differences. In addition, the medium between the emitter and the detector may contain gases or particles that either absorb, scatter, or even emit radiation, which affect the pyrometer measurements. Although the blackbody emission is isotropic, the emissivity of real materials is sometimes highly directionally specific, and the geometry of the emitter, detector, as well as the surroundings, will alter the radiation exchange.

In spite of these complicating factors, there are some reliable pyrometric detectors for temperature measurements after calibrations of the signals are made. A total-radiation pyrometer is shown in Figure 2.20. This is based on Stefan–Boltzmann's law and is designed to capture radiation emitted in the normal direction from the surface, whose temperature is to be measured. As shown in Figure 2.20, the geometry of the optical opening of the pyrometer is designed to capture the normal component of the emission emanating from the spot at which the pyrometer is pointed. The device is called a total-radiation pyrometer because it measures all of the spectral content of the emitted radiation, i.e., emissive power integrated over all wavelength range. For this reason, a lens that transmits in both visible and infrared is used to collect the incident radiation onto a thermopile detector (a set of thermocouples connected in series to detect the increase in temperature due to the incident radiation). Total-radiation pyrometers achieve good accuracy for materials with high emissivity, which needs to be known. Even with the known emissivity, it is customary to calibrate the pyrometer signal with known surface temperatures to ensure accurate measurements. In Figure 2.20, the thermopile detector housing is maintained at a constant temperature because thermopile detectors rely on the temperature increase due to incident thermal radiation. For the lens material, calcium fluoride can be used, which has good transmissivity over the 0.3–10 μm range. Other infrared optical materials include Pyrex glass (0.3–2.7 μm) and fused silica (0.3–3.8 μm). As shown by the level of thermal radiation in Figure 2.19, pyrometers are effective for high temperatures, although there are some commercial models with sensitivity below 100°C.

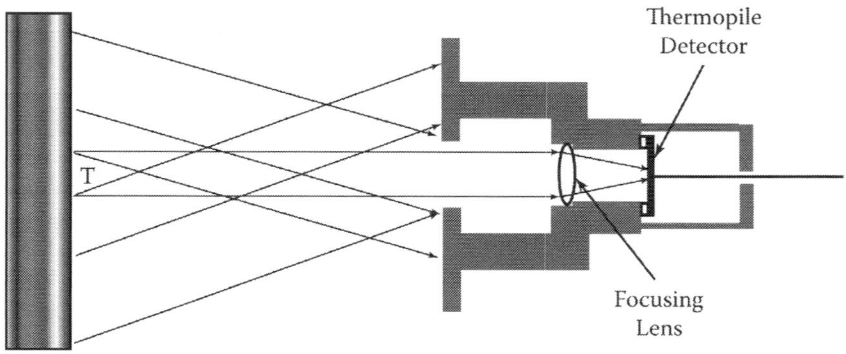

FIGURE 2.20 Total-radiation pyrometer.

An alternate method of using radiation for temperature measurements is to use two or more spectral bands and compare their ratios against a spectral radiation curve, which is the blackbody radiation curve in Figure 2.19 modified by the emissivity of the material. If we multiply the blackbody spectral intensity (Equation 2.32) by the spectral emissivity at a given wavelength, we have

$$I_\lambda = \varepsilon_\lambda I_{b\lambda} = \varepsilon_\lambda \frac{2\pi hc^2}{\lambda^5} \frac{1}{\exp\left(\frac{hc}{k\lambda T}\right) - 1} \approx C_1 \lambda^{-5} \varepsilon_\lambda \exp\left(-\frac{C_2}{\lambda T}\right) \tag{2.35}$$

$C_1 = 2\pi hc^2$
$C_2 = hc/k$

In the foregoing equation, we omit the −1 term in the denominator, which gives less than 1% error for radiation in the visible range for temperatures below 4000 K. If we select two reasonably separated wavelengths λ_A and λ_B, with respective bandwidths, $\Delta\lambda_A$ and $\Delta\lambda_B$, then the ratio of the radiation intensities at the two wavelengths is

$$R = \frac{I_{\lambda A}\Delta\lambda_A}{I_{\lambda B}\Delta\lambda_B} = \frac{\varepsilon_{\lambda A}}{\varepsilon_{\lambda B}} \frac{\Delta\lambda_A}{\Delta\lambda_B} \frac{\lambda_B^5}{\lambda_A^5} \exp\left(-\frac{C_2}{T}\left(\frac{1}{\lambda_A} - \frac{1}{\lambda_B}\right)\right) \tag{2.36}$$

By measuring $I_{\lambda A}$ and $I_{\lambda B}$, we obtain sufficient data to use Equation 2.36 to solve for the temperature, T. In contrast to total-radiation pyrometry, this approach is called two- or multicolor pyrometry (if more than two wavelength components are used) and requires the use of optical filters to separate the different spectral components. Again, even if accurate spectral emissivity data were available to use in Equation 2.36, calibration is still recommended to compensate for the wavelength-dependent responses of the detectors and optical components. The optical configuration is similar to that of the total-radiation pyrometer in Figure 2.20, except that the signal is split into two different detectors that view the signal in separate bandpass wavelengths, as shown in Figure 2.21.

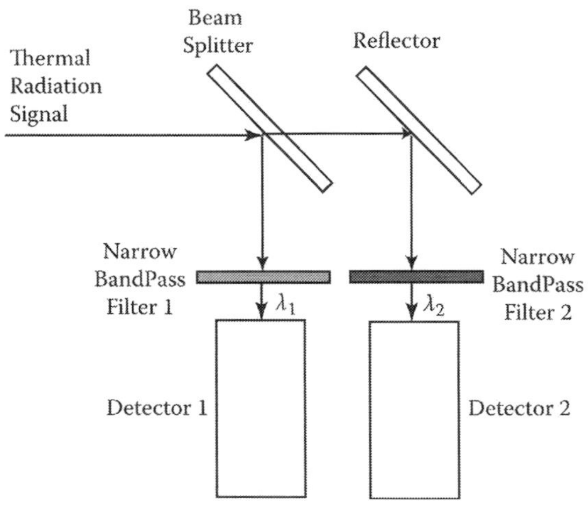

FIGURE 2.21 A two-color pyrometer setup.

There are so-called thermal imaging devices that provide two-dimensional images of thermal emission. These are similar to digital or CCD (charge-coupled device) cameras that produce an array of electrical signals in response to incident photon energy, which can be converted into an image format for display. Because most of the thermal emission is in the infrared for a large range of temperatures encountered in practice (again referring to Figure 2.19), thermal imaging devices use infrared-sensitive material. All the optical components need to be transparent to infrared radiation (0.75 to 20–100 μm or more depending on the usage), meaning that glass (mostly transparent in the visible range) or quartz (visible to ultraviolet) optical components are replaced with more expensive infrared optics (e.g., calcium fluoride, sapphire). However, for the same reasons that direct measurements of temperature are difficult in pyrometers, thermal imaging devices typically provide qualitative spatial distributions of temperature unless calibration is provided for the image data (some examples were shown in Chapter 1, Figure 1.3). Multicolor pyrometry can also be applied using infrared imaging devices by viewing separate spectral components of thermal radiation.

2.3 STRAIN GAUGES

Strain gauges are used in diaphragm-based pressure transducers, as discussed previously. The use of strain gauges is also relevant for force measurements in flows. For example, aerodynamic and fluid forces on surfaces are determined by measuring the strain on solid objects supporting the surfaces.

The strain, ε, is defined as the fractional change (δL) in the length (L) of a solid.

$$\text{Strain} = \varepsilon = \frac{\delta L}{L} \tag{2.37}$$

Such a strain is mechanically caused by an applied force on the object. The applied force (F) divided by the area (A) upon which the force acts is the *stress* (σ).

$$\text{Stress} = \sigma = \frac{F}{A} \tag{2.38}$$

The normal component of the stress is referred to as pressure in fluid flows, whereas the stress component parallel to the surface is the *shear stress*. Figure 2.22 shows the typical stress–strain relationship of solid materials. In the so-called elastic region (see Figure 2.22), there is a linear relationship between stress and strain, with the proportionality constant being Young's modulus of elasticity, E. Young's modulus is an intrinsic property of a solid and is defined as

$$\sigma = E\varepsilon \tag{2.39}$$

E = Young's modulus of elasticity

A strain in a given (longitudinal) direction in a solid of a given volume will be accompanied by a strain in the perpendicular (lateral) direction. The ratio of the lateral to the longitudinal, or axial strain, is called Poisson's ratio (ν), and is again an intrinsic property of a given solid.

$$\text{Poisson's ratio} = \nu = \frac{\varepsilon_L}{\varepsilon_a} \tag{2.40}$$

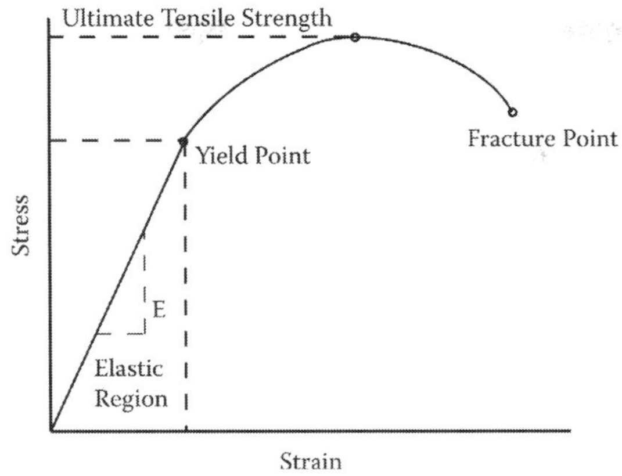

FIGURE 2.22 Stress and strain.

ε_L = lateral strain
ε_a = axial or longitudinal strain

A typical strain gauge is based on the change in resistance of a metal strip undergoing strain due to an applied stress. The resistance of a metal strip depends on the resistivity of the metal (ρ_e), the length (L), and the cross-sectional area (A) of the strip.

$$R = \frac{\rho_e L}{A} \tag{2.41}$$

Under a strain, all of the parameters on the right-hand side of the foregoing equation will, in principle, change. In differential form, this change in resistance can be written as

$$dR = \frac{d\rho_e L}{A} + \frac{\rho_e dL}{A} - \frac{\rho_e dA}{A^2} \tag{2.42}$$

Dividing by the resistance, we get

$$\frac{dR}{R} = \frac{d\rho_e}{\rho_e} + \frac{dL}{L} - \frac{dA}{A} \tag{2.43}$$

For a metal strip with a rectangular area with initial lengths X and Y, the fractional change in the cross-sectional area, dA/A, can be expressed in terms of the lateral strain and, therefore, Poisson's ratio, v.

$$\frac{dA}{A} = \frac{(X - dX)(Y - dY) - XY}{XY} \approx -\left(\frac{dX}{X} + \frac{dY}{Y}\right) = -2\varepsilon_L = -2v\varepsilon_a = -2v\frac{dL}{L} \tag{2.44}$$

Equation 3.42 can be simplified to

$$\frac{dR}{R} = \frac{dL}{L}(1 + 2v) + \frac{d\rho_e}{\rho_e} \tag{2.45}$$

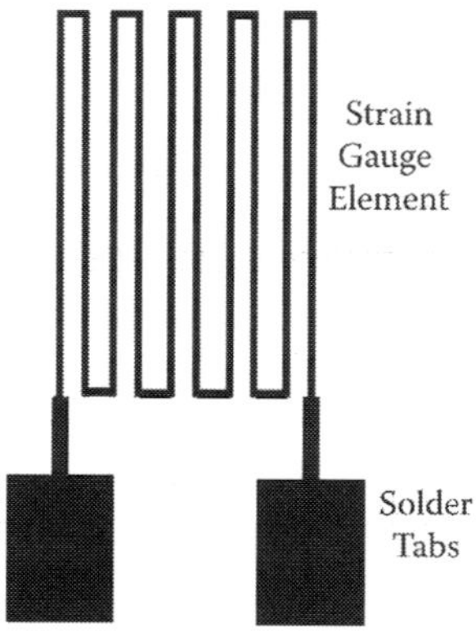

FIGURE 2.23 A typical strain gauge.

Moreover, the change in resistivity, ρ_e, is referred to as the piezoresistive effect, and a parameter called the *piezoresistance coefficient*, π_e, relates the last term in Equation 2.43 to the strain.

$$\pi_e = \frac{\frac{d\rho_e}{\rho_e}}{E\frac{dL}{L}} \tag{2.46}$$

E = Young's modulus of elasticity

We can combine these results to determine the fractional change in resistance with respect to strain, or essentially the sensitivity of the strain gauge (usually referred to as the gauge factor, GF).

$$\text{Gauge Factor} = \text{GF} = \frac{dR / R}{dL / L} = 1 + 2v + \pi_e E \tag{2.47}$$

In the practical implementation of strain gauges, a thin metal strip is bonded to the specimen so that the metal strip is subject to an identical strain as the specimen. Also, for a typical metal, an appreciable length L is needed so that the baseline resistance is in the 10–100 Ω range. This requires multiple bends and paths on the metal strip, as shown in Figure 2.23. The resistance change of strain gauges can be read using Wheatstone-type circuits similar to RTDs.

2.4 FLOW RATE MEASUREMENTS

In principle, flow rates can be measured by determining the velocity across the flow channel. Measurement of precise velocity in the flow is the subject of Chapter 4 in this book. Thus, in this section, we will discuss more basic methods by which the bulk volumetric or mass flow rates are measured without resorting to detailed pointwise velocity measurements.

For the control and characterization of flow processes, it is easy to imagine that the measurement of flow rates is important. Many thermal and flow processes depend on flow speed or flow rates, such as convective cooling, mixing processes, aerodynamic forces, and chemically reactive processes. As noted previously, the most accurate measurements of flow rates are accomplished by determining the velocity profiles across the flow section using laser Doppler velocimetry, particle image velocimetry, or similar methods. However, even this approach is not without difficulties (e.g., near-wall measurements), and obtaining velocity profiles involves relatively expensive equipment and can be time-consuming. For this reason, a number of alternative methods are used for bulk flow rate measurements. One simple method is to measure the weight or the volume of the fluid transported in a known time interval. This approach requires the extraction of the fluid and cannot be used for continuous monitoring. There is also a class of "flowmeters" that use various interactions of the flow with the internal geometry. The difference between the static and total pressure gives the dynamic pressure, which is proportional to the square of the flow velocity (as described in Section 2.1.4), and this represents a probe-based flow velocity measurement method. Through proper image analysis, flow visualization methods can be used to extract flow rate information. Similar to detailed velocity measurement methods, the approaches based on flow visualization are discussed elsewhere in Chapter 3 of this book. In the following, therefore, we will look mostly at the flowmeter- and probe-based methods for flow rate measurements.

2.4.1 Obstruction Flowmeters

Obstruction flowmeters operate by creating a change in the flow area (an obstruction) so that the local flow speed increases due to the change in flow area, and this in turn causes a change in the static pressure, which can be measured. This principle is graphically illustrated in Figure 2.24. The mathematical relationship between pressure and velocity is the conservation of momentum, one version of which is Bernoulli's equation:

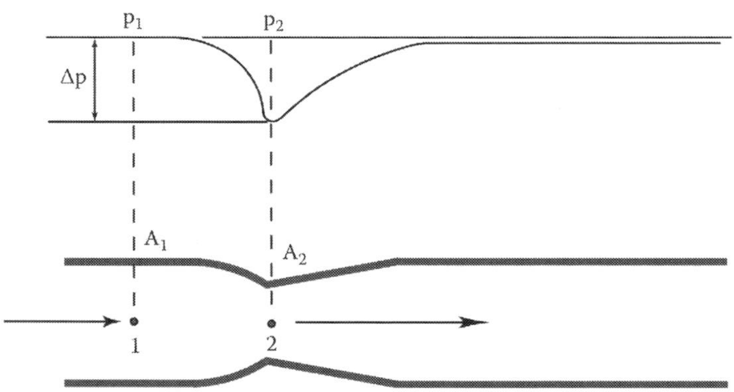

FIGURE 2.24 Flowmeters based on obstruction of the flow and a corresponding change in static pressure.

$$p_1 + \frac{1}{2}\rho V_1^2 = p_2 + \frac{1}{2}\rho V_2^2 + \Delta p_{\text{friction}} \tag{2.48}$$

$\Delta p_{\text{friction}}$ = pressure loss due to flow friction

In flows through practical devices, there is a frictional pressure loss, which is included in Equation 2.48. The volumetric flow rate, Q, is V_2A_2, and using the conservation of mass, we have $V_1A_1 = V_2A_2$ (the density is assumed to be constant at 1 and 2 for now). Thus, $V_1 = V_2A_2/A_1$, and substituting this into Equation 2.48, we obtain

$$Q = V_2A_2 = \frac{A_2}{\left[1 - \frac{A_2^2}{A_1^2}\right]^{1/2}} \sqrt{\frac{2(p_1 - p_2 - \Delta p_{\text{friction}})}{\rho}} \tag{2.49}$$

In actual applications, the pressure loss through friction is incorporated into Equation 2.49 through the flow discharge coefficient (C_D). Any change in the density that may occur in the flow from 1 to 2 is also treated with a "correction" factor called the *expansion factor* Y. Finally, the area factor in front of the square root sign in Equation 2.47 is also written in a compact coefficient form (E), so that the combined result is

$$Q = C_D E Y A_2 \sqrt{\frac{2(p_1 - p_2)}{\rho_1}} \tag{2.50}$$

C_D = discharge coefficient = ($Q_{\text{actual}}/Q_{\text{ideal}}$) = (Actual flow rate)/(Ideal flow rate with zero frictional loss)

E = factor involving the area ratio $= \dfrac{1}{\sqrt{1 - \left(\dfrac{A_2}{A_1}\right)^2}}$

Y = expansion coefficient (the effect of density change)

There are three commonly used designs for obstruction flowmeters (Venturi, nozzle, and orifice), as shown in Figure 2.25. By measuring the pressures through ports at "1" and "2," we obtain the required pressures, p_1 and p_2, in Equation 2.50. For small flow rates, the flow is assumed to be incompressible, and the expansion coefficient, Y, is taken to be 1. The discharge coefficient, C_D, is a function of the obstruction geometry and the flow Reynolds number, and for standard geometry, data are available in an ASME publication (see References).

Obstruction flowmeters based on pressure differences are useful for measuring relatively low flow rates, where the compressibility effect is small. Any change in density due to the compressibility effect needs to be corrected by using the expansion coefficient Y, as in Equation 2.50. Conversely, the compressibility effect can be used as a way of metering the flow. From gas dynamics, as the pressure difference across a nozzle (a kind of obstruction, as shown in Figure 2.25) increases, the flow rate will increase until the flow speed at the "throat" of the nozzle reaches its maximum—the sonic velocity or the speed of sound. At and beyond this point, the flow

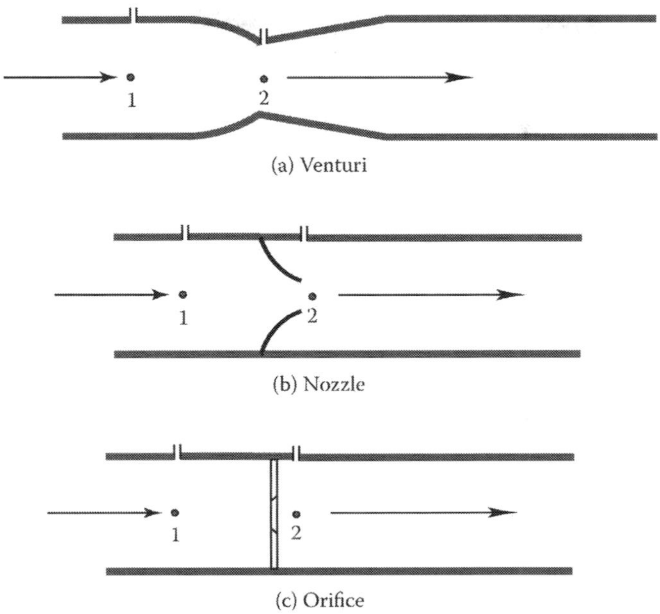

FIGURE 2.25 Commonly used designs for obstruction flowmeters.

becomes "choked," with the flow speed reaching the speed of sound at the throat, and the flow rate no longer depends on the pressure difference across the nozzle but only on the upstream pressure. The reason is that the pressure downstream cannot be sensed by the flow upstream because the flow speed is at or above the sonic velocity downstream of the throat. The choked flow condition is achieved when the upstream pressure, p_1, reaches a ratio with respect to the pressure p_0, at the obstruction

$$\frac{p_1}{p_0} = \left(\frac{\gamma+1}{2}\right)^{\frac{\gamma}{\gamma-1}} \tag{2.51}$$

p_1 = upstream pressure
p_0 = pressure at the obstruction or at the throat γ = ratio of specific heats = c_p/c_v

For air at standard conditions, γ is 1.4, and thus a pressure ratio of 1.89 is needed before the flow becomes sonic at the throat. Equation 2.51 is obtained by setting the Mach number, M_0, equal to 1 in the following gas-dynamic relationships for the stagnation temperature and pressure.

$$\frac{T_1}{T_0} = 1 + \frac{\gamma-1}{2} M_0^2 \tag{2.52a}$$

$$\frac{p_1}{p_0} = \left(1 + \frac{\gamma-1}{2} M_0^2\right)^{\frac{\gamma}{\gamma-1}} \tag{2.52b}$$

The assumption is that the upstream flow speed is small or known so that the stagnation conditions, T_1 and p_1, can be determined, which then give the mass flow rate under choked conditions:

$$\dot{m} = \frac{p_1 A_0}{\sqrt{RT_1}} \sqrt{\gamma} \left(\frac{\gamma + 1}{2} \right)^{\frac{\gamma+1}{2(\gamma-1)}} \tag{2.53}$$

R = specific gas constant of the fluid
A_0 = area at the throat

It is noted that this mass flow rate is obtained by substituting $M_0 = 1$ in

$$\dot{m} = \rho_0 U_0 A_0 = \frac{p_0 A_0}{RT_0} M_0 a_0 = \frac{p_0 A_0}{RT_0} M_0 \sqrt{\gamma RT_0} = \frac{p_1 A_0}{\sqrt{RT_1}} \frac{\sqrt{\gamma} M_0}{\left(1 + \frac{\gamma - 1}{2} M_0^2 \right)^{\frac{\gamma+1}{2(\gamma-1)}}} \tag{2.54}$$

a_0 = sonic velocity at the throat

The flow rate can be controlled by changing the upstream pressure (and temperature), or it can be determined from the upstream conditions using Equation 2.53.

2.4.2 Rotameters

Rotameters are widely used for gas or liquid flow rate measurements, and they provide flow rate data with errors of ±1% to 10%. Figure 2.26 shows the operating principle of rotameters. A "float" is subject to the forces of gravity (W), aerodynamic drag due to the flow through the tube (F_D), and buoyancy (F_B). During operation, the

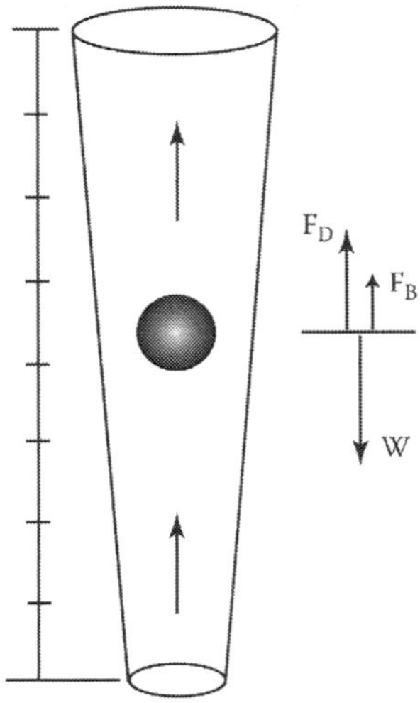

FIGURE 2.26 The basic principle of rotameter operation.

float comes to rest at a point in the rotameter where these forces balance one another (the sum of the forces equals zero, resulting in equilibrium; i.e., the position of the float is stationary).

$$\sum F_y = 0 = -W + F_D + F_B \tag{2.55}$$

For a tapered tube used in rotameters, the aerodynamic force depends on the annular area between the float and the tube wall because the average flow velocity, U, depends on the annular area. Thus, a force equilibrium is reached at a point where the aerodynamic force, due to the flow velocity, matches the weight minus the buoyancy force.

$$F_D = \frac{1}{2}\rho_g U^2 C_D A_f = W - F_B = \rho_f g V_f - \rho_g g V_f = g V_f \left(\rho_f - \rho_g\right) \tag{2.56}$$

ρ_g = fluid density
U = average fluid velocity at a vertical position in the rotameter
C_D = drag coefficient of the float in the fluid
A_f = cross-sectional area of the float
ρ_f = density of the float material
g = gravitational acceleration
V_f = volume of the float

At the condition corresponding to Equation 2.56, the flow rate, Q, is

$$Q = UA_a = A_a C \sqrt{\frac{2gV_f\left(\rho_f - \rho_g\right)}{A_f \rho_g}} \tag{2.57}$$

A_a = the annular area between the float and the rotameter wall
C = constant related to the square root of C_D.

We can see from Equation 2.57 that both A_a and U are functions of the vertical location of the rotameter tube, so a correlation between the flow rate and the vertical position can be set up for a given fluid, float geometry, material, and rotameter tube design. Moreover, the float geometry, material, and tube design can be selected so that they result in a linear relationship between the flow rate and the float vertical's position.

2.4.3 Turbine Flowmeters

Turbine flowmeters are relatively simple devices where fans or rotors are inserted into a typically circular flow section. The aerodynamic or hydrodynamic force on the rotor element is converted to angular momentum or rotation that can be picked up by an internal or external sensor. For example, an external pickup coil sensor can detect the frequency of passage of the magnetic rotor element, which can be directly converted to the rotational speed. Because the aerodynamic force acting on the rotor element depends on the density, velocity, and swirl, the effects of density and swirl need to be compensated for or eliminated.

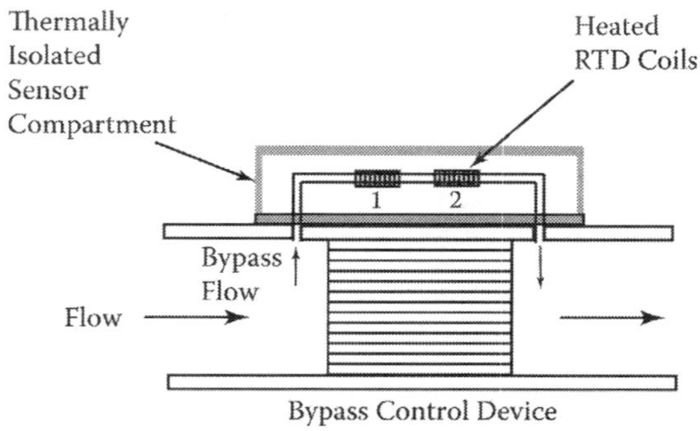

FIGURE 2.27 Operating principle of a thermal mass flowmeter.

2.4.4 Thermal Mass Flowmeters

Thermal mass flowmeters are based on the first law of thermodynamics and Newton's law of cooling for convective heat transfer. Figure 2.27 shows a schematic of a thermal mass flowmeter. A small, fixed portion of the flow is bypassed to the sensor tube by a control device, as shown in Figure 2.27. Along the sensor tube, there are RTD coil elements that are used both to add heat to the flow and to measure the temperature at each point. With a constant current passing through the RTD coil, the resistance heating will add heat to the flow. The RTD elements also measure the resistance change due to the temperature difference arising from the heat addition. The temperature difference between the two RTDs depends on the mass flow rate of the fluid passing through the two points.

The energy balance for this process illustrates this point.

$$\dot{m}c_p(T_2 - T_1) = \dot{Q}_1 \tag{2.58}$$

$\dot{m}$ = mass flow rate
c_p = specific heat of the fluid
Q_1 = rate of heat transfer to the fluid at 1

For a fixed current and temperature at the RTD at 1, the heat transfer rate will be constant because the heat dissipation by electrical resistance is IR^2, where I is the current. Also, for a given fluid at a constant temperature, the specific heat is constant. The mass flow rate can be directly related to the temperature difference as read by the RTDs at 1 and 2. The question is, what happens if the temperature of the fluid entering the sensor tube changes, or if a different fluid is used in the flowmeter, which would change the specific heat? Because the RTD at 1 can measure the fluid temperature, the temperature effect can be digitally or electrically compensated, whereas the variation in the specific heat needs to be compensated for or calibrated for each fluid used in thermal mass flowmeters. Once these effects are accounted for, thermal mass flowmeters are capable of good accuracy down to ±0.5%. In addition, the output signal can easily be interfaced with digital data acquisition and control units.

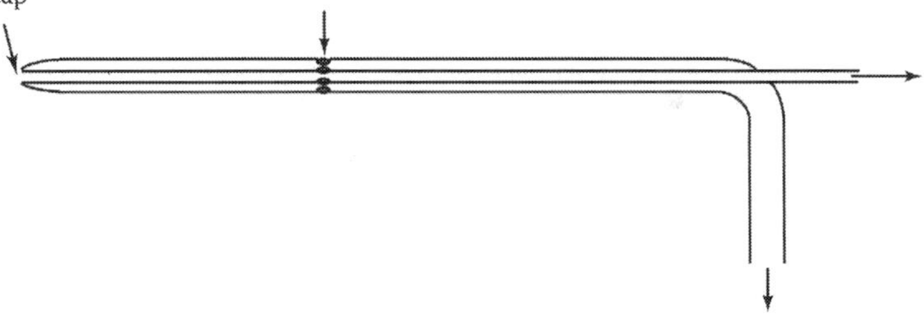

FIGURE 2.28 A Pitot-static tube for measurements of total and static pressures.

2.4.5 Flow Velocity Measurements Using Pitot and Static Pressure Probes

As discussed in Section 2.1.4, pressure probes measure either the static or the total pressure. By using two probes to measure both the static and the total pressure, the flow velocity can be retrieved. For incompressible flows, the relationship between pressure and velocity is

$$p_t = p + \frac{1}{2}\rho V^2 \rightarrow V = \sqrt{\frac{2(p_t - p)}{\rho}} \tag{2.59}$$

The limit of applicability for Equation 2.59 is usually taken as M~0.3 or below, where M is the Mach number (M = V/a, where a is the speed of sound). For high-speed compressible flows (M > 0.3), the isentropic compression effect at the Pitot tube tip needs to be accounted for. From gas dynamics, the correction for isentropic compression leads to the following expression for the velocity in terms of the static and total pressures.

$$V = \sqrt{2\frac{\gamma}{\gamma - 1}\left(\frac{p_t}{p}\right)\left[\left(\frac{p_t}{p}\right)^{\frac{\gamma-1}{\gamma}} - 1\right]} \tag{2.60}$$

Usually, the total pressure probe (the Pitot tube) is combined with static pressure ports in a single probe housing, as shown in Figure 2.28.

PROBLEMS

2.1. Convert the following gauge pressures into absolute pressure in the standard unit of pressure (kPa), if the atmospheric pressure is 101.325 kPa.
   a. 2500 psi
   b. 25.7 bars
   c. 776 mmHg
   d. 589.17 inches of H_2O

2.2. For a well-type manometer using water inclined at 45°, the liquid-column position increases by 2.5 cm on the inclined plane. What is the change in pressure?

2.3. A pressure-transducer diaphragm is made of a metal plate with a Young's modulus of 212 GPa and a Poisson's ratio of 0.25. It is designed to measure a maximum absolute pressure of 10 MPa, using a diaphragm thickness of 0.25 mm and a radius of 10 mm.

a. Determine the maximum deflection of the diaphragm.
b. For this pressure-transducer diaphragm, what is the natural frequency?
c. What happens to the natural frequency when the diaphragm thickness is increased by a factor of two?
d. What happens to the natural frequency when the diaphragm radius is decreased by a factor of two?

2.4. For a piezoelectric pressure transducer, the voltage output is given by Equation 2.13.

$$E = \frac{q}{C} = \frac{K_c pt}{c\varepsilon} = K_p pt$$

$$K_p = \frac{K_c}{c\varepsilon} = \text{ voltage sensitivity } \left[\text{Vm/N}\right]$$

If it is desired to have a voltage output of 25 mV for a pressure change of 10 kPa, using a 1 mm thickness piezoelectric element, what must be the voltage sensitivity, K_p?

2.5. What is the volumetric change of an initial 150 cm^3 of mercury when subject to a 100°C temperature increase? The volumetric expansion coefficient of mercury is $\beta = 18 \times 10^{-5}$/°C.

2.6. a. For a J-type thermocouple, what is the temperature if the voltage output is 57 mV?

b. Write a MATLAB program to convert the mV output of an R-type thermocouple to temperature in Celsius.

2.7. For a thermocouple probe, the heat transfer coefficient (h) is estimated to be 2.7 W/m^{2}K. For the surrounding temperature of $T_w = 650$ K, and thermocouple readout temperature of 1550 K, what is the radiation-corrected temperature? The emissivity of the thermocouple probe is 0.02.

2.8. In a high-speed flow of air, a thermocouple is used to measure the static temperature. For a recovery factor, *r*, of 0.75, plot the static temperature as a function of the flow speed from 25 to 100 m/s, for a thermocouple readout temperature of 1050 K.

$$T_{TC} = T_\infty + \frac{rU^2}{2c_p} = T_t + \frac{(1-r)U^2}{2c_p}$$

2.9. For a platinum RTD with ceramic coating, plot the resistance as a function of temperature from −10°C to 500°C.

$$R(T) = R_o\left[1 + \alpha(T - T_o)\right]$$

2.10. For a thermistor, the temperature coefficient, γ, is given as 3750 [K] for a reference resistance and temperature of 55 Ω at 100°C. Plot the thermistor resistance as a function of temperature from −10°C to 500°C.

$$R = R_o \exp\left[\gamma\left(\frac{1}{T} - \frac{1}{T_o}\right)\right]$$

2.11. The ratio of intensities at two different wavelengths in a two-color pyrometer is given by Equation 2.34. If the same bandwidths are used at two wavelengths of 1.5 and 2.5 μm with respective surface emissivity of 0.37 and 0.40, what is the temperature based on the intensity ratio of 1.17?

$$R = \frac{I_{\lambda A}\Delta\lambda_A}{I_{\lambda B}\Delta\lambda_B} = \frac{\varepsilon_{\lambda A}}{\varepsilon_{\lambda B}}\frac{\Delta\lambda_A}{\Delta\lambda_B}\frac{\lambda_B^5}{\lambda_A^5}\exp\left(-\frac{C_2}{T}\left(\frac{1}{\lambda_A} - \frac{1}{\lambda_B}\right)\right) \tag{2.34}$$

2.12. In a nozzle-type obstruction flow meter with a discharge coefficient of 0.95, a pressure drop of 20 kPa is observed for an area ratio of $A_2/A_1 = 0.25$ and a nozzle diameter of $D_2 = 10$ mm. The density effect may be neglected. What is the flow rate for an air flow at $T = 300$ K and $p = 1$ atm?

2.13. In a sonic orifice flowmeter, the upstream pressure is fixed at 2 MPa at a temperature of 300 K. For choked air flow through an orifice with a diameter of 0.75 mm, determine the flow rate using Equation 2.51.

$$\dot{m} = \frac{p_1 A_0}{\sqrt{RT_1}}\sqrt{\gamma}\left(\frac{\gamma+1}{2}\right)^{\frac{\gamma+1}{2(\gamma-1)}} \tag{2.51}$$

2.14. A rotameter flow rate is given by Equation 2.55. The calibration constant, C, is estimated to be 0.175. If the rotameter tube diameter varies as a function of height as $D(y) = 10 + 0.0075y$ [mm], plot the flow rate as a function of height, y, from $y = 0$ to 200 mm. The spherical float diameter is 9 mm, with a density of 497 g/cm³, and the fluid is air at $T = 300$ K and $p = 1$ atm.

$$Q = UA_a = A_a C\sqrt{\frac{2gV_f\left(\rho_f - \rho_g\right)}{A_f\rho_g}} \tag{2.55}$$

BIBLIOGRAPHY

American Society of Mechanical Engineers, *Flow Meters: Theory and Applications*, 6th ed., ASME, New York, 1971.

Beckwith, T.C., Marangoni, R.D., and Lienhard, J.H., *Mechanical Measurements*, 5th ed., Addison-Wesley, Reading, MA, 1993.

Figliola, R.S. and Beasley, D.E., *Theory and Design for Mechanical Measurements*, 4th ed., John Wiley, New York, 2006.

3 Flow Visualization and Image Analysis

Visualization renders direct and intuitive realizations of the thermal and flow processes, as illustrated by the examples shown in Figures 3.1–3.3. Figure 3.1 shows a 1/48-scale model of an F-18 aircraft being tested in a water tunnel at the NASA Dryden Flow Visualization Facility. To simulate the Reynolds number encountered by the actual aircraft, water with a much higher density than air is used. Water also allows for the easy introduction of liquid dyes, the streak patterns of which are imaged in Figure 3.1. The aerodynamics of the flow over the aircraft body can be easily seen at a high angle of attack. Figure 3.2 shows a sequence of high-speed digital images of flame propagating in a turbulent flow. Each row of images is taken at a fixed equivalence ratio (or fuel–air ratio), and the turbulence intensity is increased from left to right. Some of the changes in the flame topology can be observed, where a higher level of turbulence causes smaller scales of flame corrugation or so-called *flame wrinkles.* Figure 3.3 illustrates a somewhat more advanced technique using *laser-induced fluorescence* (LIF) imaging. Either a fluorescent dye can be injected into the fluid or, as in Figure 3.3, nitric oxide (NO) and hydroxyl (OH) molecules that are formed in the flow are optically induced to emit a fluorescence signal. The local concentrations of these molecules, along with temperature, are imaged when a laser light tuned to the excitation wavelengths of these molecules is sent in as a light

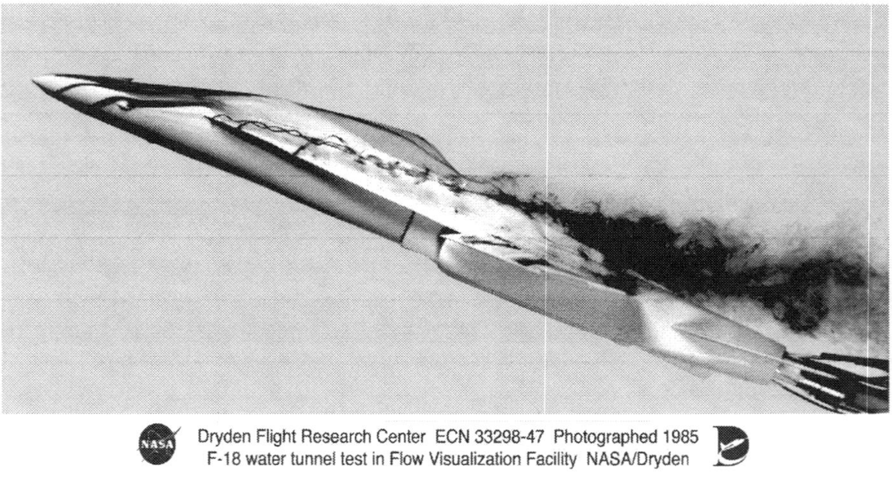

FIGURE 3.1 An example of flow visualization using liquid dye streaks. (Photo courtesy of NASA Dryden. Ref.: http://www.dfrc.nasa.gov/Gallery/Photo/FVF/HTML/ECN-33298-03.html.)

DOI: 10.1201/9781003611837-3

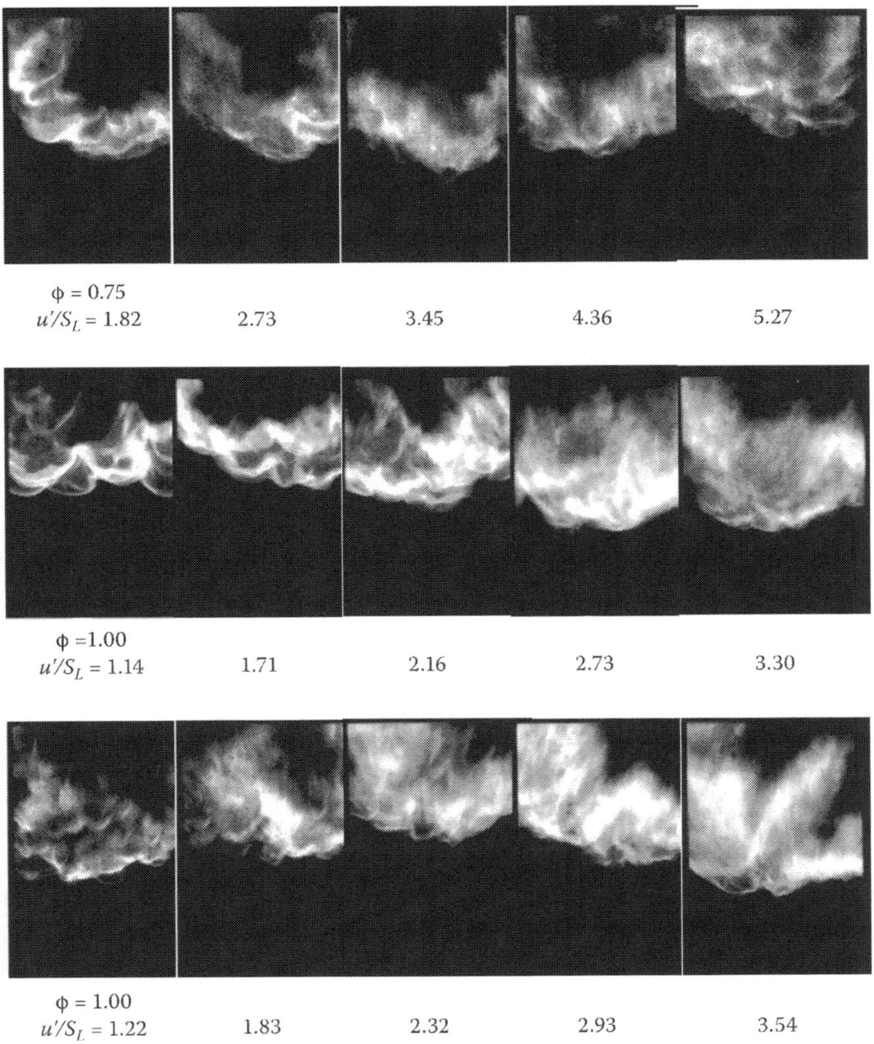

FIGURE 3.2 Photographic images of turbulent premixed flames at various conditions.

sheet. The temperature field can also be imaged from the Rayleigh scattering signal of the molecules. Details of the fluorescence and Rayleigh scattering processes for diagnostics are discussed in Chapter 5 and also briefly in this chapter.

As shown in Figures 3.1–3.3, two-dimensional views of the flow phenomena can provide immediate insight and qualitative understanding of the overall flow structure. In addition, modern image-processing and analysis tools allow quantitative data to be extracted from images if the data are acquired through appropriate means. For this reason, in many thermal-fluid experimentation and testing, the first diagnostic applied is flow visualization. In other instances, the visualization diagnostic may be the primary source of information about the process.

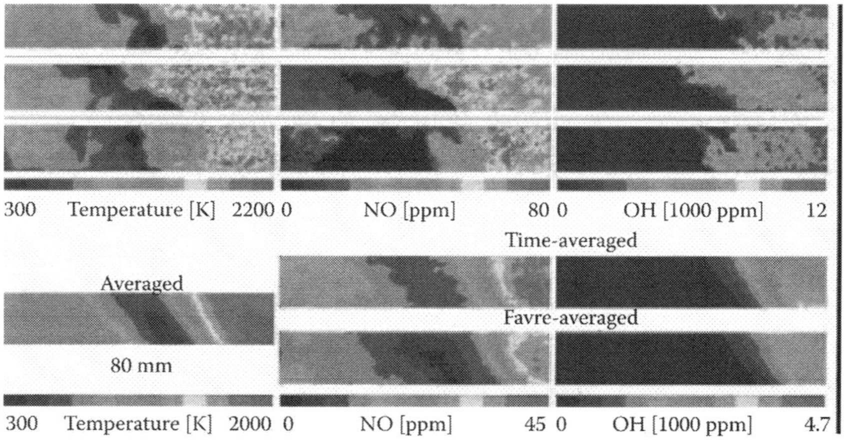

FIGURE 3.3 Species concentration imaging using laser-induced fluorescence (LIF). (From Bockle, S., Kazenwadel, J., Kunzelmann, K., and Schulz, C., Laser-diagnostic multi-species imaging in strongly swirling natural gas flames, *Applied Physics* B, 71, pp. 741–746, 2000. Images used with permission from Springer-Verlag.)

The visualization method consists of a light source, flow tracers (e.g., fluorescent dye, seeded particles), and an imaging device (camera). The tracers scatter incident light from the light source or are induced to emit light that can be detected by imaging devices. For some visualization methods, such as shadowgraph and Schlieren, the density variation in the flow produces discernible images, and thus there is no need for additional flow tracers. For remote sensing applications, satellite-derived images of the Earth's surface or atmosphere rely on solar radiation as the light source. For photographic imaging, the flow process emits light directly, as in flames (Figure 3.2). There are a large number of commercially available imaging devices that allow for subsequent image processing, analysis, and storage.

The following is a list of visualization methods commonly used for thermal and flow processes:

- Direct photography
- Shadowgraph/Schlieren imaging
- Interferometry (Mach–Zehnder interferometry, holography)
- Laser tomographic imaging using Mie-scattering, fluorescence, or other laser-induced emission

3.1 STREAMLINES, STREAKLINES, AND PATHLINES

Let us begin by examining the different flow characteristics that can be observed through flow visualization.

The *streamline* is a classical concept in fluid mechanics; it represents a line drawn tangential to the local velocity vector at a given instant, as shown in Figure 3.4a. Thus, the equation for the streamline can be obtained from the following definition:

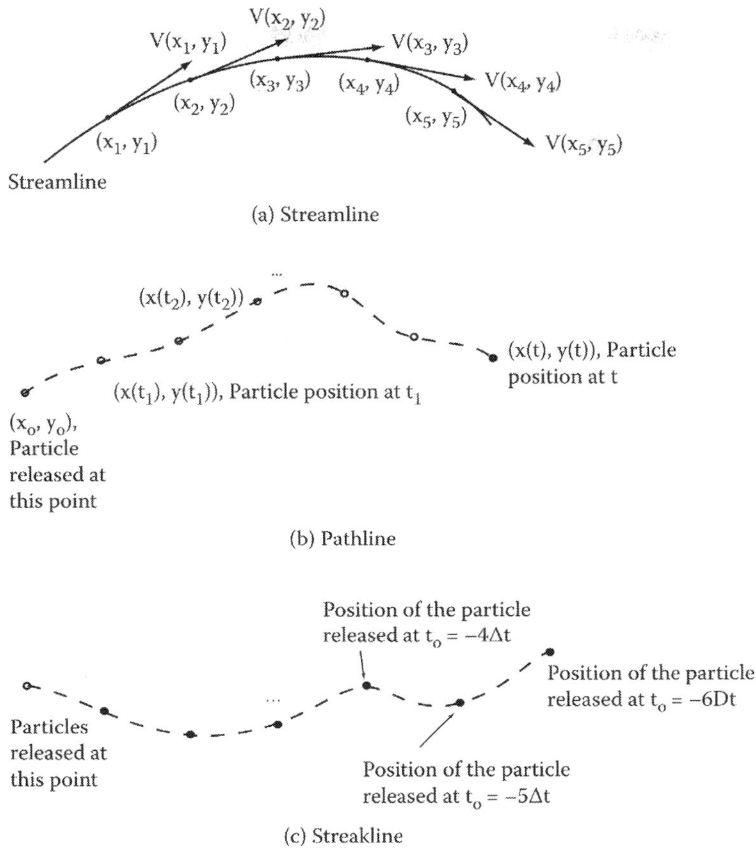

FIGURE 3.4 Streamlines, streaklines, and pathlines.

$$\frac{dx}{u} = \frac{dy}{v} = \frac{dz}{w} \tag{3.1}$$

u = x-component of the velocity vector
v = y-component of the velocity vector
w = z-component of the velocity vector

From Equation 3.1, we can proceed to write the x–y components.

$$\frac{dy}{dx} = \frac{v}{u} \tag{3.2}$$

Knowing the magnitudes of u and v from flow visualization results, the streamline can be determined using Equation 3.2. For example, Figure 3.5 shows an image of particles seeded in the flow exposed for a short duration (10 ms). Thus, each particle image appears as a line with a length corresponding to the local flow velocity times

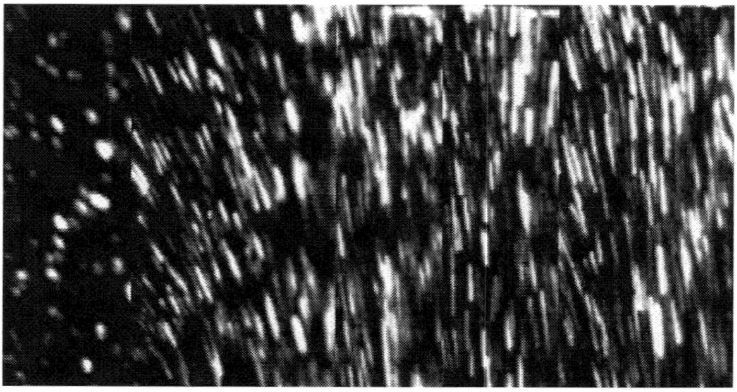

FIGURE 3.5 A flow visualization image from which streaklines may be extracted.

the time duration. The particles are small enough to follow the local flow velocity, in which case each line approximates the local streamline. Because the particle seeding density is limited, we obtained intermittent information on the streamlines, which can be remedied by taking many images and averaging them if the flow remains steady-state.

The *pathline* is defined as the line traced by a given particle released in the flow, such as a dye drop injected into the flow from a fixed point (see Figure 3.4b). Thus, pathlines can be obtained by integrating the velocity components over time.

$$
\begin{aligned}
x(t) &= x_o + \int_{t_o}^{t} u(x, y, \tau)\, d\tau \\
y(t) &= y_o + \int_{t_o}^{t} v(x, y, \tau)\, d\tau
\end{aligned}
\tag{3.3}
$$

It can be noticed that if the flow is steady-state, the flow velocity vector components are constant over time. The velocity components in the integrals in Equation 3.3 are also constant, and by taking the differential of the x- and y-component equations and dividing them, we obtain the same Equation 3.2. Therefore, for steady-state flows, the streamlines and pathlines are identical.

Streaklines are slightly different in concept. They are the loci of positions of particles sequentially released from a fixed point, as illustrated in Figure 3.4c. Thus, a streakline is something that is observed if one injects dye particles, for example, from a fixed point and takes an image of the resulting dye *streak* after some time. Again, for the same reason that streamlines and pathlines are identical for steady-state flows, streaklines, streamlines, and pathlines all become identical under such conditions.

Example 3.1

The velocity field is described by the *x*- and *y*-components of the velocity.

$$u = U = \text{const. for } y > 0$$

$$v = V\left(1 - \frac{y}{Y}\right) \quad \text{for } 0 < y < Y$$

Then, the streamline is prescribed by Equation 3.2.

$$\frac{dy}{dx} = \frac{v}{u} = \frac{V\left(1 - \frac{y}{Y}\right)}{U} \rightarrow \int_{y=y_o}^{y} \frac{dy}{1 - \frac{y}{Y}} = \frac{V}{U}\int_{x=x_o}^{x} dx$$

Evaluating the integrals, we get

$$-Y \ln\left[\frac{1 - \frac{y}{Y}}{1 - \frac{y_o}{Y}}\right] = \frac{V}{U}(x - x_o)$$

For a fluid particle passing through $(x_o, y_o) = (0, 0)$, the streamline is

$$\frac{y}{Y} = 1 - \exp\left(-\frac{V}{U}\frac{x}{Y}\right)$$

Example 3.2

Find the *x*–*y* relationship for streamline, streakline, and pathline for a fluid particle passing through $(x_o, y_o) = (1, 1)$ at $t_o = 0$. The time-dependent velocity field is

$$\vec{u} = 5\cos\frac{2\pi t}{T}\vec{i} + 1\vec{j}$$

The streamline is again determined using Equation 3.2.

$$\frac{dy}{dx} = \frac{v}{u} = \frac{1}{5\cos\frac{2\pi t}{T}}$$

The right-hand side contains neither *x* nor *y*, and the integration gives

$$y - y_o = \frac{1}{5\cos\frac{2\pi t}{T}}(x - x_o)$$

Because $(x_o, y_o) = (1, 1)$, the streamline is

$$y = 1 + \frac{1}{5\cos\frac{2\pi t}{T}}(x - 1)$$

Pathline involves integration in time, as in Equation 3.3.

$$u = \frac{dx}{dt} = 5\cos\frac{2\pi t}{T} \rightarrow x - x_o = \frac{5T}{2\pi}\left(\sin\frac{2\pi t}{T} - \sin\frac{2\pi t_o}{T}\right)$$

$$v = \frac{dy}{dt} = 1 \rightarrow y - y_o = t - t_o$$

Using $(x_o, y_o) = (1, 1)$ at $t_o = 0$ gives $y - 1 = t$, and

$$x - 1 = \frac{5T}{2\pi}\sin\frac{2\pi(y-1)}{T}$$

For streaklines, we can start from

$$x - x_o = \frac{5T}{2\pi}\left(\sin\frac{2\pi t}{T} - \sin\frac{2\pi t_o}{T}\right)$$

$$y - y_o = t - t_o$$

However, streakline is a trace of particles released at different t_o viewed at $t = 0$. Thus, the streaklines are given by

$$x - 1 = -\frac{5T}{2\pi}\sin\frac{2\pi t_o}{T}$$

$$y - 1 = -t_o$$

Combining the foregoing two equations, we obtain

$$x - 1 = -\frac{5T}{2\pi}\sin\frac{2\pi(1-y)}{T}$$

3.2 DIRECT PHOTOGRAPHY USING FLOW TRACERS

As discussed previously, what we most commonly observe are the streaklines when tracers such as dyes, smoke, or bubbles are injected continuously or intermittently into the flow. Pathlines are observed only when the individual particle images are exposed over time so that the particles appear as streaks in the image. Direct photography is used in these instances, where images can be obtained using film, or more commonly nowadays, digital imaging devices. Direct photography works if there is sufficient luminescence emanating from the object, e.g., flame luminescence as in Figure 3.2, or luminescence from injected dyes as in Figure 3.1. Bubbles, smoke particles, or other flow tracers illuminated with a light source are also used in conjunction with direct photography.

Conventional dye material can be used for flow visualization in water flows, as shown in Figure 3.1. A simple substance, such as a food dye, can be used because it is readily available and safe. However, most types of food dyes in concentrated solutions have a specific gravity (density relative to that of water) greater than one, so methanol or ethanol in small amounts is mixed with the dye solution to make it neutrally buoyant. Other common substances used as tracers in water include laundry brightener and milk. Fluorescent dyes refer to a class of materials that emit light when illuminated by a specific light source, typically a laser light sheet. Thus, a two-dimensional view of the flow structure is possible by directing the laser light sheet to the desired plane. Fluorescent dyes are commercially available, although some of them tend to have certain handling requirements due to mild toxicity. Materials such as fluorescein (green), Rhodamine-B (red), and Rhodamine-G (yellow) emit respective fluorescent colors when illuminated with the output of an argon ion laser. Figure 3.3 is an example of fluorescence imaging. Even though only the fluorescence intensity, not the color, changes as a function of the fluorescent dye concentration, a pseudo-color scale can be generated to map different fluorescence intensities, and therefore different dye concentrations, using an image-processing technique discussed later in this chapter.

Injection of dyes typically involves a hypodermic needle adapted to the test apparatus. To avoid flow disturbances from the dye injection probe, it is placed upstream of the region of interest.

Smoke can be produced in air flows by heating a thin coating of liquid materials such as paraffin, kerosene, lubricating oil, or silicone oil over an electrically heated wire. Typically, a 24-DC power supply with 0.5 to 0.8 amp sent through a stainless-steel wire of 0.1 mm diameter produces a good cloud of oil smoke particles. The thin coating of oil flows over from a reservoir using gravity or other simple means. An alternative method of producing oil smoke is by using venturi-type atomizers (these are also used in laser velocity measurements; see Figure 4.14).

Another method for producing smoke is to inject titanium tetrachloride ($TiCl_4$), which reacts with the moisture in the air to produce titanium dioxide (TiO_2) particles and HCl fumes.

$$TiCl_4 + 2H_2O \rightarrow TiO_2 + 4HCl \tag{3.4}$$

TiO_2 particles are quite visible when illuminated with a light source; however, the presence of HCl fumes in the product gas is a safety issue. For modest moisture content in ambient air, the reaction in Equation 3.4 readily occurs. If dry air is used, then the appearance of H_2O in the flow, as in combustion flows behind a flame, marks the boundary where TiO_2 particles are produced. This aspect can be used in flame imaging experiments to demarcate the flame boundary.

Hydrogen bubbles are used in water flow experiments. This method is similar to oil smoke generation using heated wires in gases. Electricity sent to a wire exposed to water produces hydrogen bubbles through electrolysis. It takes DC power of 50–300 V at a maximum current of 2 amp to produce a nearly uniform line of hydrogen bubbles over a wire length of up to 250 mm. Periodic application of voltage produces hydrogen bubble lines separated by the time interval. The evolution of such bubble lines is useful for observing the flow structure development in time and space.

3.3 SURFACE IMAGING USING THERMOCHROMATIC LIQUID CRYSTALS AND PRESSURE-SENSITIVE COATINGS

Surface temperature can be determined using a thermochromatic liquid crystal coating that changes color as a function of temperature. Although this method does not measure the temperature of the fluid directly, by setting up the experimental apparatus so that the flow is nearly two-dimensional, the temperature distribution in the flow can be inferred from surface measurements. When coated on a black surface, liquid crystals reflect different color components of the incident light in a repeatable function of temperature that can be calibrated. Therefore, thermochromatic measurements require a "white" light source, i.e., a broadband in the visible wavelength range, and a color imaging device. The measurable temperature range using commercially available liquid crystal coating is typically from –30 to about 120°C, with a resolution of 0.1°C and 0.1 mm. In that temperature range, the color change is highly nonlinear, and thus incremental temperature changes must be made during calibration, and the corresponding color (red, blue, and green intensities) needs to be recorded. Once this calibration look-up chart is made, an automated imaging analysis algorithm can convert the color data to temperature.

Pressure-sensitive coating or paint (PSP) works by emitting light whose intensity varies as a function of pressure. The electrons in the molecules of the PSP go into an excited state when illuminated with a specific wavelength from a laser or an ultraviolet lamp with a color filter. When the electrons return to the ground state, they emit photons back, following the conservation of energy. However, that downward energy transition can also occur when oxygen molecules in air collide with the PSP molecules. This process is called *quenching*, and it depends on the oxygen concentration in air, which in turn depends on pressure. The quenching subtracts from the number of PSP molecules that emit photons to return to the ground state. Therefore, the intensity of the emitted light from PSP, when illuminated with a constant level of excitation light, decreases with increasing pressure. Again, the intensity-versus- pressure relationship needs to be calibrated, and an inverse relationship is used.

$$\frac{p}{p_{\mathrm{REF}}} = A + B\frac{I_{\mathrm{REF}}}{I} \tag{3.5}$$

p_{REF}, I_{REF} = reference pressure, intensity measured at that pressure
A, B = calibration constants

The photo-luminescence process using PSP is temperature-dependent, and thus the calibration constants, A and B, in Equation 3.5 are also temperature-dependent. Care must be taken, therefore, to determine and maintain a constant temperature condition when using PSP.

There is also a class of coating material called shear-sensitive liquid crystal coating (SSLCC) that reflects different color components as a function of the surface shear stress. SSLCC is similar in molecular structure to thermochromatic liquid crystals and consists of cholesteric liquid crystal molecules with a twisted, helical structure. The "pitch" of the helical bends depends on temperature for thermochromatic liquid crystals, and on shear stress in the case of SSLCC.

3.4 SHADOWGRAPH AND SCHLIEREN IMAGING

Light rays become refracted at an interface where the refractive index changes (see Appendix C on Optics). This refraction occurs anywhere there is a change or gradient in the refractive index. When the density of the gas, such as air, changes, there is a small but finite change in the refractive index, as given by Equation 3.6. This is evident in the shimmering of light over hot surfaces or fire that we can observe with the naked eyes. Shadowgraph and Schlieren imaging work by the light refractions caused by such density variations in the flow. Flows with density variations, such as combustion flows, compressible flows with or without shocks, and convection flows with appreciable temperature gradients, can be visualized using these methods. More specifically, shadowgraph and Schlieren imaging are frequently used to determine the structure of compression, expansion, and shock waves in transonic and supersonic flows, flows with thermally driven convection, and flames where temperature change creates density gradients. Schlieren imaging, in particular, is most frequently used for visualizing shock waves in compressible flows at supersonic speeds. Because the images from shadowgraph and Schlieren originate from the second and first derivatives, respectively, of the density gradients, quantitative analyses of these images are, in principle, possible. In particular, Schlieren images can be digitally obtained using a method called *background-oriented Schlieren*, which yields spatial patterns of density gradient data. Integration of such density gradient data allows the density field to be extracted from the image.

As noted earlier, a change in the density of the fluid at a given point results in a change in the local refractive index. This change in the refractive index can be written as

$$n = n_o\left[1+\beta\left(\rho/\rho_o\right)\right] \tag{3.6}$$

β = coefficient for the refractive index change as a function of density
ρ = density of the medium
n_o = refractive index at the reference density
ρ_o = density of the medium at a reference condition (0°C, 1 std. atm.)

Some values of β are given in Table 3.1. Equation 3.6 leads to the following linear relationship between the density and refractive index gradients:

$$\frac{\partial \rho}{\partial y} = \frac{\rho_o}{\beta}\frac{\partial n}{\partial y} \tag{3.7}$$

Now we can examine how the density gradient and, therefore, the refractive index gradient can cause a change in the illumination beam path, resulting in an imaging effect. Figure 3.6 shows that each light ray traverses a distance, Δs, corresponding to the local speed of light during a time interval Δt.

$$\Delta s = c\Delta t = \frac{c_o}{n}\Delta t \tag{3.8}$$

TABLE 3.1
Coefficient for the Refractive Index Change (β)

Gas	β
Air	0.000292
Nitrogen	0.000297
Oxygen	0.000271
Helium	0.000036
Water vapor	0.000254
Carbon dioxide	0.000451

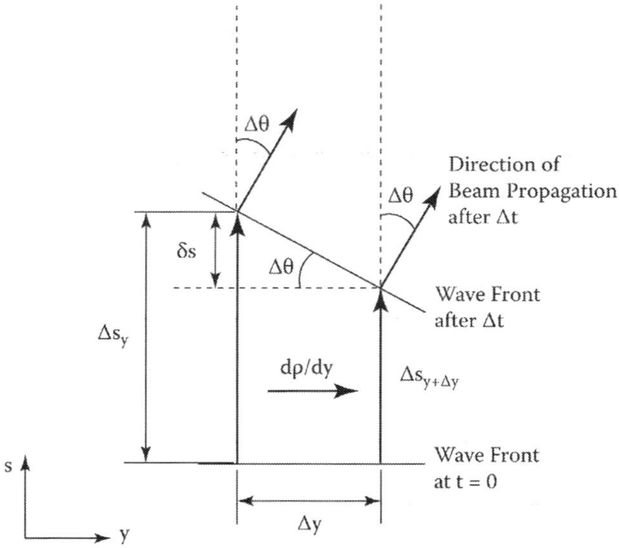

FIGURE 3.6 Effect of the density and refractive index gradient on the light paths.

$n = c_o/c$ = refractive index
c_o = speed of light in vacuum
c = local speed of light

If there is a density gradient in the y direction, and therefore a refractive index gradient, then n will vary as a function of y, as will the speed of light, c. This will result in different distances traversed; in other words, Δs will also vary with y. This incremental change can be written, after using Equation 3.8 for the last part of the equality, as

$$\delta s = \Delta s_y - \Delta s_{y+\Delta y} = -\frac{\partial(\Delta s)}{\partial y}\Delta y = -c_o \frac{\partial}{\partial y}\left(\frac{1}{n}\right)\Delta t\,\Delta y = -n\frac{\partial}{\partial y}\left(\frac{1}{n}\right)\Delta s\,\Delta y \tag{3.9}$$

Geometrically, the increment δs can be related to the angular deflection $\Delta\theta$.

$$\Delta\theta = \frac{\delta s}{\Delta y} \tag{3.10}$$

Substituting Equation 3.9 into Equation 3.10 gives us

$$\Delta\theta = -n\frac{\partial}{\partial y}\left(\frac{1}{n}\right)\Delta s = \frac{1}{n}\frac{\partial n}{\partial y}\Delta s = \frac{\partial(\ln n)}{\partial y}\Delta s \tag{3.11}$$

Equation 3.11 gives the rate of change of the light ray angle with respect to its traverse distance. For small increments, Equation 3.12 takes the differential form.

$$\frac{d\theta}{ds} = \frac{1}{n}\frac{\partial n}{\partial y} = \frac{\partial(\ln n)}{\partial y} \tag{3.12}$$

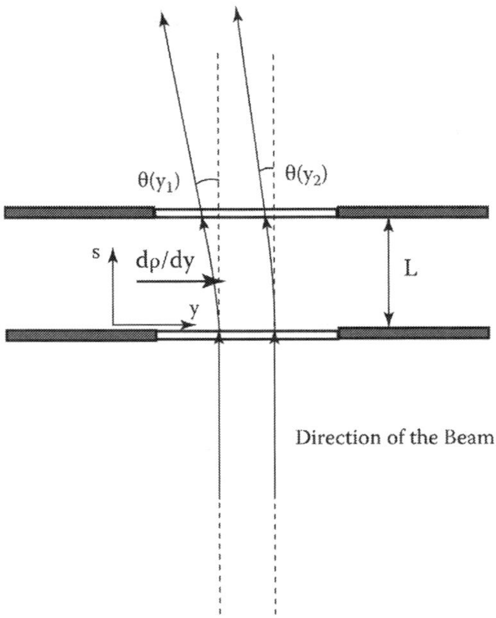

FIGURE 3.7 The light deflection for a test section of width L.

For a light beam traversing a test section of width L, as shown in Figure 3.7, the total beam deflection is calculated by integrating Equation 3.12.

$$\theta = \int_{s=0}^{L} \frac{1}{n}\frac{\partial n}{\partial y} ds \approx \frac{1}{n_o}\int_{s=0}^{L} \frac{\partial n}{\partial y} ds = \frac{1}{n_o}\frac{\beta}{\rho_o}\int_{s=0}^{L} \frac{\partial \rho}{\partial y} ds \tag{3.13}$$

n_o = refractive index of the air outside the test section

In Equation 3.13, we assume that the term $1/n$ is nearly constant and equal to $1/n_0$. That is, although the gradient in the vertical direction may be significant, we assume that the variation across a given light path is negligible. In fact, if we consider a purely two-dimensional flow where the gradient exists only in the y direction, we get $n_o \approx n$. Then,

$$\theta = \frac{\beta L}{n_o \rho_o}\frac{\partial \rho}{\partial y} \tag{3.14}$$

We use the foregoing aspects of light deflection due to transverse density gradients to analyze the principle of Schlieren imaging. Schlieren imaging is a method that produces an intensity pattern/image that depends on the local, integrated density gradient, according to Equation 3.13. A schematic of the Schlieren imaging setup is shown in Figure 3.8a. A light source is collimated and sent through the test section. On the other side of the test section, the light is converged and passes through a knife edge.

The resulting image intensity on the screen or imaging device depends on the transverse (to the direction of propagation) density gradients for the following reason.

Figure 3.8b shows the rectangular light beam from the source that has converged at the knife edge. The intensity of the light on the screen depends on how much of the light beam is intercepted by the knife edge.

$$I_R = \frac{Y}{Y_o} I_o \tag{3.15}$$

I_R = light intensity reaching the screen with no density gradient
I_o = undisturbed light intensity

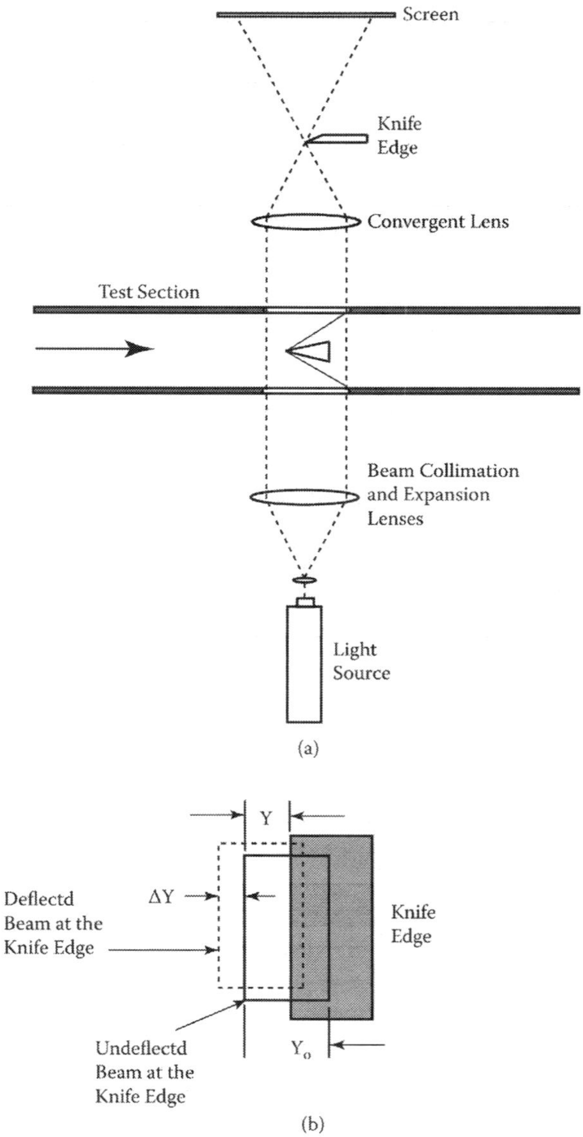

FIGURE 3.8 A schematic of the Schlieren imaging setup.

The height Y represents the portion of the light beam that passes over the knife edge when there is no light deflection due to density gradients. If there is a density gradient resulting in beam deflection, there will be a change in Y by an amount ΔY, as shown in Figure 3.8b. This will change the intensity reaching the screen.

$$I_D = \frac{Y + \Delta Y}{Y} I_R = \left(1 + \frac{\Delta Y}{Y}\right) I_R \tag{3.16}$$

I_D = light intensity reaching the screen under the density gradient
ΔY = the shift in the light beams due to the density gradient

Now, the shift ΔY is directly related to the total angular deflection, θ, and the focal length of the convergent lens, f_2.

$$\Delta Y = \theta f_2 \tag{3.17}$$

This gives the contrast of the image.

$$\text{Contrast} \equiv \frac{I_D - I_R}{I_R} = \frac{\Delta Y}{Y} = \frac{\theta f_2}{Y} = \frac{f_2 \beta}{Y n_o \rho_o} \int_{s=0}^{L} \frac{\partial \rho}{\partial y} ds \tag{3.18}$$

Depending on the sign of the density gradient, the contrast can be negative or positive. It should also be noted that, for a horizontal knife edge, the contrast follows the vertical density gradient, whereas for a vertical knife edge, the density gradient in the flow direction is resolved. In practice, the optimum image contrast is found by adjusting Y (by moving the knife edge by small amounts). Because large-aperture lenses, as shown in Figure 3.8a tend to be expensive, the lenses are often replaced with mirrors, as shown in Figure 3.9. An example of a Schlieren image is shown in Figure 3.10.

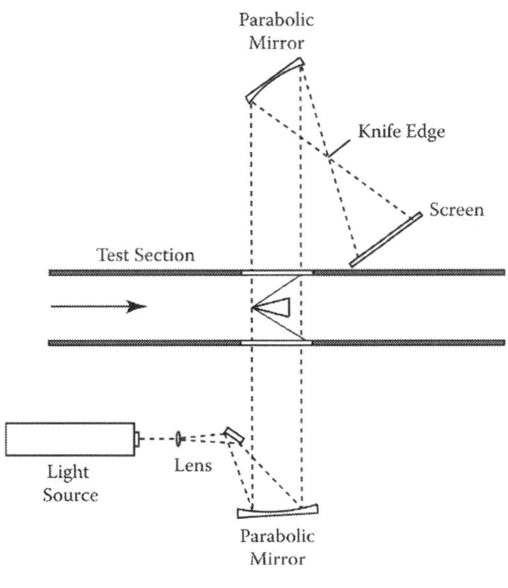

FIGURE 3.9 A Schlieren imaging setup using mirrors.

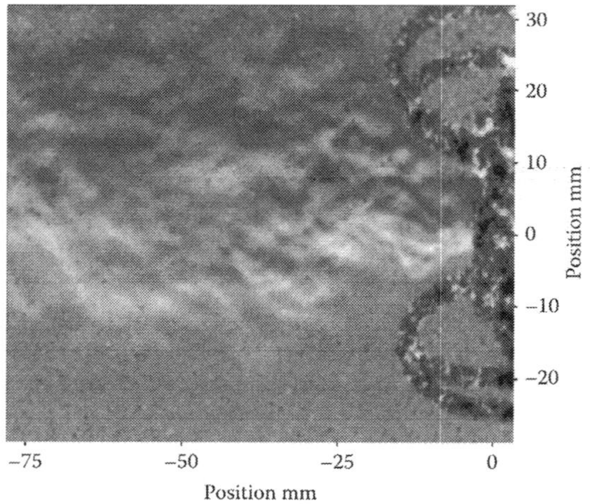

FIGURE 3.10 A Schlieren image of a helium jet injected into air.

The shadowgraph is a related technique, requiring a simpler optical setup, as shown in Figure 3.11a. Essentially, only a collimated light source and an imaging device are needed. A lens combination can be used to collimate the light, or a pair of parabolic mirrors, as shown in Figure 3.11a. Either a screen at "6" can be used to image the shadowgraph, or the beam can be recollimated for direct imaging by a digital camera, as shown in Figure 3.11a. A spatial filter is sometimes used to remove noise in the laser beam. Although laser light sources are preferred, any light source that can be collimated will work. In fact, we can often visually observe the shadowgraph effect on hot surfaces, resulting in shimmering of the natural light passing through the hot region.

If the light intensity for a segment of Δy is I_o upstream of the test section, then the light reaching the screen will depend on how much the beam has diverged.

$$I_D = \frac{\Delta y}{\Delta y_D} I_o \tag{3.19}$$

Δy = beam height at the test section
Δy_D = beam height at the screen

Now, beam divergence occurs due to the difference in light deflection from one point in the test section to another (e.g., from left to right of Δy in Figure 3.11b). We can write

$$\Delta y_D = \Delta y + s\Delta\theta \tag{3.20}$$

s = distance from the test section to the screen
$\Delta\theta$ = differential beam deflection

Again, if we write the contrast in the intensity, it will be

$$\text{Contrast } = \frac{I_D - I_o}{I_o} = \frac{\Delta y}{\Delta y_D} - 1 = \frac{\Delta y_D - s\Delta\theta}{\Delta y_D} - 1 = -s\frac{\Delta\theta}{\Delta y_D} \rightarrow -s\frac{\partial\theta}{\partial y} = -\frac{s\beta}{n_o\rho_o}\frac{\partial^2\rho}{\partial y^2} \tag{3.21}$$

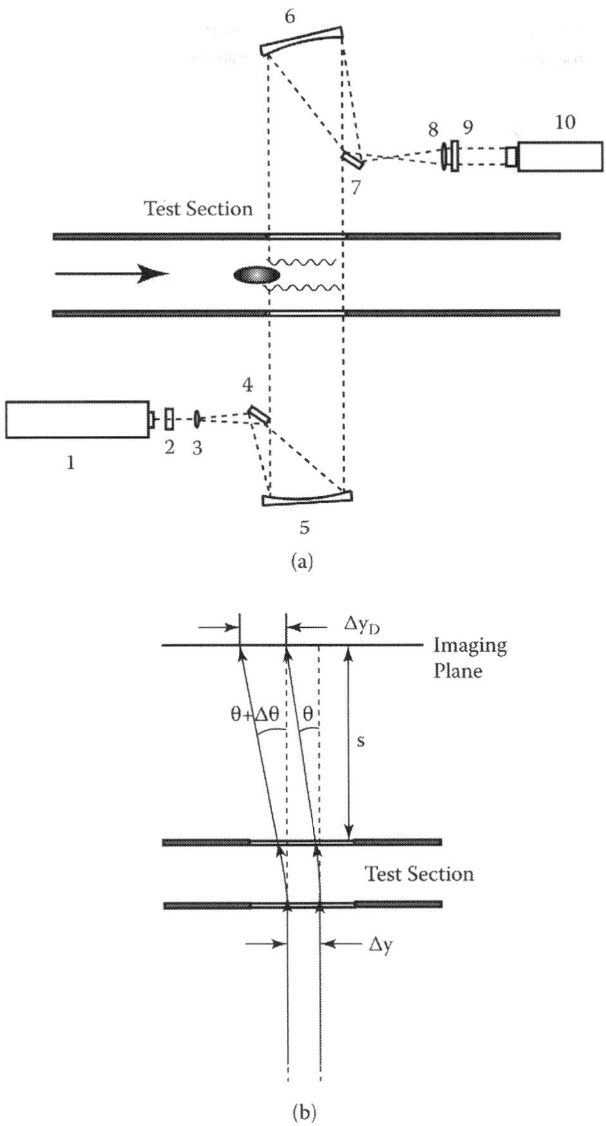

FIGURE 3.11 (a) A shadowgraph optical setup; (b) light deflection leading to image contrast on the shadowgraph imaging plane. 1: light source (typically a He-Ne or Argon Ion laser); 2: spatial filter (cleans out the beam noise); 3: lens to diverge the beam; 4: flat mirror; 5: parabolic mirror to collimate the beam; 6: second parabolic mirror to refocus the beam; 7: flat mirror; 8: collimating lens; 9: bandpass filter to reject noise; 10: digital camera.

Therefore, the shadowgraph image contrast is sensitive to the gradient of the deflection angle, or the second derivative of the density gradients. Noting that the light deflection can occur in two transverse directions perpendicular to the beam path, we have

$$\text{Contrast} = -\frac{s}{n_o}\frac{\beta}{\rho_o}\int_{s=0}^{L}\left(\frac{\partial^2 \rho}{\partial x^2}+\frac{\partial^2 \rho}{\partial y^2}\right)ds \tag{3.22}$$

Because both transverse components are impressed on shadowgraph images and the contrast is in a second-derivative form, shadowgraph images are difficult to analyze quantitatively but are nonetheless useful for qualitative visualization of small density variations in the flow.

3.5 INTERFEROMETRY

Interferometry refers to a class of imaging methods that use the interference of two coherent beams. Some mathematical treatments of the interference phenomena are given in Appendix B on Optics. One interference imaging method that is commonly used is Mach–Zehnder interferometry. A schematic of the setup is shown in Figure 3.12. As described in detail in Appendices B and C (see Equations B.34 and C.20a), propagating light beams can be mathematically prescribed in terms of the electric field amplitude, which oscillates sinusoidally in time and space.

$$\begin{aligned} E_1 &= E_{o1}\sin\frac{2\pi}{\lambda}(ct - x + \delta_1), \\ E_2 &= E_{o2}\sin\frac{2\pi}{\lambda}(ct - x + \delta_2) \end{aligned} \tag{3.23}$$

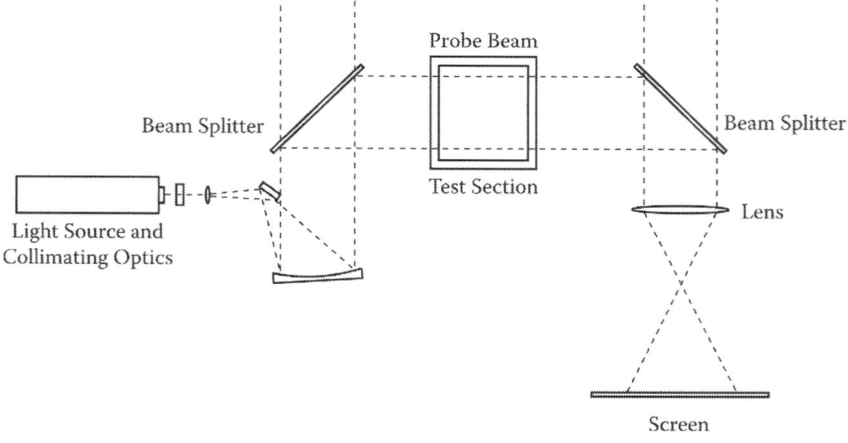

FIGURE 3.12 A setup for the Mach–Zehnder interferometry.

E_{o1}, E_{o2} = amplitudes of beams 1 and 2
λ = wavelength
c = speed of light
t = time
x = coordinate in the direction of propagation
$2\pi\delta_1/\lambda$, $2\pi\delta_2/\lambda$ = phase for beams 1 and 2

When the two beams merge, as in Figure 3.12, the electric fields are superposed.

For $\delta_1 = 0$, $\delta_2 = \delta$, and $E_{o1} = E_{o2} = E_o$, we have the combined electric field of

$$E = E_o\left[\sin\frac{2\pi}{\lambda}(ct - x) + \sin\frac{2\pi}{\lambda}(ct - x + \delta)\right] \tag{3.24}$$

Using a trigonometric identity $\sin(A + B) + \sin(A - B) = 2\sin A\cos B$, with $A = 2\pi/\lambda(ct - x + \delta/2)$ and $B = -\pi\delta/\lambda$,

$$E = 2E_o\cos\frac{\pi\delta}{\lambda}\sin\frac{2\pi}{\lambda}\left(ct - x + \frac{\delta}{2}\right) \tag{3.25}$$

The light intensity is proportional to the square of the electric field. Averaging over time will yield a constant from the sine term in Equation 3.25, so that the intensity becomes a function of the phase term, δ.

$$I \sim 4E_o^2\cos^2\frac{\pi\delta}{\lambda} = 2E_o^2\left(1 + \cos\frac{2\pi\delta}{\lambda}\right) \tag{3.26}$$

The phase term, δ, is the path length difference arising from the different light paths taken by beam 1 and beam 2 in Figure 3.12.

Now, the transit time for beams 1 and 2 will differ as they pass through different density fields and, therefore, different refractive index fields. For a total path length of L, the difference in the transit time is

$$\Delta t = \frac{L}{c_2} - \frac{L}{c_1} = \frac{L}{c_0}(n_2 - n_1) \tag{3.27}$$

There will be a corresponding difference in the optical path, which is the phase term, δ

$$\Delta x = c_0\Delta\, t = L(n_2 - n_1) \equiv \delta \tag{3.28}$$

Using Equations 3.6, 3.26, and 3.28, any difference in the density fields for beams 1 and 2 will cause fringes with a spacing of

$$\delta = L\beta\left(\frac{\rho_2 - \rho_1}{\rho_o}\right) \tag{3.29}$$

Thus, this fringe spacing is larger for a larger departure from the reference density, ρ_1. The reference beam is typically sent through a part of the test section with no density gradients. Figure 3.13 shows an example of a density field imaged using Mach–Zehnder interferometry.

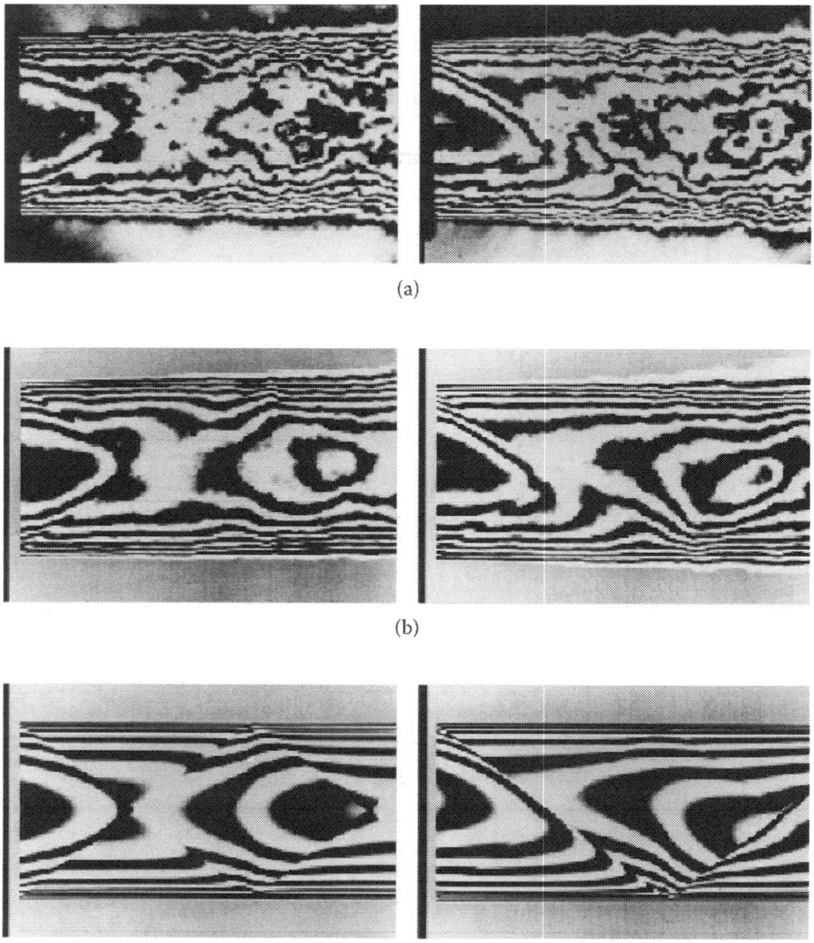

(a)

(b)

(c)

FIGURE 3.13 Density field imaged by using Mach–Zehnder interferometry. (From Dillmann, A., Wetzel, T., and Soeller, C., Interferometric measurement and tomography of the density field of supersonic jets, *Experiments in Fluids*, 25, pp. 375–387, 1998. Images used with permission from Springer-Verlag.)

3.5.1 Holographic Imaging

Phase information can be recorded on an imaging medium using holography. Holographic imaging differs from standard photography in that both amplitude and phase information are stored, allowing three-dimensional objects or structures to be recorded and reconstructed (photography records only amplitude). Holography involves two steps: (1) holographic recording using the optical arrangement shown in Figure 3.14a, and (2) holographic reconstruction using the setup shown in Figure 3.14b.

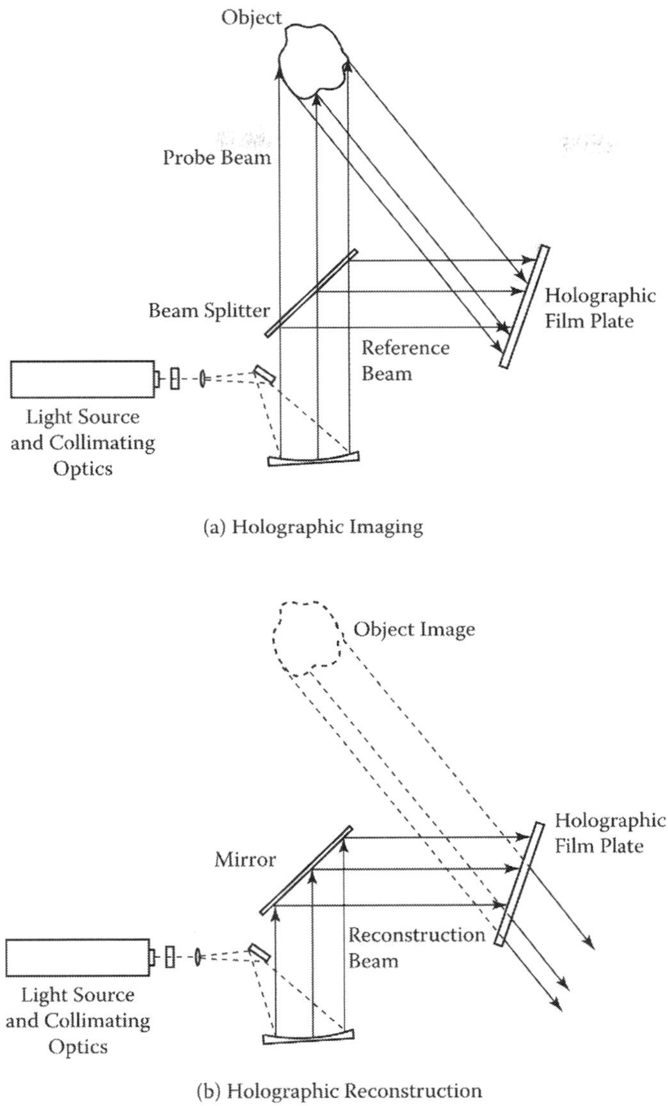

FIGURE 3.14 Optical configuration for (a) holographic recording and (b) reconstruction.

To see how this is done, we again write the electric field of light using a sinusoidal function with phase, δ.

$$E = E_o \sin \frac{2\pi}{\lambda}(ct - x + \delta) \tag{3.30}$$

Figure 3.14 shows a reference (R) and a probe (P) beam used in holographic imaging, so that we can write the electric field for each beam using

Equation 3.30, except that for holographic imaging, both the phase and amplitude information are needed.

$$E_R = E_{oR}\sin\frac{2\pi}{\lambda}(ct - x + \delta_R) = E_{oR}\mathrm{Re}\left\{e^{i\frac{i\pi}{\lambda}(ct-x+\delta_R)}\right\} \rightarrow E_{oR}e^{i\delta_R},$$
$$E_P = E_{oP}\sin\frac{2\pi}{\lambda}(ct - x + \delta_P) = E_{oP}\mathrm{Re}\left\{e^{i\frac{2\pi}{\lambda}(ct-x+\delta_P)}\right\} \rightarrow E_{oP}e^{i\delta_P} \tag{3.31}$$

In Equation 3.31, we use Euler notation to simplify the writing of the sinusoidal function. At the imaging plane, the combined electric field is again found by superposition. Then, the light intensity is calculated from the amplitude of the combined electric field.

$$I = |E_R + E_P|^2 = (E_R + E_P)(E_R^* + E_P^*) = \left(E_{oR}e^{i\delta_R} + E_{oP}e^{i\delta_P}\right)\left(E_{oR}e^{-i\delta_R} + E_{oP}e^{-i\delta_P}\right) \tag{3.32}$$

In the foregoing equation, the asterisk means the complex conjugate. Carrying out the calculation gives us the light intensity at the image plane where the hologram is recorded.

$$I = E_{oR}^2 + E_{oP}^2 + E_{oR}E_{oP}e^{i(\delta_P - \delta_R)} + E_{oR}E_{oP}e^{-i(\delta_P - \delta_R)} \tag{3.33}$$

The first two terms in Equation 3.33 represent the amplitude information, whereas the last two terms contain the phase information through $\delta_R - \delta_P$.

At the imaging plane, a special film called a holographic plate is placed. A holographic plate is a thin plate with a medium whose transmissivity varies as a function of its exposure to the incident light intensity. The final transmissivity, T, of the holographic plate is given by

$$T = T_o - \beta It \tag{3.34}$$

T_o = transmissivity of the unexposed plate
β = film sensitivity
I = intensity of the light (Equation 3.33)
t = exposure time

Using Equation 3.33 for the intensity, we get the following transmissivity recorded on the plate:

$$T = T_o - \beta t\left(E_{oR}^2 + E_{oP}^2\right) - \beta t\left(E_{oR}E_{oP}e^{i(\delta_P - \delta_R)} + E_{oR}E_{oP}e^{-i(\delta_P - \delta_R)}\right) \tag{3.35}$$

Now, the reconstruction of the original can be performed using the setup shown in Figure 3.14b. The reconstruction beam should be identical or nearly identical in wavelength and phase to the reference beam. We represent the electric field of the reconstruction beam with $E_r = E_r e^{i\delta_r}$. The reconstruction beam, in passing through the holographic plate, will be modified by its transmissivity (Equation 3.35) to give the final electric field, E_F.

$$E_F = E_r T = E_{or} e^{i\delta_r} T \tag{3.36}$$

Combining Equations 3.35 and 3.36,

$$\begin{aligned} E_F &= E_{or} T_o e^{i\delta_r} - \beta t E_{or} \left(E_{oR}^2 + E_{oP}^2 \right) e^{i\delta_r} \\ &- \beta t E_{or} \left(E_{oR} E_{oP} e^{i(\delta_P - \delta_R + \delta_r)} + E_{oR} E_{oP} e^{-i(\delta_P - \delta_R - \delta_r)} \right) \end{aligned} \tag{3.37}$$

If the phase $\delta_r \approx \delta_R$, then the third term in Equation 3.37 contains the same phase as the original beam: $-\beta t E_{or} E_{oR} E_{oP} e^{i\delta_P}$.

Thus, the effect of looking at the reconstruction beam is identical to looking at the original beam. Points at different distances will contain a different phase, δ_P, and thus distance information is intrinsically contained in holograms. By varying δ_r, we can observe different phase components or essentially different planes in the imaged object. This is the basis for the reproducibility of three-dimensional structures through holographic imaging. Figure 3.15 presents an example of holographic imaging, showing the different planes of a mixing layer.

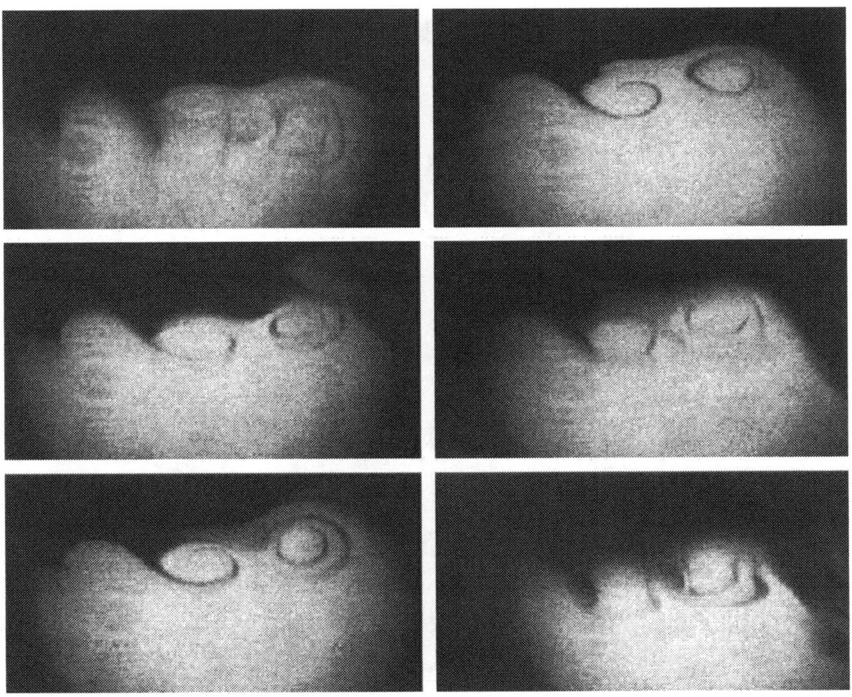

FIGURE 3.15 Holographic images of a mixing layer. (From Hermann, S., Hinrichs, H., Hinsch, K.D., and Surmann, C., Coherence concepts in holographic particle image velocimetry, *Experiments in Fluids (Suppl.)* S108–116, Springer-Verlag, New York, 2000. Images used with permission from Springer-Verlag.)

3.6 LASER TOMOGRAPHIC IMAGING

Laser tomographic imaging, also referred to as laser sheet imaging, is used to view a cross section of the flow intersected by a thin laser light sheet. The advantage of this approach is that a cross-sectional plane is interrogated at a time, so that details of the flow motion, mixing, and other characteristics can be evaluated for that plane. Figure 3.16 shows a typical setup for laser tomographic imaging. A combination of a cylindrical and a spherical lens is used to expand the beam into a sheet (cylindrical lens) while controlling the beam thickness (spherical lens). The incident laser light can be scattered by particles or other tracer molecules. Any optical or electromagnetic process that responds to the incident laser light sheet can be used for this purpose. The most common laser sheet imaging methods are based on Mie scattering and fluorescence. The light produced by the laser light sheet is captured by a digital (or analog) camera for recording and processing.

Figure 3.3, shown earlier, is an example of laser tomographic imaging, where a thin laser sheet intersects a plane in the flow, and only the fluorescence signal from that plane is imaged. The advantage of tomographic imaging is that the flow structure can be viewed plane by plane. Through digital processing, such planar images can be combined to provide a three-dimensional reconstruction of the flow field. With standard photography, where the entire volume of the flow is visualized, such cross-sectional views are not possible. An alternative, of course, is to use holographic imaging, which *is* a three-dimensional imaging technique as described in the previous section. However, in some flows, such as dense liquid sprays, the light attenuation through the liquid core may be so high as to make holographic imaging less practical.

Fine seeding particles are commonly used to seed gas flows for Mie-scattering imaging. Some of the particle seeding methods include oil smoke particles produced from venturi atomizers and titanium dioxide or similar particles generated from cyclone seeders. One interesting application of oil smoke particles is that they burn out at elevated temperatures, as found in flames, and so they have been used to

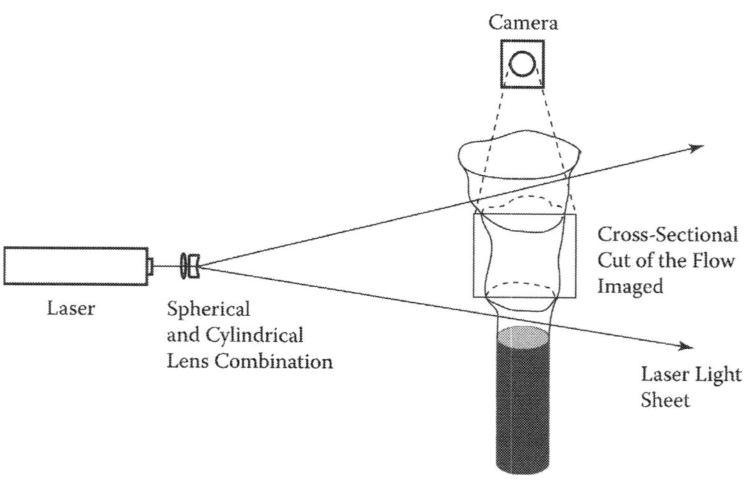

FIGURE 3.16 The optical setup for laser tomographic imaging.

demarcate the flame front in the combustion process. These particles, if produced under a controlled environment so that their diameters remain small (~1 μm), tend to follow the flow faithfully even under turbulent conditions and, therefore, can be used for flow visualization and in velocity measurement techniques to be discussed in Chapter 4.

In principle, molecular scattering phenomena, such as Rayleigh and Raman scattering, can also be used for laser tomographic imaging. However, these processes suffer from low scattering cross sections or low signal intensities and thus are easily overwhelmed by noise from Mie scattering or spurious reflections of the illumination beam. For this reason, fluorescence imaging is a preferred method when using molecular species as flow tracers. Discussions of the various optical processes, including scattering and fluorescence can be found in Chapter 5. For laser tomographic imaging, we just need to understand that these are optical phenomena that emit sensitive signals of specific wavelengths in response to the laser illumination.

Fluorescence is a process where strong signals are emitted from molecules in response to an incident laser beam. It originates from photons being emitted from molecules when they transition by an excited electronic state to a lower-energy ground state. Fluorescence imaging can be used in combustion flows where OH (hydroxyl) molecules are produced in large amounts in the flame, allowing flame characteristics to be imaged. OH molecules require a laser beam specifically tuned to the excitation wavelengths within approximately 0.01 nm. There are many excitation bands for OH molecules in the 280–312 nm ultraviolet range. The resulting fluorescence is also in a similar range, although for filtering purposes, some separation of the excitation and fluorescence bands is desirable.

For nonreacting flows, fluorescent tracers need to be seeded into the flow. Because these fluorescent tracers are molecules such as acetone or rhodamine, they have no problem following the flow. Acetone is a commonly used fluorescent tracer for flow visualization and has a high vapor pressure so that a substantial amount can easily be seeded to gas flows. It has a broad absorption spectrum in the ultraviolet range from 225 to 320 nm, which is accessible through commercially available lasers, and a strong, off-resonant (well away from the excitation wavelength) fluorescence signal that can be detected using un-intensified imaging systems. Acetone fluorescence has been used for tomographic imaging of jet flows, fuel vapor imaging, supersonic mixing layers, and internal combustion engines. Recently, 3-pentanone has been suggested as a better tracer for iso-octane (a typical representation of gasoline fuel), as it has a boiling point close to that of iso-octane, thus simulating iso-octane's evaporation characteristics. For liquid flows, there are many options for fluorescent seeders, such as fluorescein (green emission), rhodamine-B (red), and rhodamine-6G (yellow), the latter having high solubility in water and alcohol.

The principle of fluorescence is discussed in detail in Chapter 5, and the observed fluorescence signal intensity is proportional to the following factors:

$$I_f \propto I_o N \sigma(\lambda, T) L \phi(\lambda, T, p_i) \eta \tag{3.38}$$

I_f = fluorescence signal intensity
I_o = incident laser intensity

N = number density of the fluorescent molecules
σ = absorption cross section of the fluorescence molecules
λ = wavelength of the excitation laser
T = temperature
L = length of the illuminated volume
ϕ = fluorescence quantum yield (fraction of excited molecules that fluorescence)
p_i = partial pressure of the fluorescent molecules
η = collection efficiency (fraction of the emitted photons collected by the detector).

For a given fluorescence imaging system in which the incident laser energy is kept constant, along with fixed L and η, all the quantities on the right-hand side can be considered constant except N, σ, and ϕ. At a fixed laser wavelength and under isothermal conditions, the absorption cross section (σ) and the quantum yield (ϕ) are also constant, and the fluorescence signal depends only on the number density. This allows the mixing field to be mapped through fluorescence imaging if the fluorescent molecules are seeded at a constant volume percentage in the stream to be analyzed. That is, the number density of a molecular species is proportional to the fluorescence intensity.

One of the more advanced applications of fluorescence imaging in gas flows is the determination of the temperature field under isobaric (constant-pressure) conditions using fluorescence intensity. As shown in Equation 3.38, the fluorescence signal intensity has temperature dependence through the absorption cross section (σ) and the quantum yield (ϕ). Using the ideal gas equation of state, $p = NkT$ (k is the Boltzmann constant), under constant-pressure conditions, Equation 3.38 can be rewritten as follows:

$$I_f \propto \frac{I_o}{T}\sigma(\lambda,T)\phi(\lambda,T,p_i) \tag{3.39}$$

If the fluorescent molecules are uniformly seeded into the flow, then the partial pressure is proportional to the total pressure, so that we can replace p_i with p in Equation 3.39. Using overlapping laser sheets with two different wavelengths will result in different σ and ϕ because both parameters have wavelength dependence. The fluorescence signals obtained at the two wavelengths then furnish us with two equations to solve for two unknowns, σ and ϕ, which in turn depend on the temperature.

$$\frac{I_{f,\lambda 1}}{I_{o,\lambda 1}} = \sigma(\lambda_1,T)\phi(\lambda_1,T,p_i) \tag{3.40a}$$

$$\frac{I_{f,\lambda 2}}{I_{o,\lambda 2}} = \sigma(\lambda_2,T)\phi(\lambda_2,T,p_i) \tag{3.40b}$$

Therefore, if the temperature dependence of σ and ϕ is known or obtained through calibration, we can recover the temperature from Equations 3.40(a) and (b). Because σ and ϕ are multiplied, another approach is to simply take the ratio of the fluorescence signals in Equation 3.40 and match it against the ratio of the right-hand sides at a given temperature to recover the temperature.

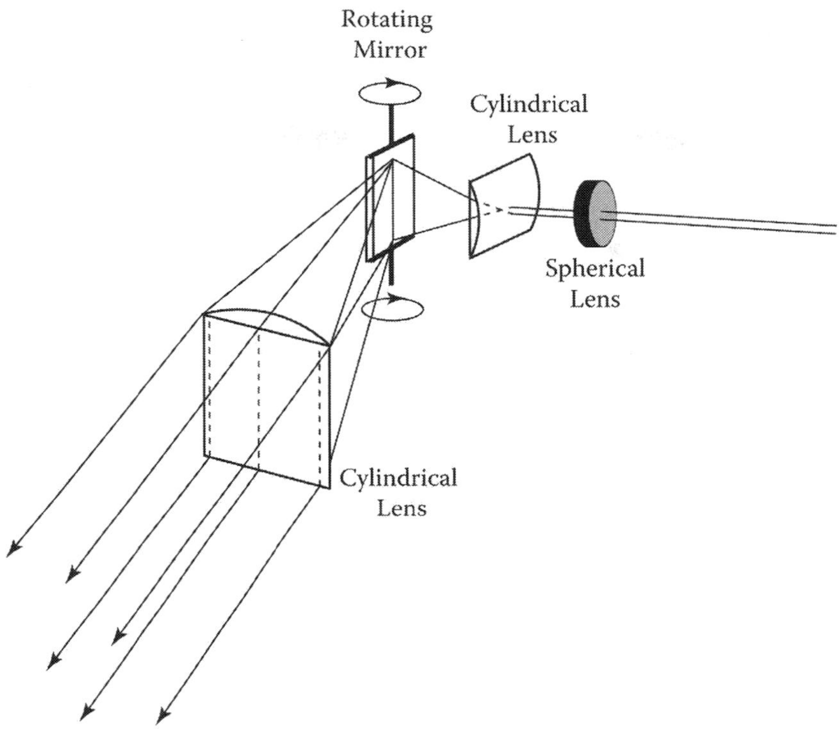

FIGURE 3.17 Rotating mirror setup for sweeping laser light sheets.

As shown in the foregoing examples, laser tomographic imaging is used to determine a variety of flow characteristics, such as flow structure, mixing of two distinct fluids, fuel vapor concentrations, and even the temperature field if the relationship between the signal spectra and scalar quantities is carefully calibrated. Two-dimensional tomographic imaging can be expanded to three-dimensional imaging or even time-dependent three-dimensional imaging, through the use of scanning beam sheets and fast image acquisition systems. A three-dimensional volume can be swept with laser light sheets using rotating mirrors, as shown in Figure 3.17. If the sweeps can be made at a frequency higher than the characteristic time scale of the flows, then four-dimensional measurements (time and space) are possible.

3.7 IMAGE PROCESSING AND ANALYSIS

Image processing is an established field with important applications in areas ranging from thermal-fluid engineering and biomedical sciences to satellite-based geophysical observations. In flow visualizations of thermal and fluid processes, important information embedded in digital images can be extracted and analyzed using image-processing methods. We will consider some of the basic elements of image processing and analysis.

Image analysis can be considered to consist of the following stages:

- Image acquisition
- Noise removal, geometric transformations, restoration
- Image enhancement
- Feature extraction, edge detection
- Storage

3.7.1 Image Acquisition

Image acquisition nowadays is facilitated by diverse hardware and software that are commercially available. Digital cameras with monochrome (black and white) or color image formats are widely available. High-resolution, high-data-rate (framing rate) cameras also allow high-speed imaging of small-time-scale processes, including direct downloading of image sequences to computing platforms (usually PCs) in convenient formats (JPG, TIF, or other image formats). Once downloaded to computing platforms, programming software such as MATLAB and other dedicated image analysis programs allow direct access to these files so that various image-processing schemes can be applied numerically. Some example MATLAB algorithms for performing simple image processing are included in this section.

There are three main characteristics of an image acquisition device: pixel density (spatial resolution), image acquisition speed (number of frames acquired per second, for temporal resolution), and dynamic range (the number of bits available for each pixel). A pixel refers to the light-detection element in an imaging device. For a typical imaging device, the number of pixels in the horizontal and vertical directions determines the imaging resolution, and it often ranges in the several millions (megapixels). For high-speed imaging, large data-rate links are required between the imaging device, memory buffer, and storage device. The taxation on the speed is dependent on the number of pixels and also on the dynamic range. Dynamic range refers to whether the light intensity is digitized into 2^8, 2^{12}, 2^{16}, etc., number of intensity levels. For example, 8-bit digitization means there are $2^8 = 256$ levels of gray, starting from 0 to 255. Digitization with a larger number of bits provides higher precision in light intensity at the expense of larger memory and longer data transfer times. Of course, red, green, and blue components can be separately imaged onto three co-located pixels to reproduce color images.

During digitization, the horizontal and vertical pixel locations are assigned to column and row numbers in an array, and the intensity then becomes the numerical value of the array at that column and row number. This is depicted in Figure 3.18. This digitization can be accomplished readily using MATLAB or other image analysis software. For example, the following MATLAB statements read a JPG file (from standard digital cameras) and convert the data into an array format.

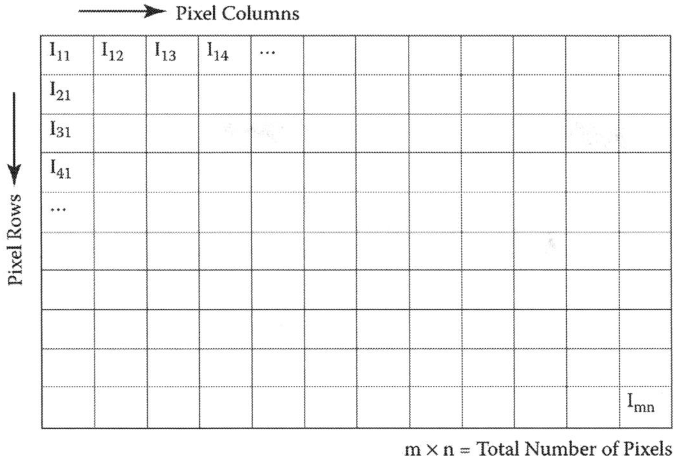

FIGURE 3.18 Data array associated with digital image format.

Example 3.3: MATLAB Algorithm to Read an Image File and Convert to a Data Array

The following command imports a JPG file, "image.jpg," into the MATLAB domain and puts the pixel intensity data into an (N, M) array format. The array size, N×M, is determined by the number of pixels in the image. MATLAB supports most image formats, such as ".jpg," ".tif," ".png," and many others.

```
I=imread('image.jpg')
```

The next command, "imshow," displays the imported image.

```
imshow(I)
```

An intensity value at a specific pixel location (i, j) can be displayed using:

```
display(I(i,j))
```

3.7.2 Basic Image Operations

Now that the image has been acquired and convert into an array format, various mathematical operations can be applied to condition these images and extract quantitative data. The arithmetic operations listed in Table 3.2 serve some useful functions. Image subtraction is useful, for example, for background subtraction. If the image data is a combination of the signal and the background noise, the background image can be taken without the signal (by shutting off the laser or other light source that produces the signals) and subtracted from the signal image as follows:

$$\text{Image Data} = \text{Signal Image} + \text{Background Image} \tag{3.41}$$

$$\text{Signal Image} = \text{Image Data} - \text{Background Image} \tag{3.42}$$

TABLE 3.2
Basic Arithmetic Operations for Image Processing

Operation	Formula	Description
Addition	$C = A + B$	Combining two images *A* and *B*, or any number of images. Used for image averaging
Subtraction	$C = A - B$	Background (*B*) subtraction from image *A*
Division	$C = A/B$	Image normalization. If image *A* has been derived from a nonuniform illumination field described by image *B*, normalization gives the true response
Inversion	$C = (2^N - 1) - A$	Inversion of image *A* with an *N*-bit digitization

Image averaging is used to generate ensemble averages. To avoid exceeding the maximum intensity scale, the following sequence of operations is typically used for image averaging, where *N* is the total number of images to be averaged.

$$I_{avg} = \frac{I_1}{N} + \frac{I_2}{N} + \cdots + \frac{I_N}{N} \tag{3.43}$$

Arithmetic image operations can be routinely performed using MATLAB subfunctions. The following are some examples.

Example 3.4: Image Arithmetic Operations Using MATLAB

If we again have an image designated with an array name, I, the following MATLAB lines subtract a constant value of 25 from the image data array and place the data in a new array, I2.

```
I2 = imsubtract(I, 25)
```

If a background image, Ib, is available, the image subtraction occurs on a pixel-by-pixel basis.

```
I3 = imsubtract(I, Ib)
```

Any of these images can be viewed, again using

```
imshow(I3)
```

Other useful image arithmetic functions include:

imadd: Adds two images or a constant.

imdivide: Divides one image by the second image or by a constant.

immultiply: Multiplies two images or a constant.

These and various other functions are described in detail on the MATLAB support homepage (http://www.mathworks.com/support/).

3.7.3 Image Enhancements

In quantitative image analysis for thermal and fluid measurements, much care and thought are needed before applying any image enhancements so as not to artificially alter the data. Some basic elements of image enhancement are nonetheless included, here as they constitute an important element in image processing and provide the basis for useful techniques such as edge detection.

Grayscale transformation: For an image that contains an intensity range from I_m to I_n, the intensity scale can be shifted, multiplied, or operated through mathematical functions. For example, for a predominantly underexposed image, adding L to the pixel intensities will result in an intensity range from $m+L$ to $n+L$, which may enhance the image. This operation is arithmetic and can be accomplished using the "imadd" function in MATLAB.

Contrast stretching: To maximize the contrast in an image (i.e., the range of intensities represented in the image), the following algorithm can be applied:

$$f(I \rightarrow J) = \begin{array}{ll} J_{\text{low}} & \text{if } I < I_{\text{low}} \\ J_{\text{low}} + \left(J_{\text{high}} - J_{\text{low}}\right)\left(I - I_{\text{low}}\right) / \left(I_{\text{high}} - I_{\text{low}}\right) & \text{if } I_{\text{low}} < I < I_{\text{high}} \\ J_{\text{high}} & \text{if } I > I_{\text{low}} \end{array} \tag{3.44}$$

This transformation converts the original intensity, I, to the new intensity, J, which is stretched from $[I_{\text{low}}, I_{\text{high}}]$ to $[J_{\text{low}}, J_{\text{high}}]$. A MATLAB function exists to perform this transformation by simply specifying the low and high intensity values. An example of contrast stretching is shown in Figure 3.19.

Example 3.5: MATLAB Algorithm for Contrast Stretching

The MATLAB function imadjust is useful for this operation. It has the following command format (notice that fractional ranges are used):

```
imadjust(I, [Ilow/Imax, Ihigh/Imax], [Jlow/Jmax, Jhigh/Jmax])
I = imread('image.jpg')
imshow(I)
J = imadjust(I,[0 0.45], [0 1])
imshow(J)
```

Figures 3.19a and b show, respectively, the original image and the enhanced image using the foregoing contrast-stretching transformation.

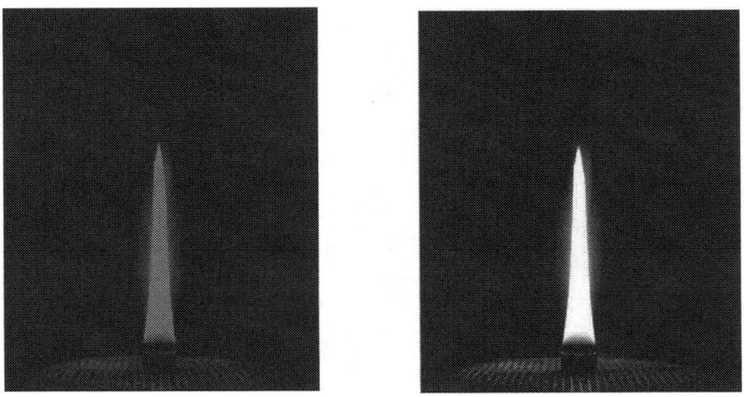

(a) (b)

FIGURE 3.19 The effect of contrast stretching on a low-contrast image.

Histogram equalization: The histogram is a probability function for the intensities present in an image. MATLAB has an internal function (imhist) that generates histograms for an input image. The histogram can be used for several useful image-processing operations, including histogram equalization, which is similar to contrast stretching except that it stretches the image optimally depending on the intensity distribution in the histogram. Another use of the histogram is to set an appropriate threshold for image binarization, as described in the following text.

Filtering: To improve image quality, image filtering may be used. However, because filtering can introduce artificial alterations to the raw data, filtering during quantitative analyses must be used with much care and judgment. A box filter involves averaging the pixel intensities in a 5×5 matrix using the following formula, where $a(m, n)$ is the input matrix and $b(m, n)$ is the output matrix:

$$b(m,n)=\frac{\left(\sum_{i=-2}^{+2}\sum_{j=-2}^{+2}a(m+i,n+j)\right)}{25} \tag{3.45}$$

A circular filter performs a similar task except that it ignores the four corner intensities in the 5×5 matrix, as follows:

$$b(m,n)=\frac{1}{21}\begin{bmatrix}0&1&1&1&0\\1&1&1&1&1\\1&1&1&1&1\\1&1&1&1&1\\0&1&1&1&0\end{bmatrix}\times a(m\pm 2,n\pm 2) \tag{3.46}$$

Another weighting matrix can be used to optimize the image filtering, as shown below:

Pyramidal filter:

$$b(m,n)=\frac{1}{81}\begin{bmatrix}1&2&3&2&1\\2&4&6&4&2\\3&6&9&6&3\\2&4&6&4&2\\1&2&3&2&1\end{bmatrix}\times a(m\pm 2,n\pm 2) \tag{3.47}$$

Gaussian filter:

$$b(m,n)=\frac{\left(\sum_{i=-2}^{+2}\sum_{j=-2}^{+2}a(m+i,n+j)g(x,y)\right)}{\sum_{i=-2}^{+2}\sum_{j=-2}^{+2}g(x,y)} \tag{3.48}$$

where

$$g(x,y)=\frac{1}{\sqrt{2\pi}\sigma}e^{-\frac{d^2}{2\sigma^2}},\quad\text{and}\quad d=\sqrt{[(x-x_c)^2+(y-y_c)^2}. \tag{3.49}$$

3.7.4 Edge Detection

For the extraction of features from images, edge detection algorithms are used. Examples of edge detection in thermal and flow image data are extractions of the boundaries of the mixing layer, flame shape, liquid droplet identification, etc. From an image that may include varying levels of contrast, and therefore a range of intensity values, one applies binarization to delineate fluid/fluid, liquid/gas, or reactant/product boundaries. From the binarized images, various edge detection schemes can be applied to obtain coordinates for the boundaries for subsequent geometrical (Euclidean or fractal) analyses.

Binarization: From an arbitrary histogram, one can set a threshold to delineate two distinct regions in the flow. For example, in an image as shown in Figure 3.19, the histogram will show that high intensities represent the flame region, whereas low intensities correspond to the background. Thus, setting an appropriate threshold value in the intermediate intensity range and applying the following binarization formula results in a clear delineation between white (object) and black (background) regions of the image.

$$\begin{aligned} b(m,n) &= I_{\text{max}} \quad \text{if } a(m,n) > I_{\text{threshold}} \\ b(m,n) &= I_{\text{min}} \quad \text{if } a(m,n) < I_{\text{threshold}} \end{aligned} \tag{3.50}$$

Edge detection algorithms: One simple method detecting the edge in a continuous outline is a circular (clockwise or counterclockwise) search for a specific intensity (maximum or minimum in binarized images), as shown in Figure 3.20. By checking the intensity value in neighboring pixels until the sought intensity is found, the pixel coordinates can be traced, which represent the contour. As shown in Figure 3.20, a search is made for "dark" cells in a fixed direction (clockwise in Figure 3.20). If it doesn't find a dark cell, it makes a right-angle turn in the clockwise direction until it finds a dark cell. In some instances, the dark cell found may be the cell the search originated from, in which case the search is continued starting from the face

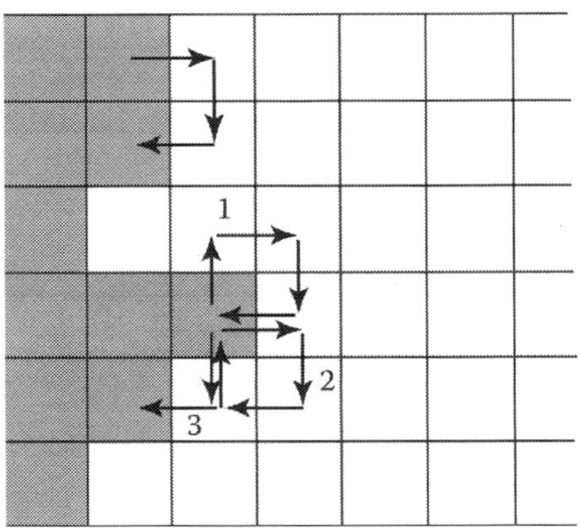

FIGURE 3.20 A circular search method for edge tracing.

FIGURE 3.21 Edge detected using two gradient-based MATLAB functions.

intersected by the search arrow. The algorithm, therefore, requires each search to start from the face last intersected by the search arrow, as shown in Figure 3.20.

There are gradient-based edge detection methods in which a threshold in the gradient of pixel intensities is used to discern the edge. Various methods are available in MATLAB, where the gradient threshold is set by the operator. The following MATLAB algorithm shows how these edge detection algorithms can be implemented.

Example 3.6: MATLAB Algorithm for Edge Detection

```
I = imread('image.jpg')
J = edge(I, 'roberts', 0.05)
imshow(J)
```

The parameter, 0.05, is the sensitivity threshold level, and in place of "roberts," other methods can be called by replacing the parameter with "prewitt," "sobel," or "canny," which refer to different methods. Figures 3.21a and b show the edges detected using the Roberts and Canny methods, respectively.

Once the edges are accurately identified by setting the appropriate threshold level, the edge coordinates can be used to find various topological parameters, such as length, perimeter, curvature, fractal dimensions, and area. The determination of length, for example, is useful in analyzing particle streak images to infer particle velocity by dividing the length by the exposure time. Perimeter, fractal dimensions, and curvature are important geometrical factors in characterizing surfaces such as iso-concentration contours and turbulent flame contours.

3.8 ADVANCED IMAGE ANALYSIS (DEEP LEARNING)

Object detection in image data is facilitated via deep-learning/machine-learning algorithms. A convenient platform for applying this technique is MATLAB®, which has many pre-programmed functions. The task of identifying different objects or patterns in the flow field has traditionally been done manually or semi-automatically.

Fast, pre-trained algorithms and functions are now fairly easily accessible in various software for advanced image analysis, or one can customize complex image analysis/pattern recognition codes using preset functions. Since deep learning requires a large throughput of labeled images for training the neural-network algorithm, a GPU (graphics processing unit) is sometimes used, although for many thermal-fluid applications a CPU (central processing unit) on PCs or laptops is sufficient. Deep-learning algorithms for object detection use so-called convolutional neural networks (CNNs or ConvNets). For example, YOLO in MATLAB® is a CNN algorithm.

Although CNN is optimally used in deep-learning algorithms, ANN (artificial neural network) is the core idea in machine learning. ANN consists of nodes that form input, hidden, and output layers. Each node is connected to other nodes with weights and a threshold to mimic how neurons are connected and exchange signals in the brain (therefore, the name artificial neural network). The weights of the connections are optimized during "training" with known inputs and outputs. CNN can be considered as a three-dimensional version of ANN, suited for the efficient optimization of a large number of two-dimensional image data, as shown in Figure 3.22. CNN consists of multiple layers: convolutional layer(s), pooling layer(s), and a fully connected layer, with each layer increasing in complexity for identifying features of the images. Feature detection is performed with a two-dimensional array of weights for the connections.

Using deep-learning algorithms, flow visualization, and image data for thermal and fluid flows can be automated for pattern and feature analysis. Some examples are:

1. Droplet, ligament, and continuous liquid core identification, and geometrical parameters (e.g., area-equivalent diameter, aspect ratio, circularity).
2. Flow regime identification in boiling flows (e.g., nucleate boiling, slug flow, film boiling).
3. Vortical patterns in turbulent flows, including vortex length scale and rotational speed.
4. Super-resolution of sub-pixel-scale features in turbulent flows.
5. Cloud formation dynamics and classification.

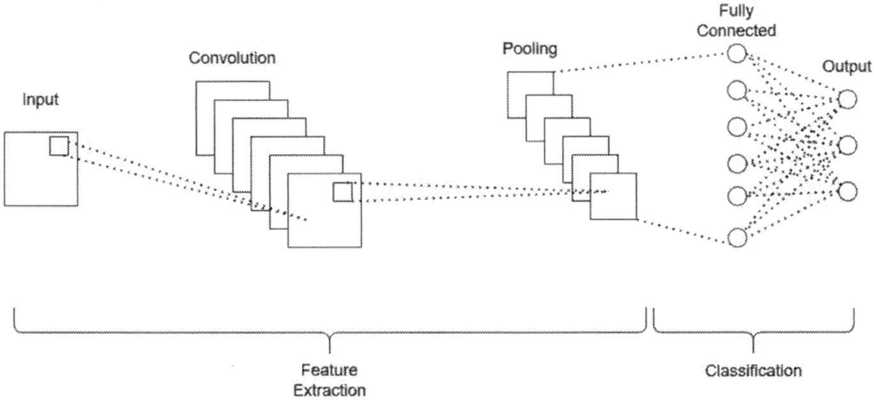

FIGURE 3.22 Schematics of CNN.

Example 3.7: Flow Feature Extraction Using MATLAB Algorithm

The objective of this example is to develop an image-processing pipeline using MATLAB to identify and analyze droplet streaks captured in spray images. Specifically, the goals are to:

- Detect and isolate streaks representing moving droplets.
- Estimate droplet velocities based on streak lengths and camera exposure time.
- Extract the drop size data.

The input images used for this analysis were captured during a spray experiment. A cross section of the spray flow was illuminated using a laser beam sheet. The camera exposure time was set to 1/1000 second, with a spatial resolution of approximately 0.05 mm/pixel. A pre-processed image (contrast adjustment based on the intensity histogram) is shown in Figure 3.23.

Then, the following steps are applied:

- Adaptive thresholding to segment bright streaks from the background.
- "Canny" edge detection to sharpen streak boundaries.
- Morphological operations to remove noise and small specks.
- Region filtering to retain only elongated streaks with a high length-to-width aspect ratio.

FIGURE 3.23 A sample image of spray flow.

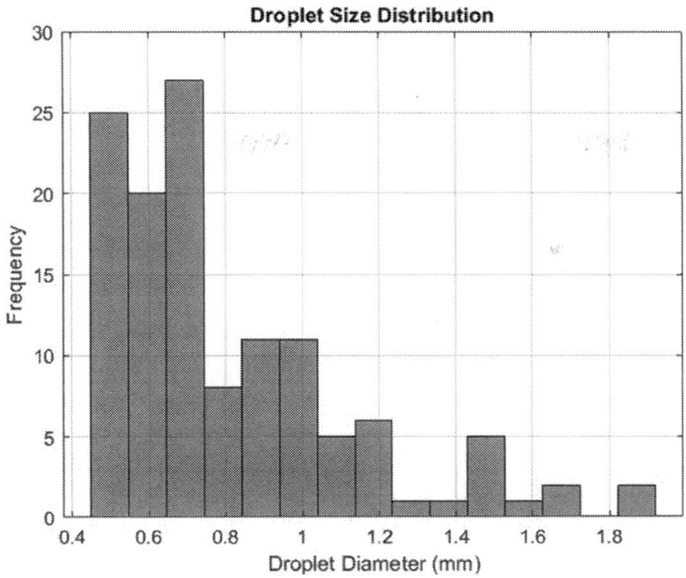

FIGURE 3.24 Drop size distribution from the spray image.

Liquid element (droplet) velocity can be determined using the detected streak lengths divided by the exposure time. In this example image, the number of streaks detected is 1134, with a mean streak length of 180.1 pixels. With an exposure time of 1 ms and a pixel scale of 0.05 mm/pixel, the average velocity is 1.67 m/s. The spray cone angle is 37.5°. For droplets with low speed, they appear as near-spherical objects, which can again be identified during image processing. Then, the drop size distribution can then be plotted (Figure 3. 24). A MATLAB code for performing the image analyses is attached.

MATLAB® code for the analyses of the spray image.

```
clc; clear; close all;

% Load the image
img = imread('CIMG0283 (1).jpg');  % Replace with your image file

% Convert the image to grayscale (if it's a color image)
gray_img = rgb2gray(img);

% Step 1: Apply a Gaussian filter to smooth the image and
reduce noise
smoothed_img = imgaussfilt(gray_img, 2);

% Step 2: Apply edge detection (Sobel operator) to highlight
streaks
edges_img = edge(smoothed_img, 'Sobel');
```

```
% Step 3: Thresholding to detect droplets (based on intensity)
threshold = 100;  % Adjust this value based on your image's
characteristics
binary_droplets = smoothed_img > threshold;  % Segmentation of
droplets

% Step 4: Remove small objects that are not droplets using
morphological operations
se = strel('disk', 5);  % Structural element (disk shape with
radius 5)
cleaned_droplets = imopen(binary_droplets, se);  % Remove
small droplets
cleaned_droplets = imclose(cleaned_droplets, se);  % Close
small gaps in droplets

% Step 5: Fill holes within droplets
cleaned_droplets = imfill(cleaned_droplets, 'holes');

% Step 6: Morphological operation to detect streaks by removing
large objects
streaks = edges_img & ~cleaned_droplets;  % Get streaks that
are not droplets

% Step 7: Combine detected droplets and streaks for visualization
final_detected_img = cleaned_droplets | streaks;  % Combine
both streaks and droplets

% Display the results
figure;
subplot(2, 3, 1), imshow(gray_img), title('Original Image');
subplot(2, 3, 2), imshow(smoothed_img), title('Smoothed
Image');
subplot(2, 3, 3), imshow(edges_img), title('Edge Detection
(Sobel)');
subplot(2, 3, 4), imshow(cleaned_droplets), title('Detected
Droplets');
subplot(2, 3, 5), imshow(streaks), title('Detected Streaks');
subplot(2, 3, 6), imshow(final_detected_img), title('Final
Detection (Droplets + Streaks)');

% Save the result if needed
imwrite(final_detected_img, 'detected _ spray _ image.jpg');



%% Spray area

% --- Compute spray area ---
pixel_scale_mm = 0.05;  % mm/pixel (adjust as per calibration)
pixel_area_mm2 = pixel_scale_mm^2;
spray_pixel_count = sum(final_detected_img(:));
spray_area_mm2 = spray_pixel_count * pixel_area_mm2;
```

```
% Overlay spray region on original grayscale image
overlay_img = imoverlay(gray_img, final_detected_img, [1 0 0]);
% red overlay

% Display annotated result
figure;
imshow(overlay_img);
title('Spray Area Overlay');

% Annotate with area value
area_text = sprintf('Spray Area: %.2f mm^2', spray_area_mm2);
position = [20, 20];  % Adjust (x, y) position if needed
text(position(1), position(2), area_text, 'Color', 'yellow',
'FontSize', 14, 'FontWeight', 'bold');



%%  Droplet size distribution

% --- Droplet Size Distribution Analysis ---
props = regionprops(cleaned_droplets, 'Area', 'Eccentricity',
'EquivDiameter');

% Filter to keep only circular features (low eccentricity)
ecc_thresh = 0.85;  % 0 = perfect circle, 1 = line
filtered_props = props([props.Eccentricity] < ecc_thresh);

% Extract diameters in pixels and convert to mm
pixel_scale_mm = 0.05;
diameters_mm = [filtered_props.EquivDiameter] *
pixel_scale_mm;

% Compute statistics
mean_diam = mean(diameters_mm);
max_diam = max(diameters_mm);
min_diam = min(diameters_mm);
std_diam = std(diameters_mm);

% Display results
fprintf('\n--- Droplet Size Distribution ---\n');
fprintf('Total detected droplets: %d\n', numel(diameters_mm));
fprintf('Mean Diameter: %.2f mm\n', mean_diam);
fprintf('Max Diameter: %.2f mm\n', max_diam);
fprintf('Min Diameter: %.2f mm\n', min_diam);
fprintf('Std Dev: %.2f mm\n', std_diam);

% Plot histogram
figure;
histogram(diameters_mm, 15);
```

```
xlabel('Droplet Diameter (mm)');
ylabel('Frequency');
title('Droplet Size Distribution');
grid on;

%% Spray Penetration Length Calculation

% --- Spray Penetration Length Calculation ---
[rows, cols] = find(final_detected_img);

if isempty(rows)
    fprintf('No spray detected – cannot compute penetration
length.\n');
else
    % Top and bottom-most Y positions
    top_y = min(rows);
    bottom_y = max(rows);

    % Calculate penetration length in pixels
    penetration_length_px = bottom_y - top_y;

    % Convert to mm
    pixel_scale_mm = 0.05;  % Adjust if needed
    penetration_length_mm = penetration_length_px *
pixel_scale_mm;

    % Display result
    fprintf('\n--- Spray Penetration Length ---\n');
    fprintf('Penetration length: %.2f mm\n',
penetration_length_mm);

    % Visualize on grayscale image
    figure;
    imshow(gray_img);
    hold on;
    line([1 size(gray_img,2)], [top_y top_y], 'Color', 'cyan',
'LineWidth', 1.5);
    line([1 size(gray_img,2)], [bottom_y bottom_y], 'Color',
'magenta', 'LineWidth', 1.5);
    plot([size(gray_img,2)*0.85], [(top_y+bottom_y)/2], 'wo');
    text(20, (top_y+bottom_y)/2,..
        sprintf('Penetration: %.2f mm', penetration_length_mm),
...
        'Color', 'yellow', 'FontSize', 14, 'FontWeight', 'bold');
    title('Spray Penetration Length');
    hold off;
end
```

PROBLEMS

3.1. A velocity field has time and spatial dependence as follows.

$$u = 5ty$$

$$v = 2t^2x^2$$

Plot the streamlines going through $(x_o, y_o) = (0, 0)$ at $t = 1$, 5 and 10 s.

3.2. For a time-dependent velocity vector given below, find the x–y relationship for streamline, streakline, and pathline for a fluid particle passing through $(x_o, y_o) = (1, 1)$ at $t_o = 0$.

$$\vec{u} = 5 \sin\frac{2\pi t}{T}\vec{i} + 2.5\vec{j}$$

3.3. The angle of refraction from one medium to another can be calculated using Snell's law (Equation B.4).

$$n_2 \sin\phi_t = n_1 \sin\phi_i \quad \text{or} \quad \frac{\sin\phi_t}{\sin\phi_i} = \frac{n_1}{n_2} \tag{B.4}$$

n_1 = refractive index of medium 1
n_2 = refractive index of medium 2

For an interface in air where there is a temperature jump from 300 to 550 K at a constant 1 atm pressure, plot the refracted angle for incident angles from 5 to 65 degrees.

3.4. In a Schlieren setup, the probe light is sent in the x-direction as shown below. The air flow inside the test section undergoes an average density change of 0.0217 kg/m^3 from $y = 0$ to 250 mm. Find the average deflection angle of the probe light for a test section width of 150 mm. The ambient is at $T = 315$ K and $p = 0.978$ atm.

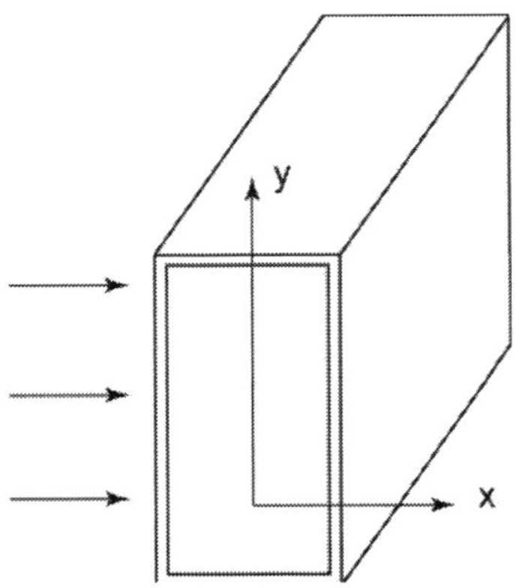

3.5. The setup in Prob. 3.4 is used now for shadowgraph imaging. Due to the heat release process, there exists an average density gradient given by $d\rho/dy = 3(y - y^2)$. Plot the resultant contrast as a function of y from $y = 0$ to 250 mm that will show up on the shadowgraph screen located 500 mm from the test section.

3.6. In Mach–Zehnder interferometric imaging, the density change is related to the fringe spacing in the image by Equation 3.29.

$$\delta = L\beta\left(\frac{\rho_2 - \rho_1}{\rho_o}\right) \tag{3.29}$$

Again, using the setup shown in Prob. 3.4, plot the spatial variation of relative density, $\rho_2 - \rho_1$, from $x=0$ to 100 mm and $y=0$ to 250 mm if the fringe spacing has a spatial distribution given by $\delta = \cos(2.5x)(y - 2y^2)$.

3.7. Scan or obtain a similar image file to the one shown below in JPG, TIF, or another image format. Write a MATLAB program to read, display, and write the image file into an intensity array. Use a MATLAB plotting function to generate a contour plot of the intensity array.

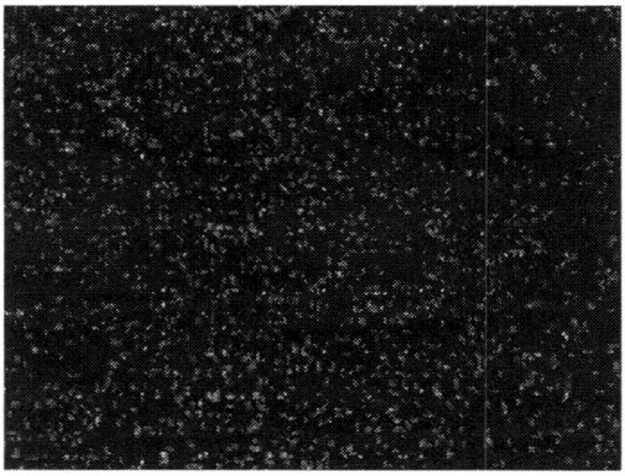

3.8. Set the intensity array file as "I_A" for the image in Problem 3.7. Generate and display new images based on the "I_A" by performing the following arithmetic operations.

a. $I_B = I_A + 50$
b. $I_c = I_A/2$
c. Invert I_A

3.9. Generate and display an artificially translated image of I_A from Problem 3.8 by performing the following operation.

$$I'_A(i,j) = I_A(i+5, j+5)$$

3.10. Using the image files from Problem 3.8, perform the following image operations and display.

$$I'_B = I_B + I_A$$

$$I'_C = I_C / I_A$$

$$I'_D = I_B - I_C$$

3.11. Perform contrast stretching to I_A and display the result.

3.12. Generate a box-filtered image of I_A.

$$I'_A(m,n) = \frac{\left(\sum_{i=-2}^{+2}\sum_{j=-2}^{+2} I_A(m+i,n+j)\right)}{25} \tag{3.45}$$

3.13. Compare the above image of Prob. 3.12 after applying the pyrantidal and Gaussian filters.

3.14. Apply an image binarization to the image of Prob. 3.7 by using a threshold intensity of 1/2, 1/3, and 1/4 of the maximum intensity found on the image.

BIBLIOGRAPHY

Bockle, S., Kazenwadel, J, Kunzelmann, K., and Schulz, C., Laser-diagnostic multispecies imaging in strongly swirling natural gas flames, *Applied Physics* B, 71, pp. 741–746, 2000.

Dillmann, A., Wetzel, T., and Soeller, C., Interferometric measurement and tomography of the density field of supersonic jets, *Experiments in Fluids*, 25, pp. 375–387, 1998.

Goldstein, R. J. (Ed.), *Fluid Mechanics Measurements*, 2nd ed., Taylor and Francis, Boca Raton, FL, 1996.

Hermann, S., Hinrichs, H., Hinsch, K.D., and Surmann, C., Coherence concepts in holographic particle image velocimetry, *Experiments in Fluids* (Suppl.) S108– 116, Springer-Verlag, New York, 2000.

Liepmann, H.W. and Roshko, A., *Elements of Gasdynamics*, John Wiley, New York, 1957.

MATLAB User Support Web site, https://www.mathworks.com/support/.

Schnars, U. and Jueptner, W., *Digital Holography*, Springer, New York, 2005.

Smits, A. J. and Lim, T. T. (Eds.), *Flow Visualization, Techniques and Examples*, Imperial College Press, London, 2000.

4 Flow Velocity Measurements

4.1 INTRODUCTION

Many interesting and useful phenomena in thermal and fluid engineering occur arise from flow processes and associated velocity fields. In Chapter 3, we saw that the flow characteristics change completely at a high angle of attack for a jet aircraft (Figure 3.1) due to flow separation from the aircraft wing–body surface. The resulting velocity field is complex and directly affects the lift and drag of the aircraft. Although flow visualization methods can provide an overall picture of the flow pattern, in many instances, we would like to determine the exact magnitude and direction of the velocity vectors to better understand and control such complex flows.

During the mixing of two different fluids, for example, the concentration field is strongly coupled to the velocity field. We can begin by looking at the governing equations for transport in a turbulent flow in a "thin layer" (thin in the y-direction, as in jet flows and boundary layers).

$$\rho\left(\bar{U}\frac{\partial\bar{U}}{\partial x}+\bar{V}\frac{\partial\bar{U}}{\partial y}\right)=-\frac{dP}{dx}+\frac{\partial}{\partial y}\left(\mu\frac{\partial\bar{U}}{\partial y}-\rho\overline{u'v'}\right),$$
$$\rho\left(\bar{U}\frac{\partial\bar{c}}{\partial x}+\bar{V}\frac{\partial\bar{c}}{\partial y}\right)=\frac{\partial}{\partial y}\left(D\frac{\partial\bar{c}}{\partial y}-\rho\overline{u'c'}\right) \tag{4.1}$$

The top equation in Equation 4.1 is the conservation of momentum equation, whereas the bottom equation is the conservation of a scalar variable, c. In this case, c can be equated to the mass concentration of jet fluid. Both equations are time-averaged to give the time mean of the velocity components and the scalar variable. The bar above each variable indicates a time-mean value, and the prime (e.g., $u'v'$) is used for the root-mean-square (rms) fluctuation about the mean. A time-series dataset for the velocity can be used to obtain the mean and the root-mean-square fluctuation about that mean, which correspond to $\bar{U}$ and u', respectively (see Equations 1.22 and 1.23 in Chapter 1). It is evident from Equation 4.1 that characterization of momentum transport requires full-field time-mean and rms velocity data. In addition, scalar transport is directly coupled to the mean velocity terms for convective transport and to the coupling of the rms velocity and scalar variable for the turbulent diffusion term $-\frac{\partial}{\partial y}\left(\rho u'c'\right)$. In turbulent flows, detailed velocity data are used to construct turbulence energy spectra, length scales, correlations, and other statistical parameters that are useful for understanding and characterizing turbulence. Moreover, velocity

DOI: 10.1201/9781003611837-4

data, if sufficient in accuracy and spatial resolution, can be used to infer other fluid dynamic parameters such as vorticity and pressure.

For flame processes, as shown in Figure 3.2 of the previous chapter, the flame topology and surface area are determined by the velocity characteristics ahead of the flame. The turbulence level and length scales affect the topology of the flame, which in turn influence flame properties such as the burning rate or the flame speed. Because the flame propagates toward the reactant, the upstream velocity field will determine the flame speed relative to the flow speed. In combustors, an imbalance between the flame and flow speeds can result in flashback (flame moving upstream toward the reactant) or blowout (flame moving downstream out of the flame zone). Thus, velocity data are a necessary component in the characterization of many thermal and flow processes.

In Chapter 2, we discussed some methods for measuring volumetric flow rates. Because the velocity field is generally not uniform, the volumetric flow rate gives only the volume-averaged flow velocity. Conversely, velocity field data can be integrated in space to provide the volumetric flow rate. In this chapter, we will discuss methods frequently used for velocity measurements, such as laser Doppler velocimetry (LDV), hot-wire anemometry (HWA), particle image velocimetry (PIV), and other techniques.

4.2 LASER DOPPLER VELOCIMETRY

4.2.1 The Operating Principle

Laser Doppler velocimetry, or LDV, uses the interference of coherent light beams to measure flow velocity. It is a pointwise measurement technique that provides time-series data for velocity, meaning that the velocity data are taken sequentially in time at one or several fixed points. The full flow-field data are obtained by traversing the LDV probe point across and along the flow. A photograph and a schematic of an LDV system are shown in Figure 4.1a and b, respectively. The chain of components starts with a coherent light source, usually a continuous-wave laser such as an argon-ion laser or, in some cases, a He-Ne laser. The laser beam is split into pairs of beams to generate an interference fringe pattern at the focal point of the beams. One or both of the split beams may be *conditioned* to improve the signals. For example, frequency shifts can be applied to the beams using "Bragg" cells, as shown in Figure 4.1b.

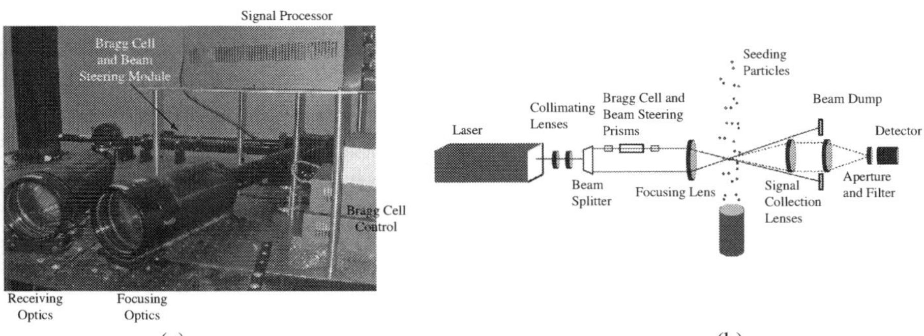

FIGURE 4.1 The laser Doppler velocimeter: (a) a photograph and (b) a schematic.

The beams are then focused using standard optical lenses. Seeded particles in the flow transit the fringe patterns at the focal point and scatter light at a frequency proportional to the velocity. If the seeded particles are sufficiently small, the particle velocity can be assumed to equal the flow velocity at that point. For multi-component velocity measurements, multi-color outputs from the laser are needed to generate two or three interference fringe patterns of different colors, so that the signals can be distinguished from one another using color filters at the detector. The detector is typically a photomultiplier tube (PMT) with a high-frequency response to capture the signal in time. Because velocity is proportional to signal frequency, the signal is sent to a combination of signal processors such as frequency filters and a frequency analyzer.

The principle of LDV operation is related to the Doppler effect, which states that a moving particle will stretch or compress the waveform depending on the relative motion between the particle and the detector. The origin of LDV lies in the application of this Doppler effect to small particles seeded in the flow. The Doppler effect refers to the change in the frequency of a source wave due to the relative motion between the source and the observer and is mathematically described as follows, where c is the speed of light.

$$f_{\text{Doppler}} = f_{\text{source}} \times \frac{1}{1 - \frac{V_{\text{rel}}}{c}} \tag{4.2}$$

$c =$ the speed of light $= 299{,}792{,}458$ m/s

If the relative velocity, V_{rel}, is positive (motion toward the detector), then the observed frequency (f_{Doppler}) is increased (the wave is "compressed") with respect to the source frequency (f_{source}). Conversely, if the relative velocity is negative (motion away from the detector), the frequency is decreased.

For LDV, the Doppler effect must be accounted for twice—once for the laser beam incident upon a moving particle, and again for the scattered light from the moving particle reaching the detector, as shown in Figure 4.2. In Figure 4.2, e_L and $e_{P\text{->}D}$ are the unit vectors in the direction of the laser toward the particle and of the scattered

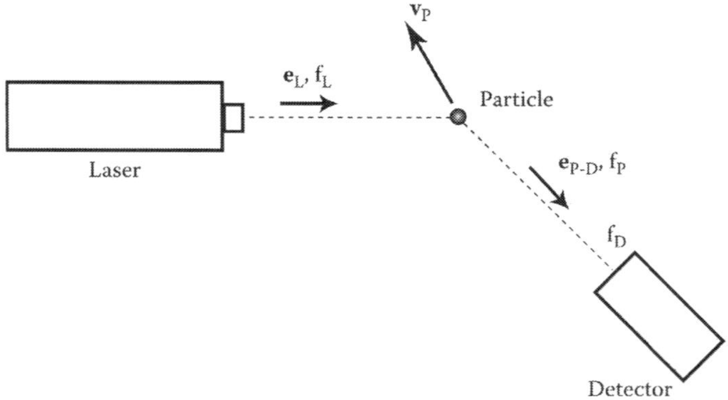

FIGURE 4.2 The Doppler effects from particle velocity.

light from the particle to the detector, respectively. Using Equation 4.2 twice, the final Doppler frequency, f_D, can be written as

$$f_D = f_P \frac{1}{1 - \dfrac{\vec{\mathbf{e}}_{P\to D} \bullet \vec{\mathbf{v}}_P}{C}} = \frac{f_L \dfrac{1}{1 - \dfrac{(-\vec{\mathbf{e}}_L) \bullet \vec{\mathbf{v}}_P}{C}}}{1 - \dfrac{\vec{\mathbf{e}}_{P\to D} \bullet \vec{\mathbf{v}}_P}{C}} \tag{4.3}$$

f_D = Doppler frequency at the detector
f_P = frequency of the light signal from the particle
f_L = frequency of the laser light
v_P = particle velocity

The first part of Equation 4.3 gives the Doppler frequency from the source frequency, which is the frequency of the light signal emanating from the particle, f_P. However, f_P itself is shifted with respect to the incident laser frequency, f_L, due to the particle velocity, v_P, as shown in the numerator of the second equality in Equation 4.3. The relative velocity between the particle and the laser is correctly obtained by taking the dot product of $-e_L$ and v_P.

Equation 4.3 can be simplified using the series expansion, $(1+x)^n \approx 1 + nx + \cdots$ for $x << 1$.

$$f_D \approx f_L \left(1 - \frac{e_L \bullet v_p}{c} + \right)\left(1 + \frac{e_{P\to D} \bullet v_p}{c} + \right) \approx f_L + f_L \frac{v_p \bullet (e_{P\to D} - e_L)}{c} \tag{4.4}$$

Using $c = f_L \lambda_L$, the frequency arriving at the detector is given by the original laser frequency (the first term) plus the Doppler shift (the second term) in Equation 4.5:

$$f_D = f_L + \frac{v_p \bullet (e_{P\to D} - e_L)}{\lambda_L} \tag{4.5}$$

λ_L = wavelength of the laser light

The foregoing equation shows that the Doppler frequency depends only on the particle velocity, the orientation of the laser and the detector, and the frequency or wavelength of the laser beam. However, for a visible-wavelength laser, the frequency f_L is of the order of 10^{14} Hz, whereas for typical flow velocities, the Doppler shift (the second term on the right-hand side of Equation 4.5) ranges from a few to 100 MHz. The frequency difference is therefore about six to eight orders of magnitude, corresponding to the signal processing equivalent of "trying to find a needle in a haystack." For this reason, optical heterodyning must be used for practical signal processing. Heterodyning is a process in which two wave components are optically or electronically mixed to produce the beat frequency, or the frequency difference between the two waves (see the following example). The original development of LDV involved optical heterodyning using a single laser beam, as shown in Figure 4.3, where the signal from the probe volume is mixed with the original laser beam at the detector. However, beam alignment for this single-beam arrangement is difficult and cannot

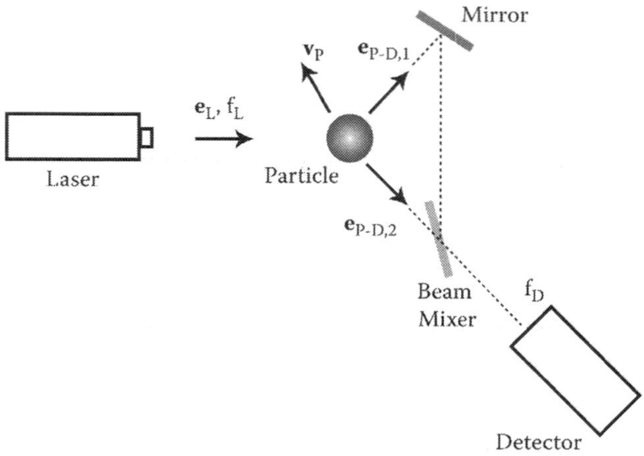

FIGURE 4.3 A single-beam LDV arrangement.

easily be corrected. All modern LDV systems therefore use a dual-beam configuration in which optical heterodyning occurs directly at the probe volume, as shown in Figure 4.4.

Example 4.1: Signal Heterodyning: The "Beat" Frequency

For two beams with frequencies f_1 and f_2, the signals can be represented as sinusoidal waves (see Appendix C).

$$E_1 = E_{o1}\sin 2\pi\left(x / \lambda - f_1 t\right)$$

$$E_2 = E_{o2}\sin 2\pi\left(x / \lambda - f_2 t\right)$$

E_{o1} and E_{o2} are the wave amplitudes, and λ is the wavelength. For frequency considerations, we only need to deal with the time-dependent terms. For simplicity, we can rewrite the foregoing equations.

$$E_1 \sim E_{o1} \sin 2\pi f_1 t$$

$$E_2 \sim E_{o2} \sin 2\pi f_2 t$$

If the two signals of equal intensity are mixed (heterodyning) with $E_{o1} = E_{o2} = E_0$, then the resulting signal is

$$E = E_{o1} \sin 2\pi f_1 t + E_{o2} \sin 2\pi f_2 t = E_o\left(\sin 2\pi f_1 t + \sin 2\pi f_2 t\right)$$

The signal "intensity" is proportional to the square of the sinusoidal function.

$$I \sim E^2 = E_o^2\left(\sin^2 2\pi f_1 t + \sin^2 2\pi f_2 t + 2\sin 2\pi f_1 t \sin 2\pi f_2 t\right)$$

If we look at the cross-product term, 2sin $(2\pi f_1 t)$ sin $(2\pi f_2 t)$, it is equal to

$$2\sin\left(2\pi f_1 t\right)\sin\left(2\pi f_2 t\right) = \cos 2\pi\left(f_1 - f_2\right)t - \cos 2\pi\left(f_1 + f_2\right)t$$

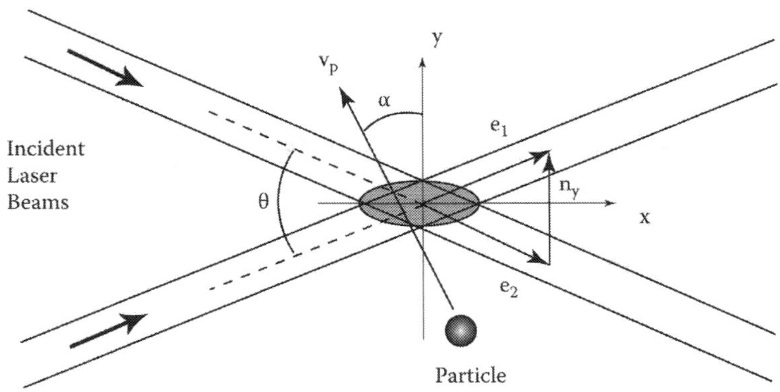

FIGURE 4.4 The dual-beam LDV configuration.

The first cosine function contains the beat frequency, $(f_1 - f_2)$. It can be observed that all of the other frequencies are higher than the frequency difference $(f_1 - f_2)$, so low-pass frequency filters can be used to extract this difference term. Heterodyning is a useful and frequently used method in signal and optical processing.

Two beams are focused to a common focal point in the dual-beam LDV configuration in Figure 4.4. Usually, a single laser is used, and the beam is split into two beams using a beam splitter (see Figure 4.lb). A focusing lens then directs both of the beams in phase to the probe volume. The Doppler frequency of each of the two incident beams can be described by Equation 4.5 using the directional unit vectors, $\vec{e}_1$, and $\vec{e}_2$, for the up-slope and down-slope beams, respectively.

$$f_1 = f_L + \frac{\vec{v}_P \bullet (\vec{e}_{P\to D} - \vec{e}_1)}{\lambda_L} \tag{4.6a}$$

$$f_2 = f_L + \frac{\vec{v}_P \bullet (\vec{e}_{P\to D} - \vec{e}_2)}{\lambda_L} \tag{4.6b}$$

The signals from the two beams mix at the detector and result in a heterodyning effect that produces the beat component, $f_{2\_}f_1$, as shown in the foregoing example. Then, the frequency of the signal at the detector is

$$f_D = f_2 - f_1 = \frac{\vec{v}_p \bullet (\vec{e}_1 - \vec{e}_2)}{\lambda_L} \tag{4.7}$$

From Figure 4.4,

$$\vec{v}_P \bullet (\vec{e}_1 - \vec{e}_2) = \vec{v}_P \bullet \vec{n}_y = 2\sin\frac{\theta}{2}|\vec{v}_P|\cos\alpha$$

Thus,

$$f_D = \frac{2\sin\frac{\theta}{2}}{\lambda_L} v_{Py}$$

$$f_D = \frac{2\sin\frac{\theta}{2}}{\lambda_L} v_{py} \tag{4.8}$$

Equation 4.8 provides the relationship between the observed frequency and the velocity component aligned in the *y*-direction.

An alternate description of the signal frequency being proportional to the particle velocity can be made by calculating the fringe spacing created by the interference of two laser beams crossing at an angle. The crossing of the beams results in constructive and destructive interference of the light waves, producing light-dark patterns in the intensity of the light called the fringe pattern. From Appendix C, a polarized, planar electromagnetic wave, such as the light output from the laser, can be described by its electric field vector, $\vec{E}$:

$$\vec{E} = \vec{e}_V E_o \cos\left(\omega t - \vec{k} \bullet \vec{r}\right) \tag{4.9}$$

Here, $\vec{e}_V$ is the unit vector in the direction of polarization (vertical or y-direction in this case), and E_o is the electric field amplitude. The propagation of the electromagnetic waves in Figure 4.4 is characterized by the circular frequency, $\omega = 2\pi f_L$, the wave vectors, $\vec{k}_1 = k\vec{e}_1 = \dfrac{2\pi}{\lambda_L}\vec{e}_1$, $\vec{k}_2 = k\vec{e}_2 = \dfrac{2\pi}{\lambda_L}\vec{e}_2$ and the position vector, $\vec{r} = \vec{e}_x x + \vec{e}_y y$. The unit vectors $\vec{e}_1$ and $\vec{e}_2$ can be decomposed into their *x* and *y* components, as shown in Figure 4.4. Then, the dot product of the wave vector and the position vector gives the sinusoidal fluctuation in the direction of wave propagation, that is:

$$\begin{aligned}
\vec{k}_1 &= \vec{e}_x \cos\frac{\theta}{2} + \vec{e}_y \sin\frac{\theta}{2}, \\
\vec{k}_2 &= \vec{e}_x \cos\frac{\theta}{2} - \vec{e}_y \sin\frac{\theta}{2}, \\
E_1 &= E_O \cos\left(\omega t - k\left(x\cos\frac{\theta}{2} + y\sin\frac{\theta}{2}\right)\right), \\
E_2 &= E_O \cos\left(\omega t - k\left(x\cos\frac{\theta}{2} - y\sin\frac{\theta}{2}\right)\right)
\end{aligned} \tag{4.10}$$

The electromagnetic wave equation is linear, and the combined electric field is obtained by simple superposition, $E = E_1 + E_2$. From Appendix C, the light intensity (I_L) of the electromagnetic wave is proportional to the square of the electric field strength averaged over time.

$$I_L \sim \langle E^2 \rangle \tag{4.11}$$

Using cos $(A+B)$+cos $(A-B)$=2 cos A cos B, and dropping the time-dependent term due to time averaging, the spatial variation of the light intensity is then given by

$$I_L \propto E_o^2 \cos^2\left(-kx \cos\frac{\theta}{2}\right)\cos^2\left(ky \sin\frac{\theta}{2}\right) \tag{4.12}$$

We are interested in the fringe pattern in the *y* (vertical) direction. Taking the y-component and using $\cos^2 A = 1 + \cos(2A)$, we get the light intensity variation:

$$I_L \propto E_o^2\left[1 + \cos\left(2ky \sin\frac{\theta}{2}\right)\right] = E_o^2\left[1 + \cos\left(\frac{4\pi}{\lambda_L} y \sin\frac{\theta}{2}\right)\right] \tag{4.13}$$

The first term inside the square bracket (*"1"*) represents the constant background intensity, referred to as the pedestal component, and the second term represents the spatial variation in the vertical direction. The spatial variation is again sinusoidal, with a spatial wavelength $\Delta y = \dfrac{\lambda_L}{2\sin\dfrac{\theta}{2}}$ referred to as the fringe spacing. For a particle crossing one fringe spacing, the transit time is $\Delta y/v_{Py}$, and for multiple fringes, the corresponding frequency is the inverse of the transit time:

$$f_D = \frac{2\sin\dfrac{\theta}{2}}{\lambda_L} v_{P_y} \tag{4.14}$$

This is exactly equivalent to the result in Equation 4.8.

Thus, a typical raw LDV signal looks like Figure 4.5a. When the particle is in the bright part of the fringe, the signal intensity is high. Conversely, the signal is

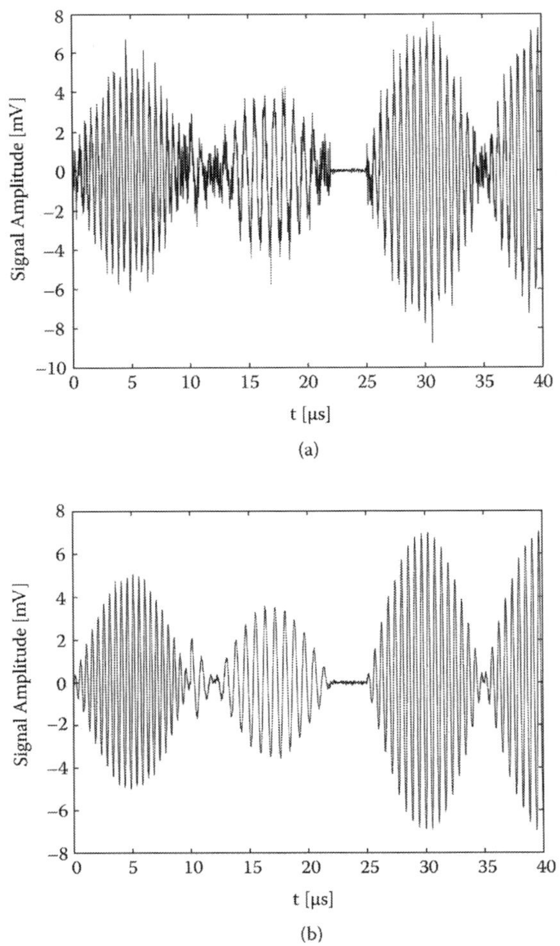

FIGURE 4.5 Typical LDV burst signals.

low when the particle is in the dark part of the fringe. The frequency components should appear reasonably clear for a properly aligned LDV system. The frequency for each signal "burst" can vary in turbulent flows, which simply means that a different velocity magnitude is registered due to intrinsic turbulent velocity fluctuations. Also, the amplitude of each signal envelope will differ depending on the exact path of the particle across the probe volume and the particle size, because the laser beam typically has a nonuniform Gaussian intensity profile. After filtering out high-frequency noise components, the signal is cleaned, as shown in Figure 4.5b, which can then be processed for zero-crossing counts for frequency analysis. For highly turbulent flows, the setting of the low- and high-end frequency filters need to be carefully adjusted to remove noise while retaining valid data frequencies. The time intervals of the zero-crossings are then validated against one another for a number of zero-crossing events set by the operator, which is then converted to frequency for the calculation of velocity using Equation 4.8 or 4.14.

4.2.2 Operational Setup

As shown earlier, an operational LDV setup is most conveniently achieved using a dual-beam configuration with a single-laser source. The most commonly used laser for LDV is the argon-ion laser, which has strong output in the green (514.5 nm) and blue (488 nm) regions, and a weaker but usable output in the violet (454 nm). These three beams can be used for full three-dimensional velocity measurements, which will be discussed later. For one-component velocity measurements, He-Ne or other lasers can also be used.

The laser output needs to be separated by color and also split into dual beams for each color component. Beam splitting is done using standard prism arrangements, as shown in Figure 4.6. Due to relatively long coherence length of gas lasers (argon-ion and He-Ne lasers), typically used for LDV, path-length equalization is not necessary through the beam-splitter optical components.

A Bragg cell is used to shift the frequency of one of the split beams. The purpose of this is to eliminate directional ambiguity. With equal laser frequencies in the split beams, the fringe pattern that is created is stationary, and we cannot determine whether the particle is moving upward or downward. We can only determine the velocity magnitude in the direction normal to the fringes. By introducing a small

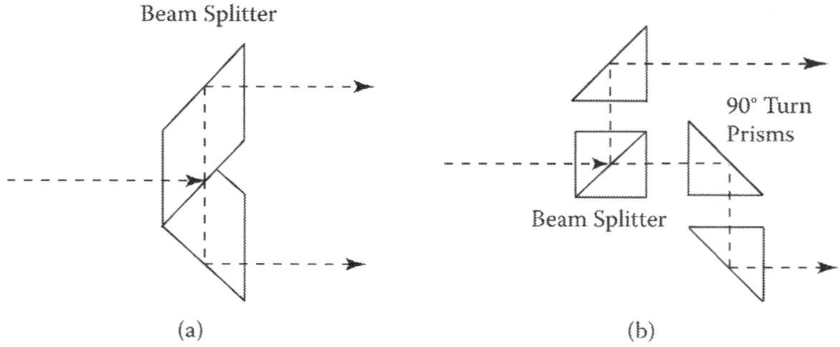

FIGURE 4.6 Beam-splitter arrangements for generating dual beams.

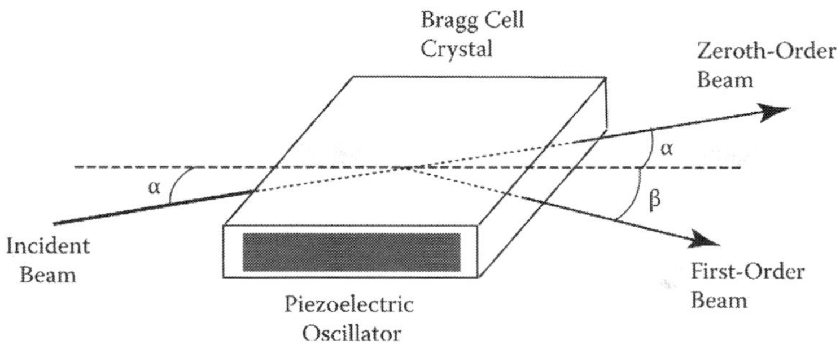

FIGURE 4.7 The Bragg cell.

frequency shift in one of the beams, there will be a frequency component in the signal for particles with zero velocity. This frequency then serves as the reference point, below which the velocity is negative. Physically, this produces moving fringes, so that any positive velocity of the particle must involve motion faster than the motion of the fringes.

The operation of the Bragg cell, shown in Figure 4.7, is based on acoustic modulation of the electromagnetic wave of the laser light. A pressure wave of a constant frequency (typically 10–40 MHz) is generated in the Bragg cell using a piezoelectric crystal. The pressure oscillation has a damping effect on the electromagnetic wave (the $\frac{\sigma}{\varepsilon}\frac{\partial E_y}{\partial t}$ term in Equation C.19b in Appendix C). The interaction of the electromagnetic wave and the pressure wave produces the first-order beam, which is shifted by the pressure oscillation frequency with respect to the frequency of the incident electromagnetic wave. This first-order beam is refracted by an angle β, as shown in Figure 4.7. The zeroth-order beam is also transmitted through the Bragg cell, unrefracted and unaffected in frequency by the acoustic modulation. Inside the Bragg cell crystal, the angles for the zeroth-order and first-order beams are related by $m\lambda = \Lambda_p(\sin\alpha - \sin\beta)$, where λ is the wavelength of the laser beam, Λ_p is the wavelength of the acoustic wave, and m is an integer. Thus, the Bragg cell not only shifts the frequency of one of the beams, which is useful for resolving directional ambiguity, but also splits the incident beam into the zeroth- and first-order beams, allowing it to serve as a beam splitter. The Bragg cell is an essential component in an LDV system with directional discrimination capability. Materials such as flint glass, fused quartz, and tellurium oxide are used depending on the wavelength range, and, as noted earlier, piezoelectric crystals are used as acoustic oscillators. Because the Bragg cell refracts the laser beam, steering prisms are used upstream and downstream of the Bragg cell to maintain parallel alignment between the Bragg-shifted and unshifted beams, as shown in Figure 4.1.

The split beams in an LDV system need to be parallel to create a proper fringe pattern at the focal point. The crossing of the two beams at the focal point forms the so-called measurement volume, as shown in Figure 4.8. From basic optics, the diameter of the beam at the focal point can be estimated as $d = 4\lambda f/(\pi D)$, where D is the initial diameter of the beam. This is the diameter of the beam's cross section, and because the beams intersect at an angle of $\theta/2$, the measurement volume diameter (d_m)and length (l_m) are given by

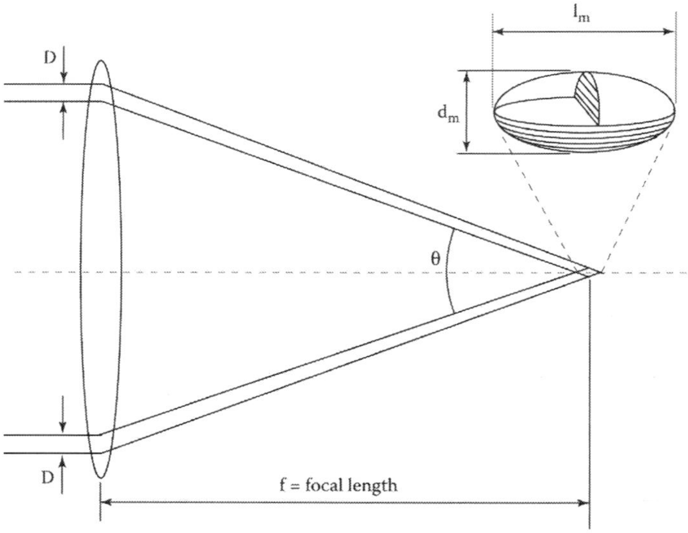

FIGURE 4.8 Focusing optics and measurement volume.

$$d = \frac{4\lambda f}{\pi D}$$
$$d_m = \frac{d}{\cos\frac{\theta}{2}} \tag{4.15}$$
$$l_m = \frac{d}{\sin\frac{\theta}{2}}$$

Example 4.2

For the following LDV specifications, determine the size of the measurement volume: laser wavelength, $\lambda = 514.5$ nm; initial beam diameter, $D = 3$ mm; total beam crossing angle, $\theta = 20°$; and focal length of the transmitting lens, $f = 250$ mm.

Using Equation 4.15,

$$d = \frac{4\left(514.5 \times 10^{-6}\text{mm}\right)(250\text{mm})}{\pi\left(3\text{mm}\right)} = 0.055\text{mm} = 55\ \mu\text{m}.$$

$$d_m = \frac{55\ \mu\text{m}}{\cos\frac{20^\circ}{2}} = 56\ \mu\text{m}.$$

$$l_m = \frac{55\ \mu\text{m}}{\sin\frac{20^\circ}{2}} = 317\ \mu\text{m}.$$

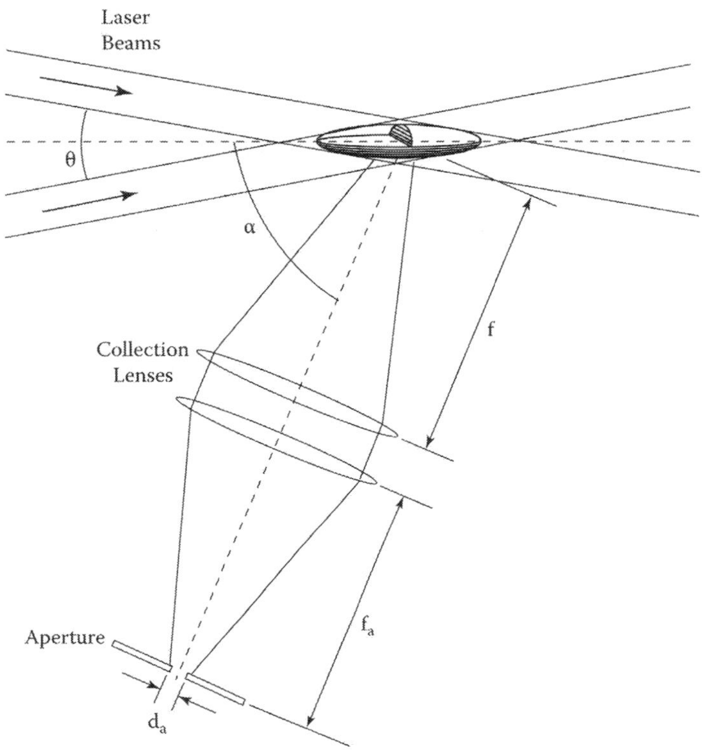

FIGURE 4.9 Probe volume.

Due to the typically small crossing angle, θ, the measurement volume is usually an elongated ellipse.

The probe volume is defined as the volume actually viewed by the detector using the receiving optics arrangement, as shown in Figure 4.9. The length of the probe volume is determined by the angle of the receiving optics, focal lengths, and the pinhole aperture diameter.

$$l_p = d_a \frac{f}{f_a \sin \alpha} \tag{4.16}$$

In Equation 4.16, d_a is the aperture diameter, and f and f_a are the focal lengths of the primary and secondary lenses, respectively, as shown in Figure 4.9. Using a large probe volume results in a higher data rate because more particles are "viewed" by the detector; however, this leads to spatial averaging of the velocity. For detailed turbulence measurements, the probe volume is optimized by making it as small as possible while still allowing for a sufficient data rate.

The LDV components are nowadays provided in compact, pre-aligned optical packages with fiber-optical interfaces between the laser, optics, and detectors, as shown in Figure 4.10. Such systems can be purchased "off the shelf," and the essential mechanics of LDV operation are identical to those of the basic system shown in Figure 4.1. In Figure 4.10, detection is achieved using the same optical components

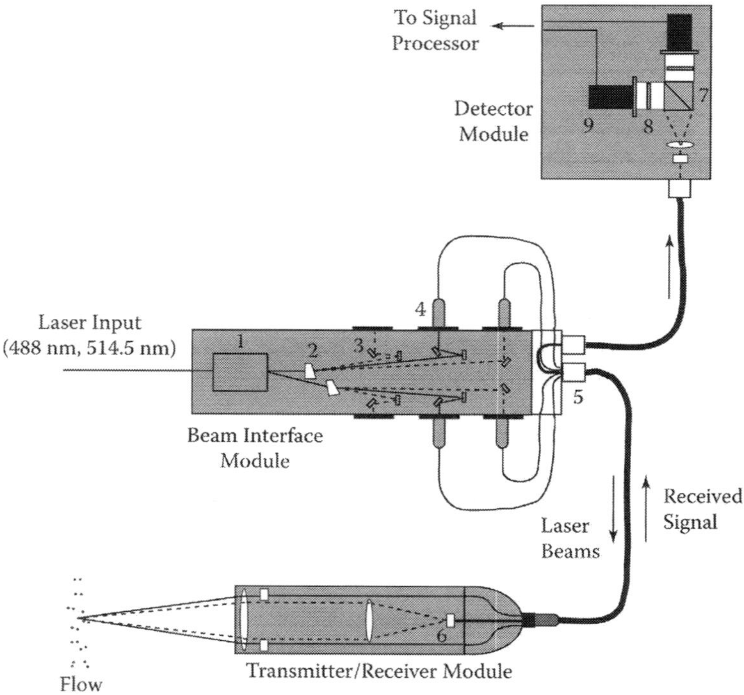

1 = Bragg cell; 2 = Color dispersion prisms; 3 = Mirrors; 4 = Fiberoptic coupling;
5 = Multiline fiberoptic link; 6 = Fiberoptic coupling for received signal;
7 = Dichroic mirror to separate color components; 8=Interference color filters;
9 = Photomultiplier tube

FIGURE 4.10 A compact LDV package with fiberoptic coupling.

as the transmitting optics in a backscatter mode, which somewhat reduces the signal intensity but has the advantage of easy alignment using a common set of optical components. Also, in the backscatter mode, where the optics are shared between laser transmission and signal collection, there is no need to align the measurement volume with the detector, allowing for a "point-and-shoot" operation.

Simultaneous measurements of two velocity components can be made using the green and blue outputs of an argon-ion laser. The beams are separated by color using dichroic mirrors or dispersion prisms (as shown in Figure 4.10), and beam splitters or mirrors (Figure 4.10) are then used to generate parallel beams. Two-component measurements can be performed using either a three-beam or four-beam configuration, as shown in Figure 4.11. The orientation of the parallel beam pair determines the component of the velocity to be measured.

For measurements close to the surface, tilting of the transmitting optical components is necessary so that the lower laser beam does not hit the surface, as shown in Figure 4.12. In such a tilted geometry, the laboratory coordinates do not coincide with the velocity coordinates obtained through the crossed probe beams. A simple coordinate transformation yields the velocity components $\bar{U}$ and $\bar{V}$ in

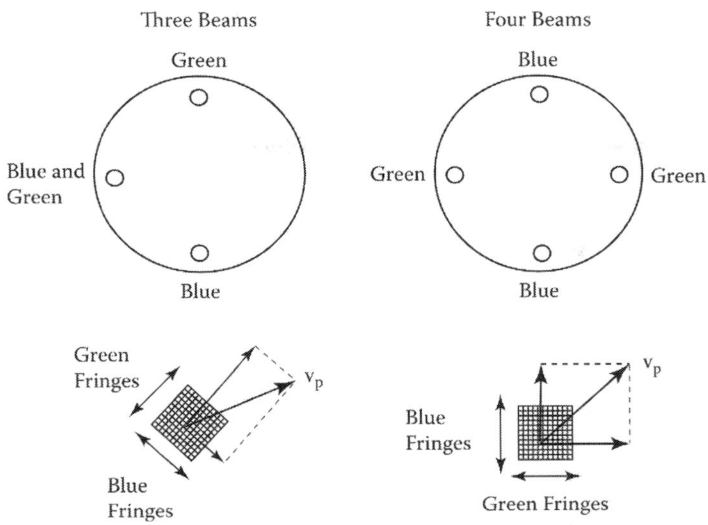

FIGURE 4.11 Beam configurations for two-component LDV.

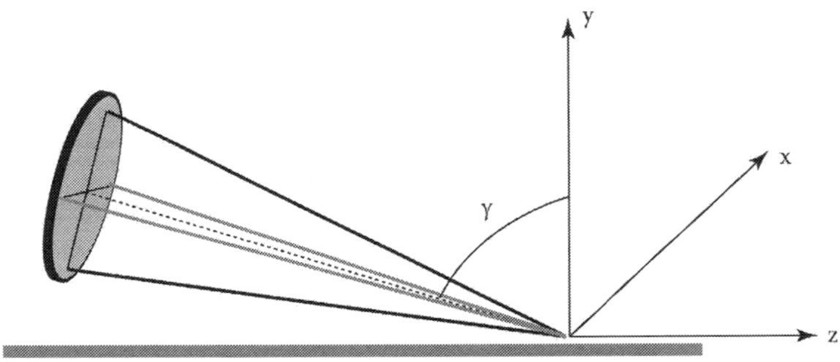

FIGURE 4.12 Tilted probe geometry.

the laboratory coordinates from the observed velocity components $\bar{U}_m$ and $\bar{V}_m$, as follows (Albrecht et al., 2002).

$$
\begin{aligned}
\bar{V} &= \frac{\bar{V}_m}{\cos\gamma} \\
\overline{u'v'} &= \frac{\overline{u'_m v'_m}}{\cos\gamma} \\
\bar{U} &= \bar{U}_m \\
u' &= u'_m \\
v' &= \frac{v'_m}{\cos\gamma}\frac{1-\overline{w'^2}\sin^2\gamma}{\sqrt{\overline{v'^2}\cos^2\gamma+\overline{w'^2}\sin^2\gamma}}
\end{aligned}
\tag{4.17}
$$

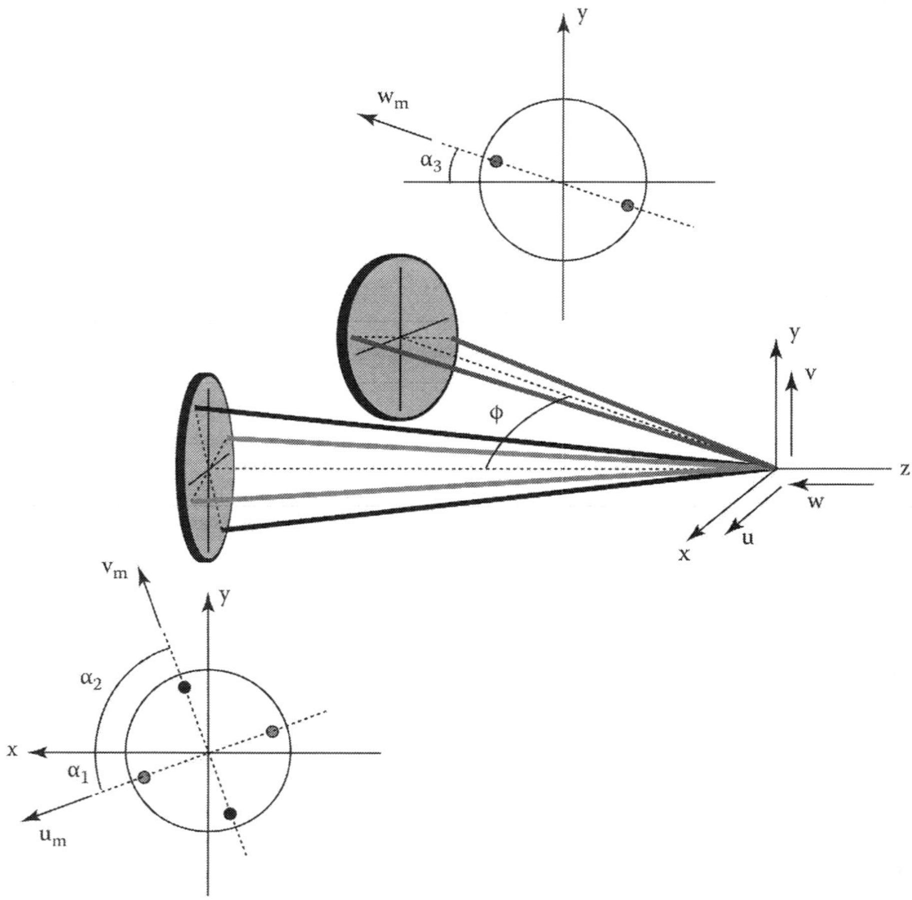

FIGURE 4.13 Three-component LDV configuration and coordinates.

In the foregoing equations, a correction for v' is needed using the w' component because fluctuations of the w-component velocity contribute to fluctuations of v'_m. Again, the barred and primed quantities indicate the time-mean and rms values of the data (see Equations 1.22 and 1.23 in Chapter 1).

For a full three-component velocity measurement, a second transmitting optical component is used, as illustrated in Figure 4.13. If the measurement coordinate system is not aligned with the laboratory coordinate frame, a full three-dimensional coordinate transformation is required (Albrecht et al., 2002).

$$\begin{bmatrix} u \\ v \\ w \end{bmatrix} = \begin{bmatrix} a_{11} & a_{12} & a_{13} \\ a_{21} & a_{22} & a_{23} \\ a_{31} & a_{32} & a_{33} \end{bmatrix} \begin{bmatrix} u_m \\ v_m \\ w_m \end{bmatrix}, \tag{4.18}$$

where

$$a_{11} = \frac{\sin\,\alpha_2}{\sin(\alpha_2 - \alpha_1)}$$

$$a_{12} = -\frac{\sin\,\alpha_1}{\sin(\alpha_2 - \alpha_1)}$$

$$a_{13} = 0$$

$$a_{21} = -\frac{\cos\,\alpha_2}{\sin(\alpha_2 - \alpha_1)}$$

$$a_{22} = \frac{\sin\,\alpha_1}{\sin(\alpha_2 - \alpha_1)}$$

$$a_{23} = 0$$

As shown here, one-, two-, or full three-dimensional components of the flow velocity can be observed using LDV. Because LDV acquires the signal over time through a series of signal bursts, as shown in Figure 4.6, the mean and rms components can be obtained using Equations 1.22b and 1.23b in Chapter 1.

$$\text{Time-mean } = \bar{U} = \frac{1}{N}\sum_{i=1}^{N} U_i \tag{1.22b}$$

$$\text{Root-mean square } = u' \equiv U_{\text{rms}} = \sqrt{\frac{1}{N}\sum_{i=1}^{N} U_i^2} \tag{1.23b}$$

N = number of LDV samples

Other statistical quantities can be evaluated, usually for detailed turbulence measurements, by manipulating the time-series LDV data. Temporal and spatial correlations, turbulence frequency spectra, and cross-correlations of the velocity components (Equation 1.26) can be obtained after some modifications to the basic LDV system and its data processing.

4.2.3 Particle Seeding

Because LDV relies on the Mie-scattering signal from seeded particles, the generation of the particles and their ability to faithfully track the flow are important issues. The motion of a particle in a flow field is described by the Basset–Boussinesq–Oseen equation, which is essentially a momentum balance equation for a particle subject to fluid forces.

$$\frac{\pi}{6} d_P^3 \rho_F \frac{d\vec{v}_P}{dt} = -3\pi\eta d_P(\vec{v}_P - \vec{v}_F) + \frac{\pi}{6} d_P^3 \frac{d\vec{v}_F}{dt} - \frac{\pi}{12} d_P^3 \rho_F \frac{d(\vec{v}_P - \vec{v}_F)}{dt} \tag{4.19}$$

vp = particle velocity
v_F = fluid velocity
ρ_P = particle density
ρ_{F} = fluid density
η = dynamic viscosity of the fluid
d_P = particle diameter

In the foregoing equation of motion for the particle, the body force and the so-called Basset history force are neglected, which is reasonable given the large density ratio between the particle and the fluid. The left-hand side of the equation represents the particle acceleration, and the terms on the right-hand side are the viscous drag force (sometimes called Stokes' drag), the pressure gradient, and the virtual mass (equal to half of the fluid mass displaced by the particle) acceleration term.

It can be immediately observed that the higher viscosity of the fluid, represented in the first term on the right-hand side of Equation 4.19, results in a larger drag force, and therefore the equilibrium between the particle and fluid velocities is achieved more quickly. The particle size requirements to track the flow are relaxed for water in comparison with air due to the larger viscosity of water. For LDV seeding, the particles must be small enough to follow the flow even through high-frequency fluctuations.

A quantitative analysis of the required particle size can be made by solving Equation 4.19 in response to sinusoidal fluid velocity oscillations. From such analyses, the particle response time, t_s, can be computed, which gives the time required for an initially stationary particle to reach 99% of the fluid velocity. The frequency cutoff, f_c, which represents the maximum oscillation frequency that a given particle can follow, can also be computed. The particle sizes corresponding to different cutoff frequencies are listed in Table 4.1. In addition, for heavy particles in air, the particle settling time may also need to be examined. It can be observed from Table 4.1 that

TABLE 4.1
Characteristics of Various LDV Seeding Particles in Air and Water

Fluid	Frequency [kHz]	Particle Diameter [μm]			
		TiO_2	Glass	Water	Silicon Oil
Air	1	1.25	1.5	2.5	2.6
	2	0.89	1.1	1.8	1.9
	5	0.56	0.69	1.1	1.1
	10	0.40	0.49	0.76	0.8
Water	1	10	13		
	2	7.2	8.9		
	5	4.6	5.6	N/A	
	10	3.2	4.0		

Source: Adapted from Albrecht et al. (2002).

the required particle size depends on the frequency of the fluid velocity fluctuations to be measured, e.g., in turbulent flows. In air, a nominal 1-μm particle can track fluctuations up to 2–5 kHz, depending on the type of particle. The maximum allowable particle diameter is much larger in water.

Liquid seeding particles are useful for LDV measurements in gas flows and can be generated using Venturi-type atomizers (see Figure 4.14). In some instances, direct injectors or Berglund–Liu droplet generators are also used. For solid particles such as glass or titanium dioxide, the cyclone-type seeder is commonly employed, as shown in Figure 4.15. For water flows using tap-water supplies, there are usually enough solid particle contaminants in the flow to serve as LDV particles. In any use of seeding particles for LDV, some elementary analysis of the particle size distribution is useful to confirm the frequency response of the particles (see Chapter 6 for some particle size measurement techniques).

LDV is one of the established techniques for flow velocity measurements in both academic and applied research. It can provide full three-component velocity data at high data rate in a non-intrusive manner. If finely tuned, it can be operated with

FIGURE 4.14 A Venturi atomizer.

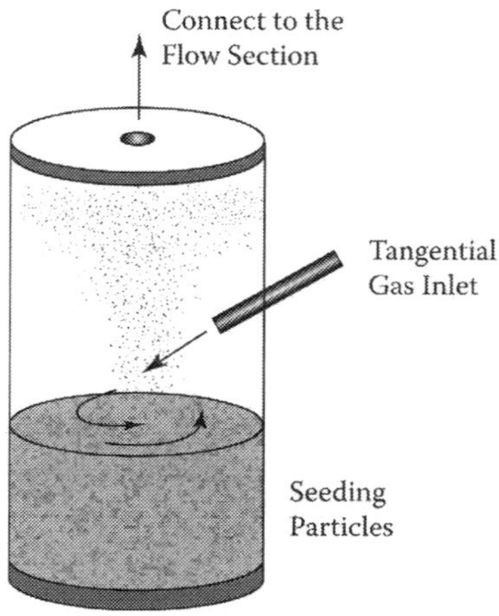

FIGURE 4.15 A cyclone seeder.

sufficient spatial and temporal resolution for detailed turbulence measurements. The technique requires particles to be seeded into the flow at appropriate diameters, depending on the characteristic time scales of the flow. A typical LDV probe, however, occupies some volume, although miniaturized fiberoptic LDV probes are available. Deployment in small-scale devices may require inventive approaches. LDV is a pointwise measurement technique, in contrast to PIV, to be discussed later, so traversing of the probe is needed for full-field flow characterization.

4.3 HOT-WIRE ANEMOMETRY

HWA measure the flow velocity by sensing the convective heat loss from a thin wire sensor exposed to the flow. The flow velocity changes the convective heat transfer rate, which lowers the sensor-wire temperature. Similar to resistance-based temperature detectors, the temperature change then leads to a resistance change that can be electronically detected. Figure 4.16 shows some typical HWA probes. A thin wire sensor (typically less than 2.5–5 μm in diameter) is supported on prong supports (designed to minimize flow disturbance), with the sensor-wire length-to-diameter ratio ranging from 100 to 600 and the sensor-wire length itself ranging from 1 to 3 mm. An electrical circuit is employed to measure the change in sensor current or to maintain a constant temperature across the sensor wire, as described in the following section.

4.3.1 Operating Principle

HWAs consist of a thin wire suspended on prongs and rely on differential convective heat loss from the wire depending on the flow velocity. The change in convective heat loss can be detected by either (1) measuring the resistance change as the sensor-wire temperature is altered by the flow velocity (constant-current method) or (2) keeping the sensor temperature constant using a feedback circuit and measuring the change in current to achieve this constant sensor temperature (constant-temperature method).

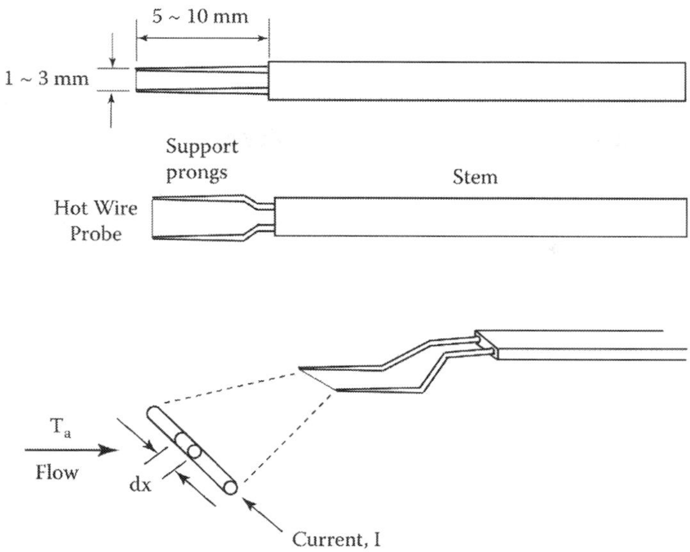

FIGURE 4.16 Hot-wire anemometer probes.

We can consider a small segment (with a length dx) on the hot-wire sensor, as shown in Figure 4.16. The temperature of this wire segment is determined by heat conduction through both ends, convective heat loss through the surface exposed to the flow, and heat generated by electrical resistance through the wire segment. This energy balance can be expressed as follows:

$$\rho_w c_w A_w \frac{\partial T}{\partial t} dx = \frac{I^2 \chi_w}{A_w} dx + k_w A_w \frac{\partial^2 T_w}{\partial x^2} dx + \pi d h (T_a - T_w) dx \tag{4.20}$$

ρ_w, c_w = density and specific heat of the wire material
I = current through the wire element
χ_w = electrical resistivity of the wire material
k_w = thermal conductivity of the wire material
d = diameter of the wire element
T_w, T_a = wire and ambient temperatures
h = heat transfer coefficient

In Equation 4.20, the left-hand side is the time rate of change of internal energy within the wire segment, and on the right-hand side are the heat generation, heat conduction, and convection terms, in that order. The electrical resistivity, a fundamental material property, is used to determine the electrical resistance (in Ω) of a wire segment with a length L and cross-sectional area A_w, and is also a function of temperature as follows:

$$R_w = \frac{\chi_w L}{A_w} = R_{\text{REF}}\left[1 + \alpha (T_w - T_{\text{REF}})\right] \quad [\Omega] \tag{4.21}$$

TABLE 4.2
Properties of HWA Materials

Material	Resistivity at 20°C [Ωm]	Temperature Coefficient of Resistivity [/°C]	Melting Point [°C]	Properties
Tungsten	5.6×10^{-8}	0.0045	3410	Oxidizes above 350°C
Platinum	10.6×10^{-8}	0.0039	1770	Soft and weak
Platinum–rhodium (90/10%)	19×10^{-8}	0.0016	1830	
Platinum–iridium (80/20%)	32×10^{-8}	0.0008	1840	
Nichrome (Ni, Fe, Cr alloy)	100×10^{-8}	0.004		

The resistance–temperature relationship is similar to that for the temperature-based temperature detectors discussed in Chapter 2. The constant α in Equation 4.21 is called the *temperature coefficient of resistivity*, and T_{REF} is typically taken as 20°C for hot-wire sensor materials. Some values of this coefficient for commonly used hot-wire materials are listed in Table 4.2. Tungsten, platinum, and platinum–iridium (80% Pt, 20% Ir) are commonly used as HWA sensor-wire materials. Tungsten wires have good strength but oxidize easily, whereas platinum is impervious to oxidation but has low strength. Platinum–iridium alloy is a compromise but has a low temperature coefficient. Thus, the choice of sensor wire depends on the experimental environment and test conditions.

Under steady-state conditions, and neglecting the small heat conduction term for thin wires in Equation 4.20, we obtain a simplified energy balance between the heat generated due to electrical resistance and the convective heat loss.

$$I^2R_w = \pi dhL\left(T_w - T_a\right) \tag{4.22}$$

To obtain the heat transfer coefficient h, we use the Nusselt number, Nu. For hot-wire anemometer sensors, it is customary to use the Nusselt number for the infinitely long cylinder because the length-to-diameter ratio of the sensor wire is 100 to 600.

$$Nu = \frac{hd}{k} \approx 0.42\mathrm{Pr}^{0.2} + 0.57\mathrm{Pr}^{0.33}\mathrm{Re}^{0.5} \tag{4.23}$$

Now, using the foregoing equation for h and the right-hand side of Equation 4.21 in Equation 4.22, we obtain

$$I^2R_w = \frac{\pi Lk}{\alpha}\frac{R_w - R_a}{R_{\mathrm{REF}}}\left(0.42\ \mathrm{Pr}^{0.2} + 0.57\ \mathrm{Pr}^{0.33}\ \mathrm{Re}^{0.5}\right) \tag{4.24}$$

where R_a is the electrical resistance of the wire at the ambient fluid temperature. Noting that the Reynolds number depends on the flow velocity, U, along with the fact that the terms on the right-hand side are nearly constant except the electrical resistance of the wire, we can write

$$\frac{I^2 R_w}{R_w - R_a} = A + BU^n \tag{4.25}$$

Or,

$$I, R_w \propto U^n \tag{4.26}$$

Thus, a relationship for the electrical current through the sensor wire and the resistance with the flow velocity is established in Equation 4.25, except that both parameters vary with the flow velocity. Therefore, HWA sensors are operated in either (1) constant-current mode ($R_w \propto U^n$); or (2) constant-temperature mode ($I \propto U^n$), using external electrical circuitry to ensure each condition. In the constant-current mode, the current supply to the sensor wire is held fixed, and the resistance change is measured using Wheatstone-type bridge circuits (see Chapter 2, Section 2.2.3). In the constant-temperature mode, a feedback electrical circuit is used to adjust the current so that a constant resistance is maintained in the HWA sensor. In both instances, the constants A and B in Equation 4.25 need to be obtained through calibration. For example, a pressurized chamber with a contoured nozzle can produce exit velocities through the nozzle that can be calculated using gas-dynamic principles (similar to the sonic orifice for flow rate measurements).

Figure 4.17 shows the HWA probe deployed in a flow. Although it is the velocity component normal to the sensor wire, U_N, that the HWA is designed to measure, the velocity vector may not always be aligned with the HWA sensor, for example, in turbulent flows or flows with steep velocity gradients. Other velocity components, as shown in Figure 4.17, will affect the HWA sensor operation through misalignment of the flow velocity and the sensor wire, as well as through interference from the sensor supports. The effective velocity extracted by the HWA sensor is written as (Bruun, 1995)

$$U_{\text{eff}} = \sqrt{U_N^2 + k_T^2 U_T^2 + k_N^2 U_{BN}^2} \tag{4.27}$$

where U_N is the velocity component to be measured, which is normal to both the sensor wire and the supports; U_T is the velocity component that is parallel to the sensor wire (responsible for the support interference); and U_{BN} is the component normal to the sensor but parallel to the supports, as shown in Figure 4.17.

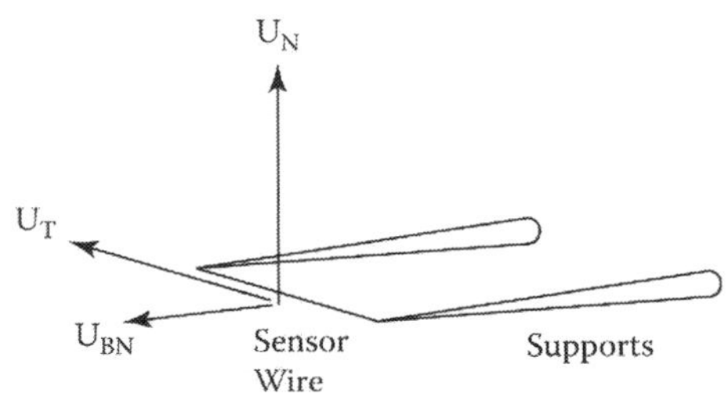

FIGURE 4.17 Angle effect for the hot-wire sensor.

The parameters k_T and k_N are empirically determined. The constant k_T decreases linearly from $k_T=0.2$ to zero as the sensor length-to-diameter ratio (L/d) increases from 200 to L/d of 600–800. The constant k_N can range from 1.0 to 1.2. The effects of k_T and k_N can be reduced by using a "plated" design at the sensor-wire ends, as shown in Figure 4.16. In Figure 4.16, the top probe is "un-plated," while the probe in the middle has the supports bent outward, which is called a plated design. Using such plated designs and proper orientation of the HWA probes, good accuracy can be maintained for single-component velocity measurements without accounting for the wire orientation effects, i.e., $U_{\text{eff}} \approx U_N$. Usually, k_T^2 is a small number, and the U_{BN} component is also small except in intense turbulence (>20% turbulence intensities).

4.3.2 Operational Considerations

Figures 4.18 and 4.19 show the electrical circuitry based on the Wheatstone bridge to operate the HWA in the constant-current mode (Figure 4.18) and constant-temperature mode (Figure 4.19). In the constant-current mode, the operation is quite similar to the resistance measurement for resistance temperature detector (RTDs) (Equations 2.25 and 2.26 in Chapter 2). The potentiometer resistance, R_3, is adjusted until the Wheatstone bridge is balanced, i.e., there is no current or voltage reading across BD ($I_{BD}=0$ and $V_{BD}=0$). The fact that V_{BD} is zero means that the voltage drop across the elements must be the same for both sides of the bridge: $V_{AB}=\mathrm{V}_{\mathrm{AD}}$ and $V_{BC}=V_{DC}$. If $V_{AB} \neq V_{AD}$, then $V_B \neq V_D$, which would violate V_{BD} being equal to zero. Using Ohm's law (V = IR), this condition gives $I_{AB}(R_1+R_w)=I_{AD}R_3$, and I_{BC} $R2=I_{DC}R_4$. The zero-current condition ($I_{BD}=0$) also means that $I_{AB}=I_{BC}$ and $I_{AD}=I_{DC}$. This sets up the Wheatstone resistance relationship for the HWA wire resistance.

$$\frac{R_w + R_1}{R_2} = \frac{R_3}{R_4} \tag{4.28}$$

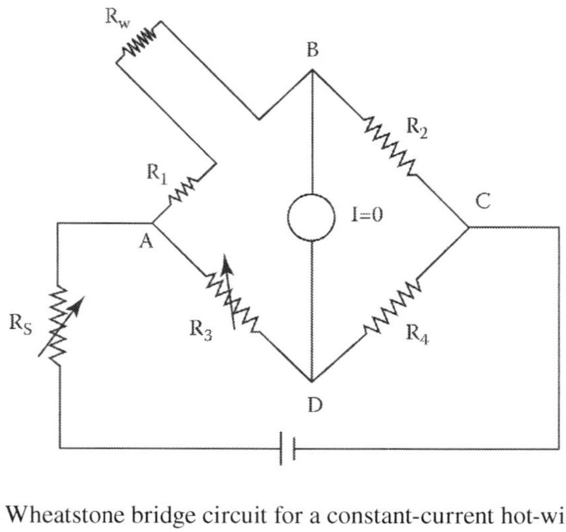

FIGURE 4.18 A Wheatstone bridge circuit for a constant-current hot-wire anemometer.

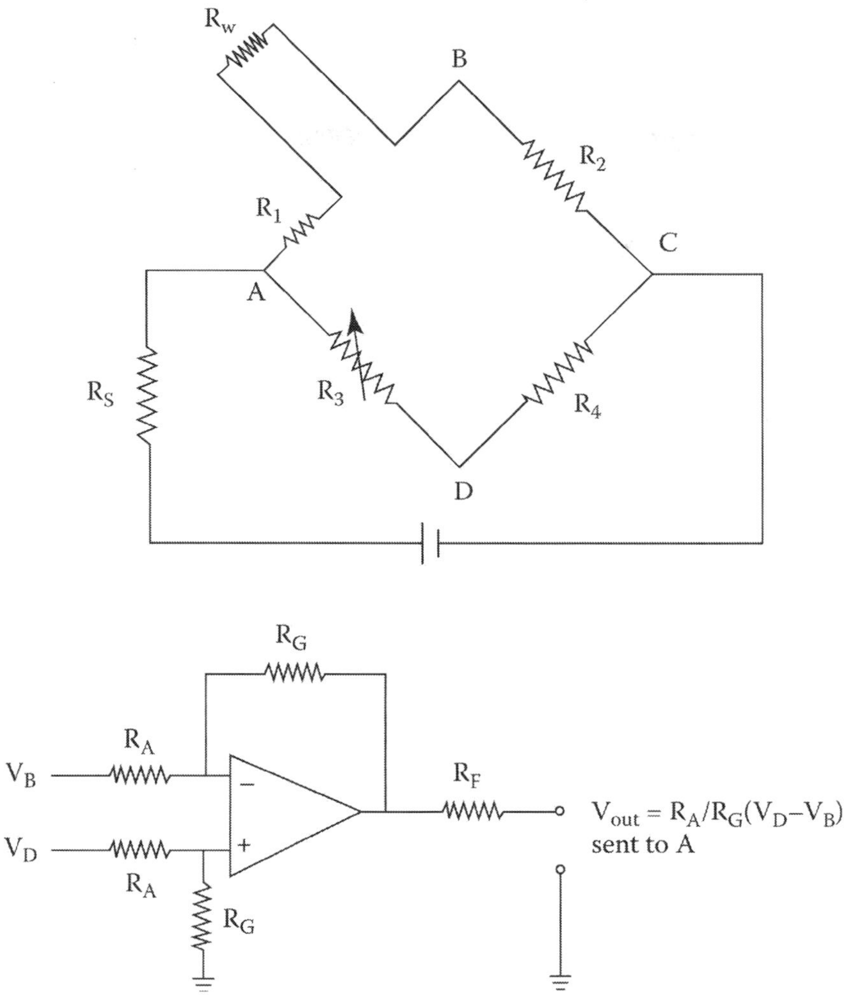

FIGURE 4.19 A feedback circuit for a constant-temperature hot-wire anemometer.

To stabilize the current going through the wire, a resistance–capacitance compensation circuit using an op-amp can be added at "A." For calibration of the constant-current HWA, the current is kept constant at different levels of known flow velocities, and the Wheatstone bridge is balanced by adjusting R_S and R_3, at which point Equation 4.27 can be used to determine R_w. Using $n=0.5$ in Equation 4.25, the calibration constants A and B are then obtained. Subsequent application involves reading R_S and R_C, from which R_w is again obtained through Equation 4.27, and, with the known calibration constants, the flow velocity is extracted. However, the fact that both R_S and R_C need to be adjusted makes the constant-current HWA cumbersome, and in most instances, the constant-temperature operation is preferred, as described in the following text.

The constant-temperature HWA uses a circuit as shown in Figure 4.19, where a fast-response differential amplifier (see Appendix A) is added to the Wheatstone bridge. The flow velocity induces a change in the wire resistance R_w, as in Equation 4.25, and this leads to a change in voltage at point B with respect to the voltage at D (off-balanced Wheatstone bridge). This voltage difference, $V_B - V_D$, is fed to the operational amplifier, which then outputs a voltage and current proportional to the voltage difference and therefore inversely proportional to R_w. This output is connected to the supply voltage to the HWA at "A" in Figure 4.19, thus lowering the current. If the resistance values in the feedback circuit are chosen properly, this feedback loop will maintain a constant R_w. With modern operational amplifiers with very rapid response, along with signal conditioning elements, a frequency response of up to 10 kHz for the feedback loop can be easily achieved.

For one-component velocity measurement using a single hot-wire probe, we can apply Equation 4.27. If we assume that the probe is oriented to measure the primary velocity component U_1, with $\bar{U}_2 = \bar{U}_3 = 0$, where a velocity component is broken into the mean and fluctuating components as in turbulent flows $\left(U = \bar{U} + u'\right)$, then the effective velocity according to Equation 4.27 is

$$U_{\text{eff}} = \sqrt{\left(\bar{U}_1 + u_1'\right)^2 + u_3'^2} \tag{4.29}$$

In the foregoing, we use the facts that $k_T{}^2$ is small and k_N is approximately one. As noted earlier, when u'/U is less than about 0.2, the contribution from u_3' is small (with the error in U_1 measurements being approximately 2% for most turbulent flows). Thus, a common practice is to simply assume that the velocity measured is that of the primary velocity component. Then, the mean and fluctuating quantities of that velocity component are obtained through appropriate statistical processing of the data.

For two-component velocity measurements, an X-shaped sensor is typically used, as shown in Figure 4.20. If sensors A and B are orthogonal and oriented with respect to the velocity components, as in Figure 4.20, then each velocity component can be extracted.

Similar to the LDV probe, offset angles of the sensor wires with respect to the laboratory coordinate axes require a transformation from the measured velocity components to those in the laboratory coordinate system. For an X-shaped HWA probe design with an offset angle α (Figure 4.20), the effective velocities of the two sensors are given through a coordinate transformation in terms of the mean (U_1, U_2, U_3) and the fluctuating (u_1' ,u_2' ,u_3') components in the laboratory coordinate frame.

$$U_{A,\text{eff}}^2 = (U_1\cos\alpha - U_2\sin\alpha)^2 + k_T^2(U_1\sin\alpha + U_2\cos\alpha)^2 + k_N^2U_3^2 \tag{4.30}$$

$$U_{B,\text{eff}}^2 = (U_1\sin\alpha + U_2\cos\alpha)^2 + k_T^2(U_1\cos\alpha - U_2\sin\alpha)^2 + k_N^2U_3^2 \tag{4.31}$$

Similar to the single sensor-wire case, we may take $k_T=0$ and $k_N=1$, and then the foregoing equations simplify somewhat as follows:

$$U_{A,\text{eff}}^2 = (U_1\cos\alpha - U_2\sin\alpha)^2 + u_3'^2 \tag{4.32}$$

$$U_{B,\text{eff}}^2 = (U_1\sin\alpha + U_2\cos\alpha)^2 + u_3'^2 \tag{4.33}$$

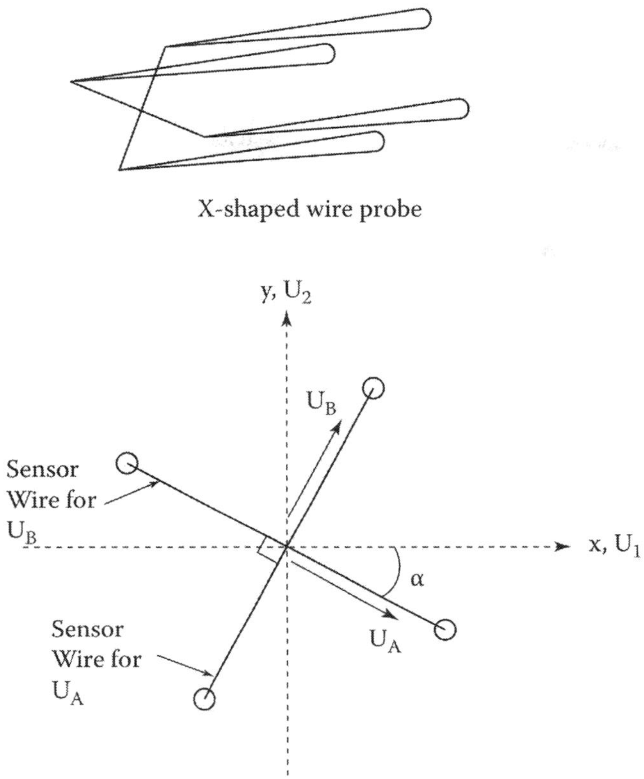

FIGURE 4.20 An X-shaped probe and velocity coordinates.

Furthermore, if we take $\alpha = 45°$ and neglect u_3', then we can obtain simple expressions for U_1 and U_2 from the foregoing two equations.

$$U_1 = \frac{1}{\sqrt{2}}\left(U_{A,\text{eff}} + U_{B,\text{eff}}\right) \tag{4.34}$$

$$U_2 = \frac{1}{\sqrt{2}}\left(U_{A,\text{eff}} + U_{B,\text{eff}}\right) \tag{4.35}$$

This approach can be generalized to the X probe with arbitrary offset angles with respect to the laboratory coordinate axes, as shown in Figure 4.21. The transformation from the measured velocity components (U_A, U_B) to (U, V) is more complex, but the results can be verified through some algebraic work. Again, the barred and primed quantities represent time-averaged and rms values.

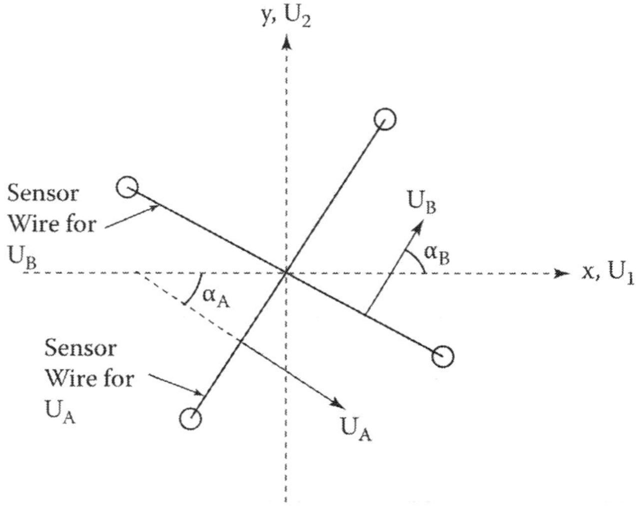

FIGURE 4.21 An X-shaped probe with arbitrary orientations.

$$
\begin{aligned}
\bar{U} &= \frac{\bar{U}_A \tan\,\alpha_B + \bar{U}_A \tan\,\alpha_B}{\tan\,\alpha_A + \tan\,\alpha_B} \\
\bar{V} &= \frac{\bar{U}_B - \bar{U}_A}{\tan\,\alpha_A + \tan\,\alpha_B} \\
\overline{u'^2} &= \frac{\overline{v_A'^2}\tan^2\alpha_B + 2\overline{v_A' v_B'}\tan\,\alpha_A \tan\,\alpha_B + \overline{v_B'^2}\tan^2\alpha_A}{\left(\tan\alpha_A + \tan\alpha_B\right)^2} \qquad 4.36) \\
\overline{v'^2} &= \frac{\overline{v_A'^2} - 2\overline{v_A' v_B'} + \overline{v_B'^2}}{\left(\tan\,\alpha_A + \tan\,\alpha_B\right)^2} \\
\overline{u'v'} &= \frac{\overline{v_B'^2}\tan^2\alpha_A + \overline{v_A' v_B'}\left(\tan\,\alpha_B - \tan\,\alpha_A\right) - \overline{v_A'^2}\tan^2\alpha_B}{\left(\tan\,\alpha_A + \tan\,\alpha_B\right)^2}
\end{aligned}
$$

A multiwire HWA probe geometry, as shown in Figure 4.22, can be used for simultaneous measurement of three-dimensional velocity components and for spatial correlations of the flow velocities. HWAs can be designed for accurate, high-frequency measurements over a wide range of velocities with very good spatial resolution, and thus have been applied in detailed measurements of turbulence properties including spatial correlations, cross-correlations, and turbulence spectra.

The HWA sensor is probe-based and therefore can produce flow perturbations and corresponding errors due to flow interference. For example, the thermal wakes

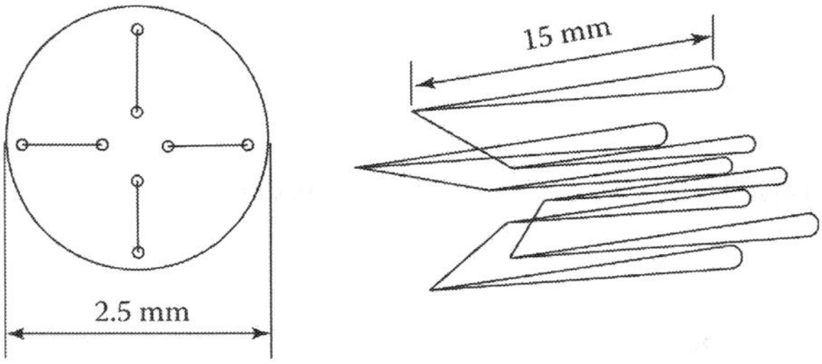

FIGURE 4.22 A four-wire sensor.

caused by multiple sensor wires can alter the sensor-wire response in downstream sensors. Also, the prong wake can be a source of interference. The following sensor component dimensions are recommended to keep these errors below 2% in most flow configurations (Comte–Bellot, 1977).

Stem diameter: 4 mm

Support prong spacing: 3 mm

Support prong length: 20 mm

Support prong diameters: 0.4 mm (base); 0.2 mm (tip)

Wire spacing: 1 mm

Wire length: 1 mm

Finally, HWA is essentially an isothermal technique, and any drift in the fluid temperature needs to be compensated for in HWA measurements. Also, HWA is sensitive to oxidation or other contaminations, in addition to being relatively fragile. These aspects preclude the use of HWA in heat transfer or combustion flow measurements.

4.4 PARTICLE IMAGE VELOCIMETRY

PIV operates on the simple principle that velocity is equal to distance divided by the time interval. In PIV measurements, particles are seeded to follow the flow (similar to LDV), and a sequence of images (usually two) of particles is taken under short-pulse illumination. This pair of images of the same particles displays the distance corresponding to the velocity multiplied by the time interval, as shown in Figure 4.23. The analysis of this displacement distance is facilitated by digital image processing (see Section 3.8 in Chapter 3). The illumination needs to be precisely controlled in time, and usually pulsed Nd:YAG lasers are used to generate two pulses with preset time intervals between the pulses. If each laser pulse has a very short duration (typically 10 ns or less using an Nd:YAG laser), then the particle motion is essentially frozen at the instant of the given laser pulse. If the second laser pulse is applied after a preset time interval (on the order of a few to a few hundred microseconds) following the initial pulse, then the flow velocity is simply the distance traveled by the particles divided by the time interval. The laser light sheet is generated through a combination of spherical and cylindrical lenses, resulting in a light sheet thickness of approximately 1 mm. The thickness of the laser light sheet sets the spatial resolution in the

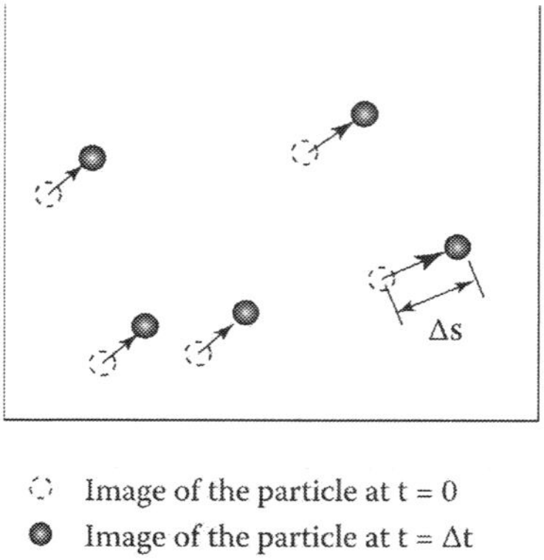

FIGURE 4.23 The basic principle of particle image velocimetry.

direction perpendicular to the light sheet, whereas the pixel density of the digital imaging system sets the spatial resolution in the imaging plane. However, because more than one particle is typically used during post-processing, the spatial resolution is much less than the pixel resolution and is selected by the operator as the best compromise between spatial resolution and accuracy. Because PIV is an image-based detection system, it is a powerful technique for quantitatively visualizing spatial flow patterns and gradient-based parameters (e.g., velocity gradients, vorticity), but it is not well-suited for acquiring detailed time-series data such as turbulence statistics.

The distance between the pair of particle images is analyzed using a spatial correlation technique in PIV. Using image acquisition methods (see Section 3.8 in Chapter 3), the particle images are placed in two M-by-N two-dimensional arrays, where the column and row numbers correspond to spatial coordinates, and the intensity data correspond to the particle or background. For a sequence of image data stored in M-by-N arrays, I_1 and I_2, the spatial correlation can be defined as

$$\Phi(i,j)=\sum_{k=1}^{M}\sum_{l=1}^{N}I_1(k,l)I_2(k+i,l+j) \tag{4.37}$$

Here, k, l, $k+1$, and $k+j$ are the matrix indices that correspond to incremental spatial coordinate locations. In actual implementations, the foregoing cross-correlation function is refined through image normalization to maximize contrast, as follows:

$$\Phi(i,j)=\frac{\sum_{k=1}^{M}\Sigma_{l=1}^{N}\left(I_1(k,l)-\bar{I}_1\right)\left(I_2(k+i,l+j)-\bar{I}_2\right)}{\left[\Sigma_{k=1}^{M}\Sigma_{l=1}^{N}\left(I_1(k,l)-\bar{I}_2\right)^2\Sigma_{k=1}^{M}\Sigma_{l=1}^{N}\left(I_2(k+i,l+j)-\bar{I}_2\right)^2\right]^{1/2}},$$

$$\bar{I}=\frac{1}{MN}\sum_{k=1}^{M}\sum_{l=1}^{N}I(k,l) \tag{4.38}$$

Equations 4.37 and 4.38 produce the cross-correlation between the images I_1 and I_2 in space, as indicated by the values of the indices i and j. The maximum in the

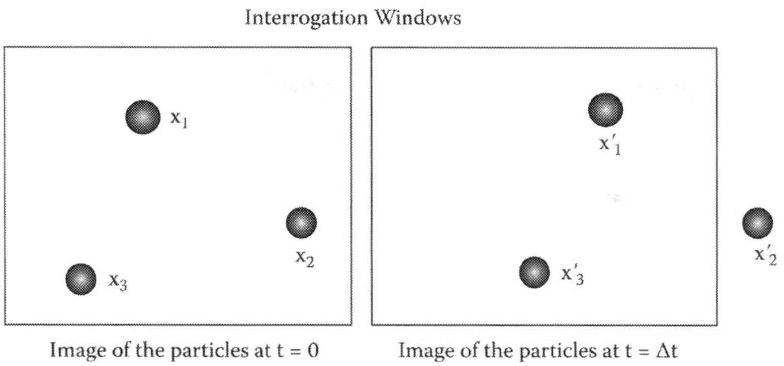

FIGURE 4.24 A Sketch showing the particle motion during a time interval between laser pulses.

cross-correlation will be found at a spatial location that corresponds to the position of the particle image on I_2 $(k+i,\ l+j)$ that has moved from the position of the particle image on I_1 $(k,\ l)$. If there were no particles at $(k,\ j)$, then the intensity or value of the array I_1 at that point will be zero, and no cross-correlation will be produced anywhere.

We can consider a simple example. Figure 4.24 depicts an interrogation spot, where a small subsection of the PIV image is analyzed. Here, the assumption is that the interrogation spot is small enough so that all the particles in it move at the same velocity. The choice of the size of the interrogation spot depends on the particle seeding density, the level of velocity gradients in the flow, and the desired degree of computational speed and accuracy. In the first image in Figure 4.24, we have placed three particle images. If the particles are displaced after a time Δt, the second image will appear as shown on the second image of Figure 4.24. Notice that x_2' is not located within the interrogation spot, leading to a loss of pair, which can very well happen during processing. Also, the cross-correlation is applied to all possible image pairs, including those between identical and different particles, such as $x_1' - x_1$, $x_1' - x_2$, and $x_3' - x_1$ (the correlations between different particles are incidentally referred to as *convolutions*). However, for an appropriately selected interrogation spot, as in Figure 4.24, the correlation between identical particles, $x_1' - x_1$ and $x_3' - x_3$, occur twice, and this leads to the tallest peak in the cross-correlation function. As noted earlier, the algorithm works by searching for the maximum in the correlation function as a function of the separation distance (i, j), and that separation distance divided by Δt gives the velocity vector for that interrogation spot.

Example 4.3

We can consider the simplest array data given below.

$$I_1 = \begin{bmatrix} 1 & 0 \\ 1 & 0 \end{bmatrix}, \quad I_2 = \begin{bmatrix} 0 & 1 \\ 0 & 1 \end{bmatrix}$$

The sequential array data simply indicate that the particles represented by intensity 1 are moving across in the horizontal direction.

It can be verified that the only correlation that is not equal to zero, and therefore the maximum, is

$$\Phi(0,1)=\sum_{k=1}^{2}\sum_{l=1}^{2}I_1(k,l)I_2(k+0,l+1)$$

$$=I_1(1,1)I_2(1,2)+I_1(2,1)I_2(2,2)+I_1(1,2)I_2(1,3)+I_1(2,2)I_2(2,3)$$

$$=(1)(1)+(1)(1)+(0)(0)+(0)(0)=2$$

In the foregoing calculations, any index outside the 2-by-2 window region is ignored. The maximum correlation thus occurs for $i=0$ and $j=1$, and this furnishes the direction and magnitude of the velocity vector for this array data.

During actual calculations, the primary peak will appear from the correlation of identical pairs, and secondary peaks are due to convolutions between different particles and other fluctuation noise. The ratio of the primary and secondary peak heights is then a measure of the reliability of velocity measurement for the interrogation spot.

4.4.1 Particle Image Density

The particle number density captured in the image has an effect on the choice of the interrogation spot size, as noted earlier. If the particle number density (the number of particles per unit volume) is N, and the laser light sheet thickness is δz_o, then the source density (the number of particle images per unit area of the image), N_s, can be written as (Goldstein, 1996)

$$N_s = N\delta z_o \frac{\pi d_I^2}{4M^2} \tag{4.39}$$

M = image magnification due to camera optics
d_I = diameter of the particle image

The diameter of the particle image is also a function of the imaging optics and is given by (Goldstein, 1996)

$$d_I = \left(M^2 d_p^2 + d_s^2\right)^{1/2} \tag{4.40}$$

M = image magnification
d_p = actual particle diameter
d_s = diffraction diameter

The *diffraction diameter* represents the effects of diffraction during the imaging of small particles. For a very small particle (a point particle), the particle image will be formed with a diameter given by d_s.

$$d_s = 2.44(1+M)f\#\lambda \tag{4.41}$$

where $f\#$ is the f-number of the imaging optics (the focal length divided by the lens aperture), and λ is the wavelength of the illumination light.

For small N_s, as given by Equation 4.39, the number of image pairs in the interrogation spot may be small and may also suffer from frequent loss of pairs. At the other extreme, if $N_s >> 1$, this leads to the so-called "speckle" condition, where the particles are so close to one another that they form interference patterns, or speckles, on the images. In fact, speckle velocimetry was a precursor to PIV; however, because it is difficult to maintain such large particle density in gas or liquid flows, speckle velocimetry is nowadays reserved for the detection of solid surface motion.

A common practice is thus to maintain an intermediate N_s and choose a magnification so that velocity gradients can be captured in 32-by-32 or 64-by-64 pixel array interrogation spots. Then, using the cross-correlation function, one velocity vector is produced for each interrogation spot. For example, with a camera pixel density of 1280-by-1024, the foregoing interrogation spot sizes yield 1280 velocity vectors for a 32-by-32 interrogation spot size or 320 vectors for the 64-by-64 spot.

An example of a PIV-generated velocity vector field is shown on the cover of this book. It is a simultaneous PIV-PLIF (planar laser-induced fluorescence) image; the latter is the topic of the next chapter.

4.4.2 Stereoscopic PIV for Three-Component Velocity Measurements

PIV is a planar imaging technique, and thus two of the velocity components are always resolved in vector form. The third velocity component in the direction normal to the imaging plane can also be resolved to some extent using a stereoscopic method. Figure 4.25 shows the arrangement involving two cameras at offset angles. A velocity vector, as shown in Figure 4.25, will appear to have different lengths when viewed by cameras L and R, because the cameras acquire the velocity vector images (through cross-correlations) projected onto the object plane, O-O′. The vector lengths depend on the magnitude of the velocity component w in the normal direction, and

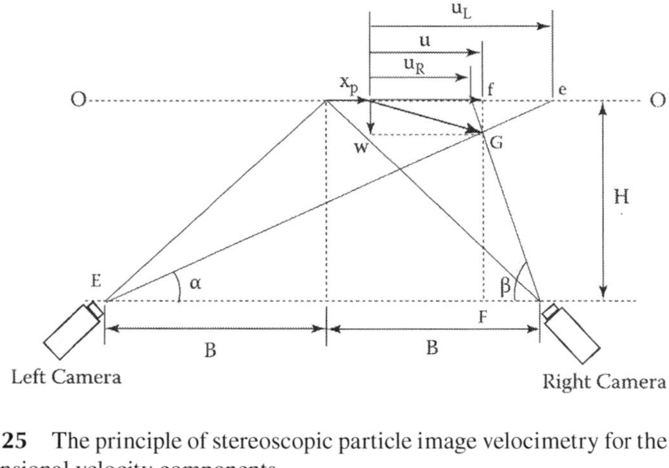

FIGURE 4.25 The principle of stereoscopic particle image velocimetry for the resolution of three-dimensional velocity components.

also on the position of the particle image, x_p. These components can be obtained through geometrical relationships (see the following example).

$$u_L = \frac{Hu + w(B + x_p)}{H - w}$$
$$u_R = \frac{Hu - w(B - x_p)}{H - w} \tag{4.42}$$

The velocity component v, normal to the page, appears to have the same length for both cameras, but because the cameras are not perpendicular to the object plane, v now depends on w and the position of the particle image, y_p.

$$v_L = v_R = \frac{Hv + y_p w}{H - w} \tag{4.43}$$

We can subtract the u_L and u_R equations in Equation 4.42 and then solve for w. Other "true" components can also be found from Equations 4.42 and 4.43 (see the following example).

$$u = \frac{B(u_L + u_R) + x_p(u_L - u_R)}{2B + u_R - u_L}$$
$$v = \frac{y_p(u_L - u_R) + 2Bv_L}{2B + u_R - u_L} \tag{4.44}$$
$$w = \frac{H(u_R - u_L)}{2B + u_R - u_L}$$

Example 4.4

In Figure 4.25, B is the distance from a reference point in the image (e.g., center) to the camera, and x_p is the position of a particular particle. In Figure 4.25, we can see that the triangles EFG and efG are similar. The tangent of the angle for each of these triangles is the same. For the triangle efG, the lengths of the sides are e–f ($\Delta u_L = u_L - u$) and f–G (w).

$$\tan\alpha = \frac{w}{\Delta u_L}$$

For the triangle EFG, the sides have the lengths of $B + x_p + u$ and $H - w$.

$$\tan\alpha = \frac{H - w}{B + x_p + u}$$

Using these, we get u_L.

$$u_L = u + \Delta u_L = u + \frac{w}{\tan\alpha} = u + w\frac{B + x_p + u}{H - w} = \frac{Hu + w(B + x_p)}{H - w}$$

We can perform a similar calculation for the angle β made by the right camera.

$$\tan\beta = \frac{w}{\Delta u_R}$$

$$\tan\beta = \frac{H - w}{B - x_p - u}$$

From these, we get u_R.

$$u_R = u - \Delta u_R = u - \frac{w}{\tan\beta} = u - w\frac{B - x_p - u}{H - w} = \frac{Hu - w(B - x_p)}{H - w}$$

We can rewrite the equations for u_L and u_R to get u and w.

$$(H - w)u_L = Hu + w(B + x_p)$$

$$(H - w)u_R = Hu - w(B - x_p)$$

Subtraction of the foregoing equations eliminates u, and we can solve for w.

$$w = \frac{H(u_R - u_L)}{2B + u_R - u_L}$$

Adding the two foregoing equations and using w, we can get u.

$$u = \frac{B(u_L + u_R) + x_p(u_L - u_R)}{2B + u_R - u_L}$$

4.5 OTHER IMAGE-BASED METHODS

There is a class of other image-based methods to determine flow velocity. This includes scalar imaging velocimetry, molecular tagging velocimetry, and others. They can be considered scalar analogs of PIV, where instead of Lagrangian particles, the concentration field of a conserved scalar, such as fluorescent or phosphorescent dyes, is imaged and processed to determine the velocity field.

Molecular tagging velocimetry (MTV) involves the recovery of flow velocity information from the displacement of phosphorescent or fluorescent dyes. Both phosphorescence and fluorescence refer to light emission from molecules when they transition from a higher energy state to a lower energy state. Fluorescence corresponds to a quantum-mechanically "allowed" transition between molecular electronic energy states (see Appendix D on Quantum Mechanics), and for this reason has a short lifetime with emissions lasting typically a few nanoseconds. Phosphorescence, on the other hand, is due to "forbidden" transitions between electronic energy states, and therefore is associated with a very low probability and a long emission lifetime on the order of a few milliseconds. For MTV, phosphorescent supra-molecules, such as 1-Bromon-naphthalene (10^{-5M} concentration in aqueous solution) bound to glucosyl-modified cyclo-dextran (Gb-CD) molecule (10^{-3M} concentration), are used with alcohol added to this mixture. The presence of alcohol leads to a phosphorescence signal in the range 480–650 nm with a phosphorescence lifetime of 0.1–5 ms, which is excitable using an XeCl-excimer laser at a wavelength of 308 nm. An alternative tagging dye for MTV is a so-called caged dye or photoactivated fluorophore.

These are normally fluorescent molecules that have been rendered non-fluorescent through *chemical caging*, or attachment to a chemical group that suppresses fluorescence. The tagging laser cleaves the caging chemical group, leaving only the exposed molecules susceptible to fluorescence in the subsequent interrogation through planar laser-induced fluorescence imaging.

The sequence of events for MTV using caged dye is shown in Figure 4.26. A laser "line" illumination on the dye "tags" the fluid, the phosphorescence signal of which is detected using an imaging device. To determine the velocity vectors, a grid illumination is used, as shown in Figure 4.27. Movements of a given grid point during a known time interval between tagging and imaging provide information on the velocity of the fluid at that point.

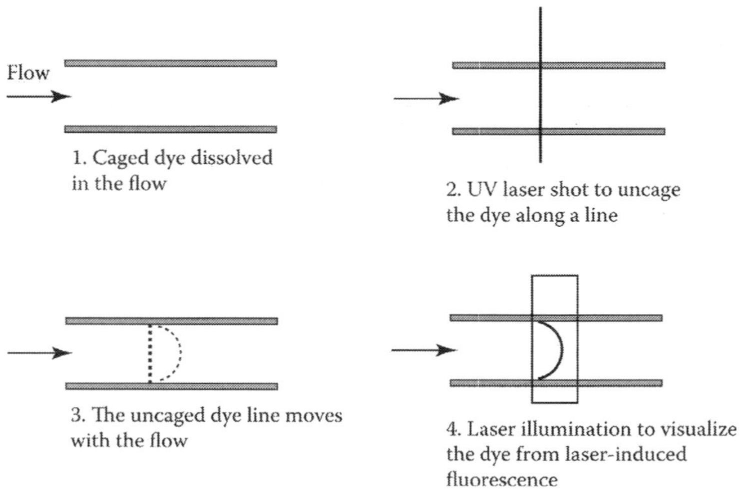

FIGURE 4.26 The operating principle of the molecular tagging velocimetry: use of caged dye.

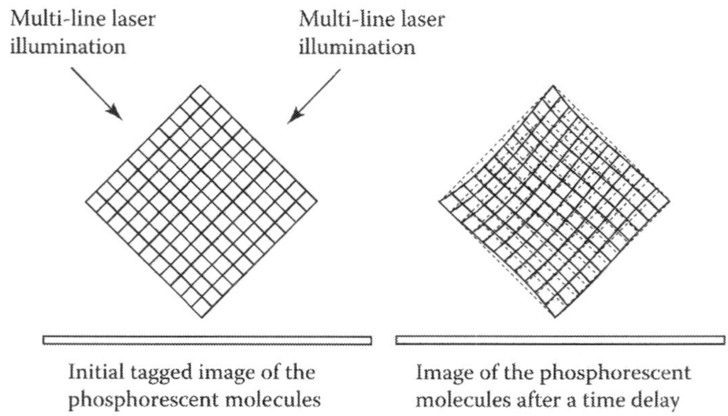

FIGURE 4.27 Grid illumination for two-dimensional velocity resolution in molecular tagging velocimetry.

TABLE 4.3
Types of Molecular Tagging Velocimetry

Name	Molecule	Excitation Method
RELIEF (Raman excitation +	O_2 laser-induced electronic fluorescence	Raman pair + UV
APART (air photolysis and recombination tracking)	O_2	Three UV
VENOM (vibrationally excited NO monitoring)	NO_2	Three visible, UV
OTV (ozone tagging velocimetry)	O_3	Three UV
HTV (hydroxyl tagging velocimetry)	OH	Three UV
KTV (krypton tagging velocimetry)	Kr	Three visible, UV
Acetone MTV (acetone molecular tracking velocimetry)	Acetone	UV
PHANTOMM (photoactivated non-intrusive tracking of molecular motion)	PAF	Three visible, UV
FLEET (femtosecond laser electronic *excitation tagging*)	N_2	UV to NIR (near infrared)

Fairly large variations of MTV have been implemented, as shown in Table 4.3. Some involve a combination of laser breakdown and excitation schemes. For example, in OTV (oxygen tracking velocimetry), tag line(s) of ozone (O_3) are generated from O_2 in air using a high-intensity pulsed ArF (argon fluoride) excimer laser at 193 nm (UV). Then, a second excimer laser (KrF) operating at 248 nm (UV) causes photo-dissociation of O_3 back to O_2, accompanied by a fluorescence signal. This second beam can be sent as a laser sheet so that two-dimensional tracking of molecules is enabled. Methods such as FLEET (femtosecond laser electronic excitation tracking) require a single-laser source and are preferred in some applications (e.g., high-speed flows discussed in the next section).

An interesting technique called *scalar imaging velocimetry* has been suggested by Dahm and coworkers (Dahm et al., 1992). Although the method is based on the premise that high-resolution spatial and temporal data on scalar concentration are available and that the Schmidt number is large, it is an interesting example of using scalar data to retrieve velocity information and will thus be discussed briefly.

The scalar concentration, ζ, in a flow field obeys a conservation equation that states that the time rate of change of the scalar concentration is balanced by convection and diffusion:

$$\frac{\partial \zeta}{\partial t} + \vec{u} \bullet \nabla \zeta = D\nabla^2 \zeta \tag{4.45}$$

D is the molecular diffusivity. The convection term $\vec{u} \bullet \nabla \zeta$ can be interpreted as the velocity component aligned in the direction of the concentration gradient (Dahm et al., 1992). That is,

$$\vec{u} \bullet \nabla \zeta = u_{//} |\nabla \zeta| \tag{4.46}$$

Thus, this particular velocity component can be obtained from Equation 4.46, provided that the scalar concentration data in both space and time are available at sufficient resolution and accuracy to permit the second-derivative calculations included in Equation 4.45.

$$\overrightarrow{u_{//}} = \frac{D\nabla^2\zeta - \dfrac{\partial\zeta}{\partial t}}{|\nabla\zeta|} \tag{4.47}$$

The objective is to determine the actual velocity vector, $\vec{u}$ and to do so, the following equation for the spatial variation of $\overrightarrow{u_{//}}$ is used using the chain rule for differentiation.

$$\nabla\overrightarrow{u_{//}} = \vec{u} \bullet \nabla\overrightarrow{e_{\nabla\zeta}} + \nabla\vec{u} \bullet \overrightarrow{e_{\nabla\zeta}} \tag{4.48}$$

where $\vec{e}_{\nabla\zeta} = \dfrac{\nabla\zeta}{|\nabla\zeta|}$

The foregoing equation states that the change in $\vec{u}_{//}$ is due to both changes in $\vec{u}$ itself and in the concentration gradient, represented by a unit directional vector in the direction of the concentration gradient, $\overrightarrow{e_{\nabla\zeta}} = \dfrac{\nabla\zeta}{|\nabla\zeta|}$. This is illustrated in Figure 4.28, where the local velocity vector and its projection onto the $\nabla\zeta$ vector are shown from one point to another.

Equation 4.48 contains not only the unknown flow velocity vector at a given point but also the unknown velocity components at adjacent points because the $\nabla\vec{u}$ term also appears in the equation. Thus, an iterative solution procedure is required where the velocity terms are solved based on an initial estimate of the flow velocities.

Solution Procedure:

1. Obtain an initial estimate of $\vec{u}$ from $\vec{u}_{\text{initial}} \approx \overrightarrow{u_{//}} = \vec{u} \bullet \overrightarrow{e_{\nabla\zeta}} = \dfrac{D\nabla^2\zeta - \dfrac{\partial\zeta}{\partial t}}{|\nabla\zeta|}$
2. Iterate until the following equation converges:

$$\nabla\overrightarrow{u_{//}} = \vec{u}_k \bullet \nabla\overrightarrow{e_{\nabla\zeta}} + \nabla\vec{u}_{k-1} \bullet \overrightarrow{e_{\nabla\zeta}} \tag{4.49}$$

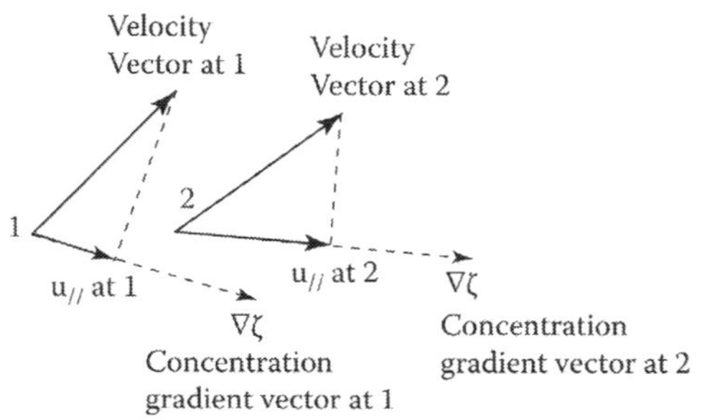

FIGURE 4.28 The vector relationship between the velocity and scalar gradients for the scalar imaging velocimetry.

In step (2), $\nabla\vec{u}_{//}$ on the left-hand side is available from $\vec{u}_{//}$ (Equation 4.47), whereas $\vec{e}_{\nabla\zeta}$ and $\nabla\vec{e}_{\nabla\zeta}$ are obtained directly from the measured scalar field. Thus, from the initial estimates of the flow field based on $\vec{u}_{//}$, the velocity field is updated until the left- and right-hand sides of Equation 4.49 match.

4.6 HIGH-SPEED VELOCIMETRY

Due to interest in hypersonic vehicles (Mach number > 5), velocimetry techniques are adapted for high-speed flows. Some are applications of methods described in the previous sections, but with important modifications to account for aero-optical effects and high-temperature conditions. In addition to the increased velocity, high-speed flows involve a large pressure and temperature range, accompanied at times by chemical reactions and high heat flux. Measurements of the scalar variables are discussed in the next chapter. Testing high-speed flows also requires complex and more restrictive test facilities, e.g., shock tubes and high-speed wind tunnels.

Doppler global velocimetry (DGV) uses the same principle as the LDV discussed in Section 4.2, and as prescribed by Equation 4.5. For supersonic flows, a typical Doppler shift may be in the hundreds of MHz. Instead of two-beam optical heterodyning, direct measurements of the Doppler shift are made at the detector in DGV, thus in principle making use of the arrangements in Figures 4.2 and 4.3. This approach renders possible two-dimensional measurements or imaging of the velocity field, the challenge being the detection of MHz MHz-level frequency shifts from GHz-level signals. Beam mixing at the detector side (Figure 4.3) can be used. Alternatively, interferometers (Fabry–Perot, Michelson) or an iodine absorption cell can be used to discern the frequency shift. Basic principles of interferometers are discussed in Appendix B. Fabry–Perot interferometers make use of multiple reflections between closely spaced semi-reflective layers, as shown in Figure 4.29. By tuning the incident light angle (α in Figure 4.29), the transmittance wavelength (λ) can be set, according to

$$2d\cos\alpha = m\lambda \tag{4.50}$$

In this way, the precise wavelength corresponding to the Doppler-shifted frequency can be sequentially scanned. The frequency shift can be converted to the flow velocity using Equation 4.5.

The iodine cell method makes use of the finite absorption bandwidth of iodine vapor, as shown in Figure 4.30. Due to the random motion of atoms, the iodine absorption band is broadened, leading to a finite bandwidth. Depending on the frequency shift, different amounts will be absorbed. By selecting the absorption wavelength range, the Doppler shift can be measured using the differential absorption. This requires careful calibration, but then three-dimensional components of the flow velocity vector can be imaged.

Application of PIV in high-speed flows requires fast-framing cameras, along with appropriate selection and distribution of seeding particles. Both are necessitated by the short time scales of the flow under supersonic or hypersonic conditions. Other challenges include different drag due to polydisperse seeding particles. Due to the high velocities involved, even small variations in the seeding particle diameter lead to broadening of the observed velocities. Also, if a large interrogation window is selected, additional errors are introduced due to velocity amplification from the

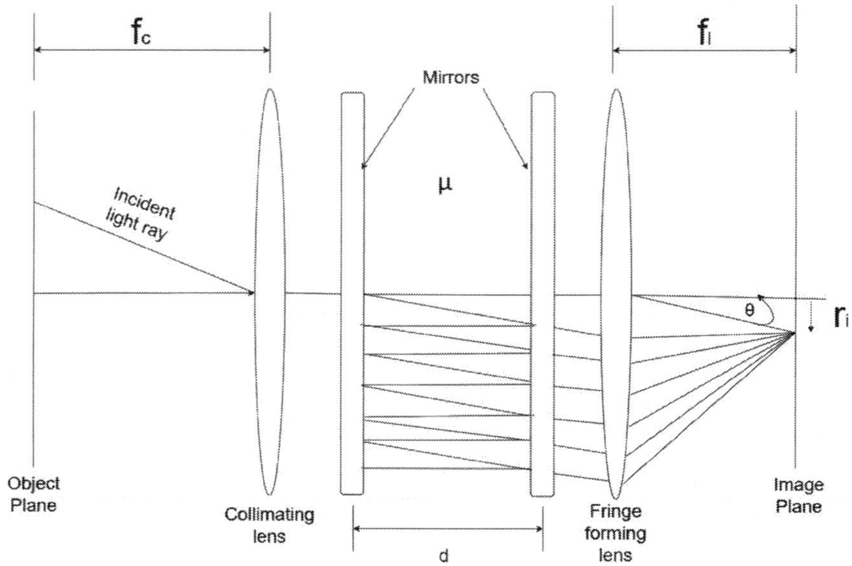

FIGURE 4.29 Operating principle for Fabry–Perot interferometer.

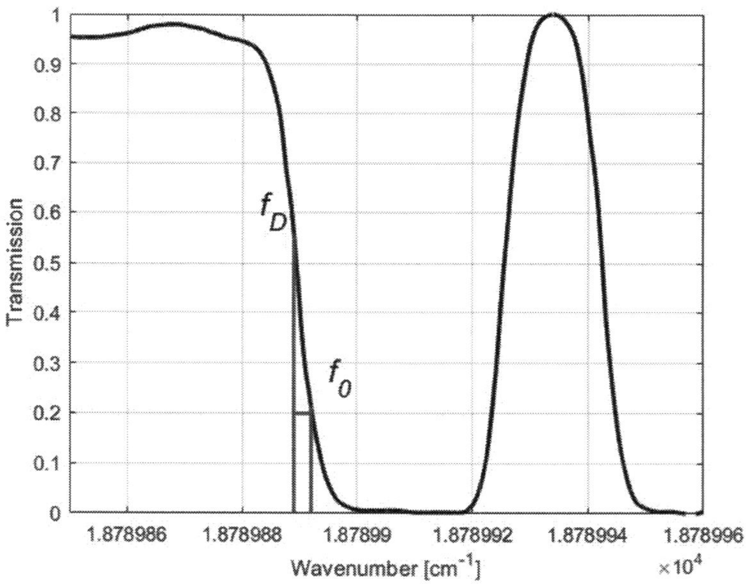

FIGURE 4.30 Iodine absorption cell method, for Doppler global velocimetry.

particle time lag. From Equation 4.19, the time lag (τ_p) for a seeding particle (diameter = d_p, density = ρ_p) can be estimated as

$$\tau_p = d_p^2 \frac{\rho_p}{16\mu} \tag{4.51}$$

Since compressible flows often involve density variations, this leads to a nonuniform refractive index field, resulting in refraction errors that impact the cross-correlation distance. In spite of these difficulties, successful PIV in high-speed flows is possible. Seeding particles such as calcium carbonate and graphite have been used, while liquids (olive oil, acetone) can be introduced as aerosols.

FLEET and PLEET (picosecond laser electronic excitation tagging) refer to a class of MTV, as shown in Figure 4.26, except very short-pulse lasers are used to excite nitrogen molecules. A femtosecond is 10^{-15} s, while a picosecond is 10^{-12} s. Ti:sapphire lasers generate 50~70 fs laser pulses at 800 nm to excite nitrogen molecules, and the resulting laser-induced fluorescence is imaged to track the flow velocity field (Figure 4.26). With frequency modulation (doubling, tripling, or quadrupling), the fundamental output wavelength (800 nm) of Ti:sapphire lasers can generate 400, 267, or 202 nm for FLEET. Nd:YAG (neodymium-doped yttrium aluminum garnet) with 1,064 nm output has been used for PLEET. While other MTV methods involve seeding of molecules (e.g., NO_2, acetone), nitrogen excitation used in FLEET and PLEET bypasses this requirement for flows involving air.

PROBLEMS

4.1. a. If the sound speed is 340.29 m/s, what is the Doppler-shifted frequency of a sound with an original frequency of 11.5 Hz when the sound source moves toward the receiver at 100 m/s?

b. The speed of light, c, in vacuum is 299,792,458 m/s. What is the Doppler-shifted wavelength of light with an original wavelength of 532 nm when the light source moves away from the receiver at 100 m/s? Note $c = f\lambda$.

c. What is the speed of a light source moving toward the observer if the Doppler shift of 0.0025 nm is observed?

4.2. Consider the following fundamental LDV configuration, with the Doppler-shifted frequency given by Equation 4.5.

$$f_D = f_L + \frac{\vec{v}_P \bullet \left(\vec{e}_{P\rightarrow D} - \vec{e}_L\right)}{\lambda_L} \tag{4.5}$$

λ_L = wavelength of the laser light

Using $c = 299{,}792{,}458$ m/s and a laser wavelength of 514 nm, determine the Doppler-shifted frequency of the signal if the magnitude of the velocity is 2.5 m/s at an angle of 25 degrees from vertical.

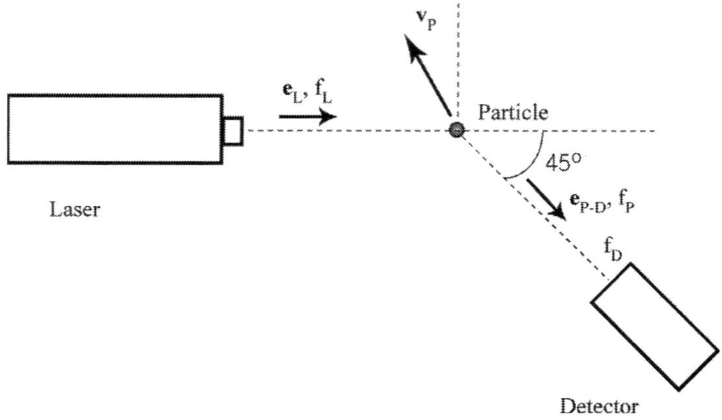

4.3. If we optically mix (heterodyning) two light signals of wavelengths 514 and 514.28 nm, find all of the wavelength components that will be observed.

4.4. Consider the dual-beam LDV configuration shown below, with the laser wavelength of 488 nm. If the laser beam cross-angle is 15 degrees and the magnitude of the velocity vector is 12.2 m/s with $\alpha = 20°$, what is the resulting frequency of the LDV signal?

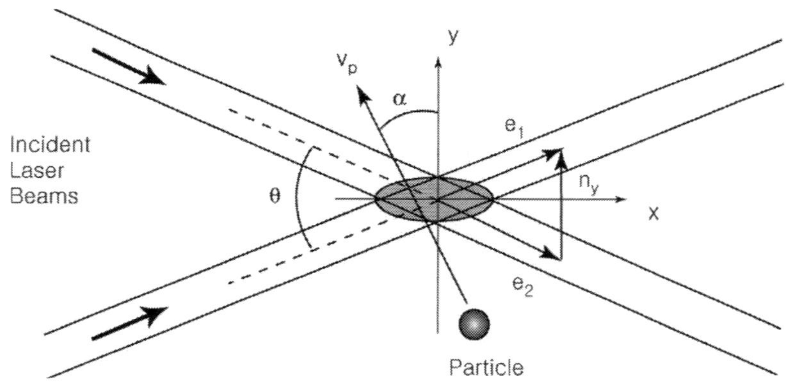

4.5. Show that the fringe spacing resulting from the crossing of two beams at an angle θ with a laser wavelength of λ_L is given by $\Delta y = \dfrac{\lambda_L}{2\sin\dfrac{\theta}{2}}$.

4.6. Determine the ideal dimensions (diameter and length) of the LDV measurement volume, for $\theta = 7.5°$ and laser wavelength of λ_L of 514 nm. The initial laser beam diameter, D, is 0.9 mm. What are experimental factors that can lead to deviations from these ideal dimensions? Use $f = 250$ mm.

4.7. Sketch a two-component (horizontal and vertical velocity components) LDV system using the same components as shown in Figure 4.1b, except now we will need two sets of optical components for the two velocity components.

The laser outputs two wavelengths of 512 and 488 nm, so beam separation optics also need to be included.

4.8. The time scale of turbulence can be estimated by using the length and velocity scales. For example, if the turbulence velocity and length scale are $u' = 1.5$ m/s and $l = 1.5$ mm, then the time scale is estimated at $l/u' = 1$ ms or a frequency of 1 kHz. For the flow frequency in air, what are the maximum diameters for the seeding particles made of titanium dioxide, glass, and silicon oil?

4.9. Using tungsten wire of 5 mm length and 200 μm diameter in air, determine the relationship (Equation 4.24) between the hot-wire anemometer current and the flow velocity. The wire is kept at a constant temperature of 125°C, and the ambient temperature is 20°C. The Prandtl number, *Pr*, can be taken to be 0.707, while *k* is the air conductivity of 0.0263 W/mK. Air density and viscosity at 1 atm and 20°C will also be needed in Equation 4.24.

$$I^2 R_w = \frac{\pi L k}{\alpha} \frac{R_w - R_a}{R_{REF}} \left(0.42\text{Pr}^{0.2} + 0.57\text{Pr}^{0.33}\text{Re}^{0.5}\right) \tag{4.24}$$

4.10. For calibration of the above hot-wire probe, a current of 275 mA is observed at a flow velocity of 1.1 m/s. What is the wire resistance?

4.11. The effective flow velocity (U_{eff}) measured by a hot-wire probe is given by Equation 4.29. The mean flow velocity (U_1) of 2.25 m/s, and the turbulence fluctuation in the primary flow direction u_1' is 10% of the mean flow velocity. Plot U_{eff}/U_1 velocity as a function of u_3' / U_1. At what u_3' / U_1 does U_{eff} deviate more than 10% from U_1.

4.12. For a two-component hot wire probe aligned at 45° from the primary flow direction, determine the flow velocity in the primary and lateral directions if the observed velocities are $U_{A,eff} = 4.5$ m/s, and $U_{B,eff} = 2.7$ m/s. u_3' / U_1 is estimated to be 0.05. What are the flow velocities if u_3' is neglected?

4.13. For the data arrays given below, calculate the following cross-correlations.

a. $\Phi(2,0)$

b. $\Phi(1,0)$

c. $\Phi(0,1)$

$$I_1 = \begin{bmatrix} 127 & 25 & 35 \\ 17 & 115 & 5 \\ 7 & 12 & 121 \end{bmatrix}, I_2 = \begin{bmatrix} 14 & 125 & 27 \\ 27 & 21 & 121 \\ 107 & 10 & 21 \end{bmatrix}$$

4.14. Write a MATLAB program to calculate the normalized cross-correlation function (Equation 4.38) and apply to the data arrays in Problem 4.13.

4.15. Using the image in Problem 3.7, generate a synthetically translated image by performing $I_2(i, j) = I_1(i-5, j-5)$.

4.16. Using the images I_1 and I_2 from Problem 4.15 and the MATLAB program from Problem 4.14, perform cross-correlation of I_1 and I_2. Display the velocity vectors.

4.17. Derive the equations in Equation 4.44.

BIBLIOGRAPHY

Albrecht, H.-E., Borys, M., Damaschke, N., and Tropea, C., *Laser Doppler and Phase Doppler Measurement Techniques*, Springer Verlag, New York, 2002.

Bruun, H.H., *Hot-Wire Anemometry, Principles and Signal Analysis*, Oxford University Press, New York, 1995.

Comte-Bellot, G., Hot-wire anemometry, *Annual Review of Fluid Mechanics*, 8, p. 209, 1977.

Dahm, W.J.A., Su, L.K., and Southerland, K.B., A scalar imaging velocimetry technique for fully resolved four-dimensional vector velocity field measurements in turbulent flows, *Physics of Fluids* A, 4(10), p. 2191, 1992.

Goldstein, R.J., *Fluid Mechanics Measurements*, Taylor & Francis, New York, 1996.

Khan, M.A. and Bruun, H.H., Signal analysis of X hot-wire probes using a look-up table method, *Dantec Information* 13, January 1994.

Raffel, M., Willert, C., and Kompenhans, J., *Particle Image Velocimetry*, Springer, New York, 1998.

5 Optical Diagnostics for Measurements of Species Concentrations and Temperature

5.1 INTRODUCTION

Light represents energy, and when this light energy is incident upon particles (gas molecules, soot particles, water droplets, etc.), there is an interaction between them that affects the energy states of both the light and the particle. This interaction of light with particles leads to many interesting phenomena involving emission, absorption, and scattering of the light by the particle. Optical diagnostic methods make use of one or a combination of these processes. As shown in Figure 5.1, the incident light may be broadband in its spectral (wavelength) content, and some of the light energy can be absorbed by the particles. Then, the transmitted light will have these absorbed components subtracted from the initial spectral content. This is the basis for so-called absorption spectroscopy, or line-of-sight methods. The light energy that is not absorbed is scattered, and the scattering characteristics depend on the particle size and optical properties. Thus, the scattering signal can be used to infer the particle characteristics. Some of the absorbed energy can be re-emitted through internal energy transitions within the molecule, leading to such effects as fluorescence or incandescence.

We can describe light–particle interactions from two viewpoints. One is a macroscopic perspective, where electromagnetic theories are applied to describe the

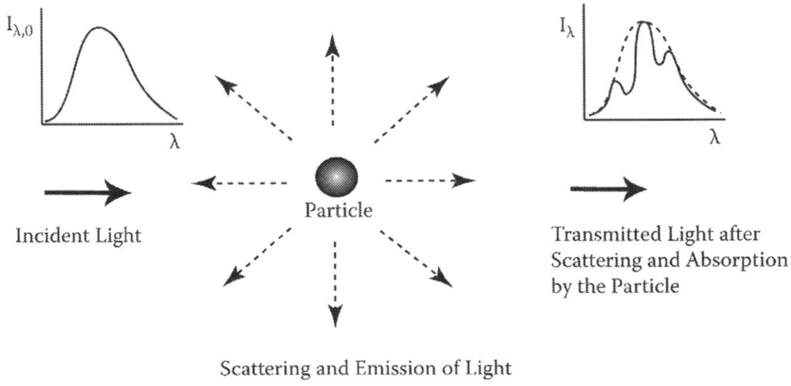

FIGURE 5.1 Light absorption, emission, and scattering.

DOI: 10.1201/9781003611837-5

changes in the electromagnetic field, resulting in light-scattering processes such as Mie, Rayleigh, and Raman scattering. The second approach is a microscopic or quantum-mechanical description, where the energy states of the molecules or atoms undergo transitions resulting in photoabsorption or emission. The basic elements of quantum mechanics are presented in Appendix D.

We will begin by describing the scattering phenomena, Mie and Rayleigh scattering, through electromagnetic theories. Discussions will follow on other spectroscopic methods, including Raman scattering, Fourier-transform infrared spectroscopy, and laser-induced fluorescence. These methods yield signals that are related to particle energy states and thus can be used to determine particle identity and concentrations, as well as other scalar properties such as temperature.

5.2 RAYLEIGH SCATTERING

5.2.1 Rayleigh Scattering Theory

As shown in Figure 5.1, some of the incident light can be scattered by particles in all directions. Solar radiation scattered by atmospheric gas molecules (oxygen and nitrogen) is a good example of Rayleigh scattering (the blue color of the sky). On the other hand, cloud particles are due to a different scattering process (Mie scattering, discussed in the next section). For particles that are small compared to the incident light wavelength, the scattering process can be described by the Rayleigh scattering theory. The molecular scattering of sunlight falls into this category. The general criterion for the applicability of the Rayleigh theory is that the particle diameter is much less than one.

$$x = \frac{\pi d}{\lambda} \ll 1 \tag{5.1}$$

d = particle diameter
λ = wavelength of the incident light

For particles not meeting the foregoing criteria (i.e., larger particles such as cloud particles), the Mie theory needs to be used, as described in the next section.

From classical electrodynamics (see Appendix C on Electromagnetics and Electromagnetic Radiation), the electric field associated with the incident light can be mathematically described as a sinusoidal wave. The sinusoidal fluctuation in the electric field can induce oscillations in the electric dipole moment of the particle. The electric dipole moment results from the nonuniform electric charge distribution in the molecules or particles through the spatial distribution and movement of the electrons. The electric dipole moment oscillates at the same frequency as the incident electric field through the electrostatic effect. This process is described by an electrostatic formula, where **p** is the induced dipole moment, **E** the electric field of the incident electromagnetic wave, and *the* polarizability (the tendency of the particle to be polarized).

$$\mathbf{p} = \alpha \mathbf{E} \tag{5.2}$$

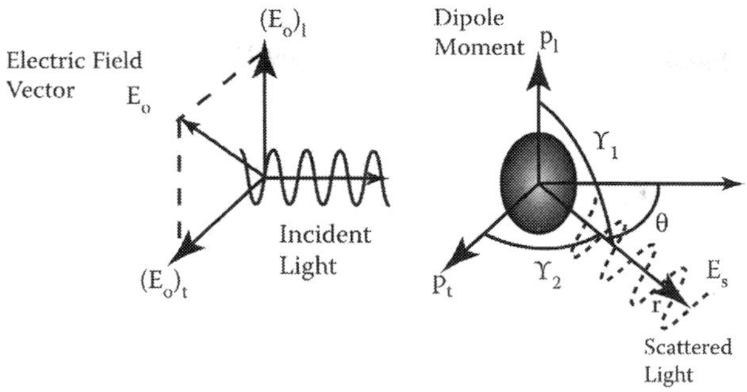

FIGURE 5.2 The Rayleigh scattering principle.

Both the induced dipole moment and the electric field are written as vectors because they can have arbitrary magnitudes and orientations in space. The polarizability, α, depends on the particle size and its molecular structure, and, as will be seen later, it can be expressed as a function of the particle diameter and the refractive index of the particle material.

The Rayleigh theory states that the scattered electric field of the light, scattered through the Rayleigh process, is proportional to the acceleration of the induced dipole moment, the sine of the angle between the incident light and the scattered directions, and inversely proportional to the distance r. This is analogous to radar or antenna processes, where the energy of moving charges results in the propagation of the electromagnetic wave, proportional to the acceleration of the dipole moments (electric dipole radiation). The Rayleigh theory is an approximation in which the scattering particle is considered "small," i.e., a point source of the scattered light.

For an electric field, as shown in Figure 5.2, with components in two orthogonal directions, the induced dipole moment can be decomposed into two orthogonal components. The angle between the incident and scattered light can then be defined using two angles, γ_1 and γ_2, as shown in the figure. The angle θ is between the direction of the incident light and that of the scattered light. Using the Rayleigh theory, the scattered electric field can be written as follows:

$$\mathbf{E}_s = \frac{1}{c^2}\frac{1}{r}\frac{\partial^2 \mathbf{p}}{\partial t^2}\sin\gamma \tag{5.3}$$

From the mathematical description of the electromagnetic wave, the incident electric field and the dipole moment are:

$$\begin{aligned} \mathbf{E} &= \mathbf{E}_o e^{-ik(r-ct)} \\ \mathbf{p} &= \alpha \mathbf{p}_o e^{-ik(r-ct)} \end{aligned} \tag{5.4}$$

$k = 2\pi/\lambda$
r = radial position or the distance along the direction of radial propagation
c = speed of light
t = time

E_o = amplitude of the incident electric field

Using Equation 5.4 in Equation 5.3, the electric field of the scattered radiation is

$$\mathbf{E}_s = -\mathbf{E}_o \frac{e^{-ik(r-ct)}}{r} k^2 \alpha \sin \gamma \tag{5.5}$$

We can consider the components of the scattered electric field in the longitudinal direction, l, and the transverse direction, t, by setting $\gamma_1 = \pi/2$ for the "l" component and $\gamma_2 = \pi/2 - \theta$ for the "t" component.

$$\begin{aligned}(E_s)_l &= -(E_o)_l \frac{e^{-ik(r-ct)}}{r} k^2 \alpha \sin \gamma_1 \\ (E_s)_t &= -(E_o)_t \frac{e^{-ik(r-ct)}}{r} k^2 \alpha \sin \gamma_2\end{aligned} \tag{5.6}$$

The intensity of the scattered radiation is proportional to the square of the electric field (see Appendix C). Thus, using $\sin \gamma_1 = 1$, $\sin \gamma_2 = \cos\theta$, and normalizing by the corresponding components of the incident light intensity, we obtain

$$\begin{aligned}\frac{(I_s)_l}{(I_o)_l} &= \frac{\left|(E_s)_l\right|^2}{\left|(E_o)_l\right|^2} = \frac{k^4 \alpha^2}{r^2} \\ \frac{(I_s)_t}{(I_o)_t} &= \frac{\left|(E_s)_t\right|^2}{\left|(E_o)_t\right|^2} = \frac{k^4 \alpha^2}{r^2} \cos^2\theta\end{aligned} \tag{5.7}$$

The total scattered light intensity is the sum of the two foregoing components.

$$I_s = (I_s)_l + (I_s)_t = \left[(I_o)_l + (I_o)_t \cos^2 \theta\right] \frac{k^4 \alpha^2}{r^2} \tag{5.8}$$

For unpolarized incident light, the longitudinal and transverse components are equally probable in occurrence and therefore equal in value, i.e., $(I_o)_l = (I_o)_t = \frac{I_o}{2}$. Also, by writing $k = 2\pi/\lambda$, we obtain an expression for the Rayleigh scattering intensity.

$$I_s = \frac{I_o}{r^2} \alpha^2 \left(\frac{2\pi}{\lambda}\right)^4 \frac{1 + \cos^2\theta}{2} \tag{5.9a}$$

The polarizability, α, for a sphere is given by $\frac{3}{4\pi} V \left|\frac{\mathbf{n}^2 - 1}{\mathbf{n}^2 + 2}\right|$, where $\mathbf{n}$ is the complex refractive index ($\mathbf{n} = n - ik$) of the particle. With $V = \pi D^3/6$, this leads to

$$I_s = \frac{I_o}{r^2} \frac{D^6}{8} \left|\frac{\mathbf{n}^2 - 1}{\mathbf{n}^2 + 2}\right|^2 \left(\frac{\pi}{\lambda}\right)^4 \left(1 + \cos^2\theta\right) \tag{5.9b}$$

First, we can extract some of the more obvious conclusions from Equation 5.9b. The Rayleigh scattering intensity depends on the incident light intensity and decreases with the square of the distance from the scatter, r. Next, polarizability has a large effect on the Rayleigh scattering intensity through the sixth power of the particle diameter and the squared terms for the refractive index. The term involving the cosine of the angle gives the angular dependence of the scattering and is referred to as the *phase function*. The fourth-power dependence on the inverse of the wavelength gives rise to interesting effects, such as the blue tint of the sky. Atmospheric gas molecules are small compared to the visible wavelength components of solar radiation and thus scatter light in the Rayleigh mode. The fourth-power inverse proportionality means that the shorter visible wavelength components (violet and blue) are scattered much more strongly than the longer visible wavelength components (red and orange), leading to the blue tint of the sky. In contrast, during sunset, the violet and blue components are predominantly scattered, so the solar radiation reaching the observer contains more of the red and orange components. For the same reason, laboratory observations of Rayleigh scattering can be made more efficient by using a shorter-wavelength light source to achieve a significantly higher scattering intensity.

For laboratory measurements of Rayleigh scattering, a laser light source with vertical polarization is typically used so that the transverse component is zero. Then, Equations 5.8 and 5.9 simplify to

$$I_s = \frac{I_o}{r^2}\frac{D^6}{8}\left|\frac{\mathbf{n}^2-1}{\mathbf{n}^2+2}\right|^2\left(\frac{\pi}{\lambda}\right)^4 \tag{5.10}$$

For vertically polarized incident light, Rayleigh scattering does not have any angular dependence (isotropic scattering) in that plane.

Before proceeding, it is useful to define some standard terms for the light-interaction properties of particles. We define the scattering cross section, C_{sca}, as the amount of scattered light in terms of radiative power P_{sca} [W] divided by the incident irradiance, I_o [W/m^2]. If the signal is observed over a wavelength band, then both P_{sca} and I_o will correspond to the respective radiative fluxes in that wavelength band. By quantifying the scattering cross section, we can calculate the scattering radiative power of a single scattering particle by multiplying it with the incident irradiance.

$$P_{\text{sca}} = C_{\text{sca}}\, I_o \quad \text{or} \quad C_{\text{sca}} = \frac{P_{\text{sca}}}{I_o} = \text{scattering cross section} \tag{5.11}$$

From Equation 5.11, the scattering cross section has the unit of area, and the scattering radiative power, P_{sca}, is found by integrating I_s (from, e.g., Equation 5.10) over the spherical region surrounding the particle.

$$P_{\text{sca}} = \int_0^{2\pi}\int_0^{\pi} I_{\text{sca}} r^2 \sin\theta d\theta d\phi = \text{scattering radiative power} \tag{5.12}$$

Here, $r^2 \sin\theta d\theta d\phi$ is the differential area element in spherical coordinates, and the surface integration is from $\theta=0$ to π and $\phi=0$ to 2π.

Example 5.1

For Rayleigh scattering with vertically polarized incident radiation, I_{sca} is given by Equation 5.10, $I_s = \frac{I_o}{r^2}\alpha^2\left(\frac{2\pi}{\lambda}\right)^4$. Integrating this over the sphere and dividing by I_o yields the Rayleigh scattering cross section for vertical polarization: $(C_{sca})_{VV} = 4\pi\alpha^2\left(\frac{2\pi}{\lambda}\right)^4$, where the additional subscript *VV* refers to the fact that it is a vertically polarized scattering signal from vertically polarized incident irradiance. The polarizability, α, for the sphere is given by $\frac{3}{4\pi}V\left|\frac{\mathbf{n}^2-1}{\mathbf{n}^2+2}\right|$, where $\mathbf{n}$ is the complex index of refraction ($\mathbf{n} = n - ik$, from Appendix C). With $V = \pi D^3/6$, this leads to an expression for the Rayleigh scattering cross section.

$$(C_{sca})_{VV} = \pi D^6\left(\frac{\pi}{\lambda}\right)^4\left|\frac{\mathbf{n}^2-1}{\mathbf{n}^2+2}\right|^2 \left[\text{cm}^2\right]$$

The unit for cross sections is typically cm². In many instances, a related term, the *differential cross section,* is used. This quantity is simply the scattered radiative power in a specific direction per solid angle (steradian) divided by the incident light intensity. For the foregoing example, where the scattering is isotropic, the differential cross section is found by dividing by 4π str, which is the solid angle of the spherical surface.

$$\text{Differential cross section} = (\sigma_{sca})_{VV} = \frac{(C_{sca})_{VV}}{4\pi} = \frac{D^6}{4}\left(\frac{\pi}{\lambda}\right)^4\left|\frac{\mathbf{n}^2-1}{\mathbf{n}^2+2}\right|^2 \left[\text{cm}^2/\text{str}\right]$$

Differential cross sections are particularly useful in measurement situations because detectors and detector optics only view a small solid angle of the total spherically scattered radiation.

Alternatively, the incident intensity, I_o, can be converted to the incident radiative power by multiplying it by the projected cross-sectional area of the particle, *A*, and taking the ratio of the scattered radiative power (P_{sca}) to the incident radiative power (P_o). This ratio defines the scattering *efficiency*, Q_{sca}, which is dimensionless.

$$\frac{P_{sca}}{P_o} = \frac{C_{sca}I_o}{AI_o} = \frac{C_{sca}}{A} = Q_{sca} = \text{scattering efficiency} \tag{5.13}$$

Both the scattering cross section and the efficiency refer to the scattering power for a single scattering particle. In many instances, there may be a number of scattering particles, in which case the total scattered power will depend on the number density of the particles. Thus, the scattering *coefficient* is defined as the scattering cross section multiplied by the number density of the particles, *N* [cm^{-3} or m^{-3}], for monodisperse particles (all particles having the same size).

$$K_{sca} = NC_{sca} = \text{ scattering coefficient } \left[\text{cm}^{-1} \text{ or } \text{m}^{-1}\right] \tag{5.14}$$

Thus, multiplying the scattering coefficient by the incident intensity yields the total scattered power per unit volume. For polydisperse particles (i.e., particles with a range of sizes), the scattering cross section will differ for different particle sizes.

For example, the Rayleigh scattering cross section varies with particle size because the polarizability, α, is a function of the particle diameter. Therefore, we need to integrate over the particle size distribution, $N(D)$, to obtain K_{sca}.

$$K_{sca} = \int_0^\infty N(D) C_{sca}(D) dD \tag{5.15}$$

Analogous definitions can be made for the amount of incident light absorbed by the particle, as well as for the sum of the absorbed and scattered light, which is referred to as light extinction.

$$C_{abs,ext} = \frac{P_{abs,ext}}{I_o} = \text{absorption or extinction cross section} \tag{5.16}$$

$$Q_{abs,ext} = \frac{C_{abs,ext}}{A} = \text{absorption or extinction efficiency} \tag{5.17}$$

$$K_{abs,ext} = \int_0^\infty N(D) C_{sca,ext}(D) dD = \text{absorption or extinction coefficient} \tag{5.18}$$

5.2.2 Applications of Rayleigh Scattering

A typical setup for laboratory measurements of Rayleigh scattering is shown in Figure 5.3. The setup is relatively straightforward, except that great care must be taken to eliminate any spurious scattering from surfaces or particles, as these can easily overwhelm the Rayleigh signal even with aggressive optical and digital filtering. The setup shown in Figure 5.3 can be used to measure fluid density because the scattering coefficient depends on the number density of the molecules (Equation 5.14). Under constant-pressure conditions, Rayleigh scattering can also be used to measure

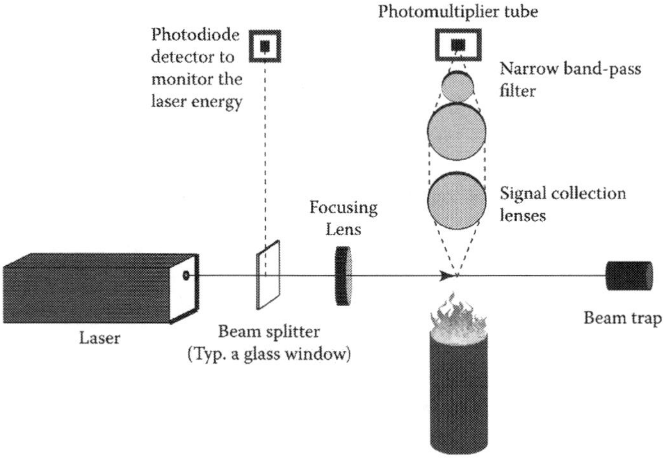

FIGURE 5.3 An experimental setup for Rayleigh scattering measurements.

temperature using the ideal gas equation of state, but only in "clean," particle-free flows. For example, in sooting flames, the Mie scattering from soot particles completely overwhelms the Rayleigh scattering signals from gas molecules.

For temperature measurements, we can start with Equation 5.14, which can be used to calculate the scattered intensity per unit volume illuminated by the incident intensity, I_o.

$$I_s = cl(\Delta\Omega)K_{\text{sca}}I_o \tag{5.19}$$

c = calibration constant of the collection optics
l = length of the probe volume
$\Delta\Omega$ = solid angle for the signal collection
K_{sca} = overall scattering coefficient of the gas mixture
I_o = incident laser intensity

Normally, multiplying the scattering coefficient by the incident intensity would give the scattered radiative power through Equations 5.11 and 5.14. However, we take the Rayleigh scattering to be isotropic and consider a solid angle (ΔQ) out of the total spherical solid angle of 4π, and absorb this and other conversion factors into the calibration constant, c, in Equation 5.19.

Within the probe volume, there may be several gas species; therefore, K_{sca} needs to be written as $K_{\text{sca}} = N\sum_{i=1}^{M} X_i (C_{\text{sca}})_i$, where there are M different gases present in the mixture, and X_i's are their individual mole fractions.

Now, for flows at a constant pressure, the ideal gas equation can be written as

$$p = \frac{N}{A_o}RT \quad \text{or} \quad N = \frac{pA_o}{RT} = \frac{p}{kT} \propto \frac{1}{T} \tag{5.20}$$

A_o = Avogadro's number = $6.02214199 \times 10^{26}$ particles/kmol
k = Boltzmann's constant = $1.3806503 \times 10^{-23}$ J/K
R = Universal gas constant = 8314 J/(kmol K)

Thus, substituting for N in Equation 5.19 and combining all of the constant terms of the experiment into a single constant C', we obtain

$$I_s = \frac{C'}{T}\left(\sum_{i=1}^{M} X_i (C_{\text{sca}})_i\right) \tag{5.21}$$

If the gas composition in the mixture is the same throughout the flow, i.e., X_i = constant, then the signal is inversely proportional to temperature. For chemically reacting systems, the gas composition changes from reactants to products, i.e., X_i will vary from one location in the flow to another. However, for most "premixed" combustion systems involving a reactant mixture of fuel with air, a large fraction of the gas mixture is nitrogen on both the reactant and product sides. Therefore, the effective Rayleigh cross section is dominated by that of nitrogen, and approximately a 10% error is associated with Rayleigh temperature measurements when assuming that the Rayleigh cross section remains constant.

Alternatively, Rayleigh scattering can be used to measure the concentrations (N_1 and N_2) of two-component gas mixtures under isothermal and constant-pressure conditions. For such flows, the overall number concentration is the same: $N = N_1 + N_2$, and Equation 5.21 can be written for two-component mixtures as follows:

$$
\begin{aligned}
I_s &= C''\left(N_1 (C_{sca})_1 + N_2 (C_{sca})_2\right) = C''\left(N_1 (C_{sca})_1 + \left(\frac{p}{kT} - N_1\right)(C_{sca})_2 \right) \\
&= C''\left(\left[(C_{sca})_1 - (C_{sca})_2\right] N_1 + \frac{p}{kT}(C_{sca})_2 \right) = A'' N_1 + B''
\end{aligned}
\tag{5.22}
$$

The signal, I_s, is proportional to N_1, from which one can obtain $N_2 = p/kT - N_1$. For good signal sensitivity, the constant A'' should be large, or a large difference between the Rayleigh cross sections of the two gas components is needed.

Rayleigh scattering is also used to characterize small particles such as flame-generated soot (Santoro et al., 1983). As noted in Section 5.2.1, the differential Rayleigh scattering cross section of a sphere for vertically polarized incident light is given by

$$
(\sigma_{\text{sca}})_{VV} = \frac{D^6}{4}\left(\frac{\pi}{\lambda}\right)^4 \left|\frac{\mathbf{n}^2 - 1}{\mathbf{n}^2 + 2}\right|^2 \tag{5.23}
$$

Here, $\mathbf{n}$ is the complex refractive index of the soot particle (assumed to be spherical), so that $\mathbf{n} = n - ik$. The complex refractive index represents the refractive index through the real part and the absorption index through the imaginary part (see Appendix C).

For a soot particle size distribution, $P(D)$, the scattering coefficient based on the differential cross section (Equation 5.23) can be written using Equation 5.15.

$$
K_{\text{sca}} = \frac{N}{4}\left(\frac{\pi}{\lambda}\right)^4 \left|\frac{\mathbf{n}^2 - 1}{\mathbf{n}^2 + 2}\right|^2 \int_0^\infty P(D) D^6 dD \tag{5.24}
$$

Similarly, the extinction coefficient can be written as

$$
K_{\text{ext}} = \frac{\pi^2 N}{\lambda} \left| \text{Im}\left\{\frac{\mathbf{n}^2 - 1}{\mathbf{n}^2 + 2}\right\} \right| \int_0^\infty P(D) D^3 dD \tag{5.25}
$$

A generalized mean diameter is useful for the current analysis, as defined by

$$
D_{pq} = \left[\frac{\int_0^\infty P(D) D^p dD}{\int_0^\infty P(D) D^q dD} \right]^{\frac{1}{p-q}} \tag{5.26}
$$

The particle volume fraction, f_v, is then given by $\frac{\pi}{6} N D_{30}^3$, where D_{30}^3 is simply the mean cubed diameter and also the integral contained in Equation 5.25. Thus, the particle volume fraction, f_v, is determined from the extinction measurement.

$$f_v = \frac{\lambda K_{\text{ext}}}{6\pi \left| \text{Im}\left\{ \frac{n^2-1}{n^2+2} \right\} \right|} \tag{5.27}$$

A measure of the soot particle diameter can be obtained from the sixth-to-third moment ratio, D_{63}, which is the ratio of the integrals contained in Equations 5.24 and 5.25, respectively.

$$D_{63} = \lambda \left(\frac{4}{\pi^2} \frac{\left| \text{Im}\left\{ \frac{n^2-1}{n^2+2} \right\} \right|}{\text{Im}\left| \frac{n^2-1}{n^2+2} \right|} \frac{K_{\text{sca}}}{K_{\text{ext}}} \right)^{1/3} \tag{5.28}$$

Finally, the number density can also be estimated using f_v and D_{63}.

$$N = \frac{12 f_v}{\pi D_{63}^3} \tag{5.29}$$

Thus, this approach requires both the scattering and absorption coefficients from measurements of the scattering intensity and light extinction. A schematic of the experimental setup used to obtain these parameters is shown in Figure 5.4.

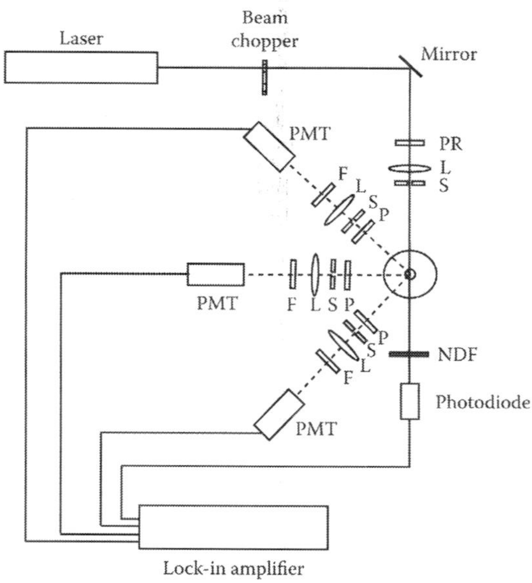

PR, polarization rotator; L, lens; S, spatial filter (pinhole aperture);
F, narrow-band filter; P, polarization filter; NDF, neutral-density filter;
PMT, photomultiplier tube.

FIGURE 5.4 Soot measurements using Rayleigh scattering. (Adapted from Santoro et al., 1983.)

5.3 MIE SCATTERING

Mie-scattering signals from small particles are used in laser Doppler velocimetry and particle image velocimetry, as described in Chapter 4. For these applications, the details of the Mie-scattering processes are not necessary, other than the requirement that these particles generate observable signals. However, Mie scattering is responsible for the scattering of light in many physical and engineering processes and is therefore an essential component in the diagnostics of these phenomena.

We can start from the electromagnetic wave equation developed in Appendix C.

$$\frac{\partial^2 \mathbf{E}}{\partial t^2} = c^2 \nabla^2 \mathbf{E} \tag{5.30}$$

E = electric field vector
c = speed of light in the medium

If we write the electric field vector, **E**, as $\mathbf{E} = \mathbf{E}_o \exp(i\mathbf{k} \bullet \mathbf{x} - i\omega t)$ (as in Appendix C) and use it on the left-hand side of Equation 5.30, the second derivative with respect to time becomes $-\omega^2\mathbf{E}$. Here, **k** is the wave vector and ω the circular frequency. Then, the wave equation becomes

$$\nabla^2 \mathbf{E} + \frac{\omega^2}{c^2} \mathbf{E} = 0 \tag{5.31}$$

Using $k = 2\pi/\lambda$, $\omega = 2\pi v$, and $c = \lambda v$, where λ and v are the wavelength and frequency, respectively, we get $\omega/c = k$. The electromagnetic wave contains both electric and magnetic fields; thus, including the magnetic component, the vector equations for the electromagnetic wave can be written as

$$\begin{aligned} \nabla^2 \mathbf{E} + k^2 \mathbf{E} &= 0 \\ \nabla^2 \mathbf{H} + k^2 \mathbf{H} &= 0 \end{aligned} \tag{5.32}$$

The electric and magnetic field vectors, in general, have three components in each orthogonal direction. To solve for these components for Mie scattering of spherical particles, a scalar wave equation is first written using Equation 5.32 in the spherical coordinate system. Using the Laplacian in spherical coordinates, the scalar wave equation is

$$\frac{1}{r^2}\frac{\partial}{\partial r}\left(r^2 \frac{\partial \psi}{\partial r}\right) + \frac{1}{r^2 \sin\theta}\frac{\partial}{\partial \theta}\left(\sin\theta \frac{\partial \psi}{\partial \theta}\right) + \frac{1}{r^2 \sin\theta}\frac{\partial^2 \psi}{\partial \phi^2} + k^2 \psi = 0 \tag{5.33}$$

The individual components of the vectors in Equation 5.32 do not, in general, satisfy the scalar wave equation, such as Equation 5.33, except in the case where the vectors are in the Cartesian coordinate system. This is because $\nabla^2 \mathbf{E} = \nabla \bullet \nabla \mathbf{E} \neq \nabla(\nabla \bullet \mathbf{E})$. However, the solution of the scalar wave equation can be used to generate vectors that do satisfy the vector wave equation. Through extensive vector algebra, both the vectors **M** and **N** can be shown to satisfy Equation 5.32 if Ψ satisfies Equation 5.33 (Bohren and Huffman, 1983).

$$\mathbf{M} = \nabla \times (\mathbf{r}\psi),$$
$$k\mathbf{N} = \nabla \times \mathbf{M} \tag{5.34}$$
$$\mathbf{r} = r\mathbf{e_r}$$

Thus, the remaining task (mathematically intense) is to write the incident, internal, and scattered electric and magnetic field components in terms of the so-called spherical harmonic vectors in Equation 5.34, which are generated from the scalar solution, Ψ, and apply the boundary condition at the spherical particle surface. The boundary condition at the particle surface ($r=a$) is that the incident field is decomposed into internal and scattered fields.

$$\begin{aligned}(\mathbf{E}_\mathrm{i} + \mathbf{E}_\mathrm{s} - \mathbf{E}_\mathrm{I}) \times \mathbf{e}_r &= 0\\ (\mathbf{H}_\mathrm{i} + \mathbf{H}_\mathrm{s} - \mathbf{H}_\mathrm{I}) \times \mathbf{e}_r &= 0\end{aligned} \tag{5.35}$$

A classic monograph on this topic by Bohren and Huffman (1983) devotes approximately seventeen pages of text to the solution procedure, and that only shows the main steps. We do not need to repeat this mathematical adventure but only need to be able to use such results. To start, the scattered electric field, in terms of spherical harmonics, is determined to be

$$\mathbf{E}_\mathrm{s} = E_o \sum_{n=1}^{\infty} i^n \frac{2n+1}{n(n+1)}(ia_n\mathbf{N} - b_n\mathbf{M}) \tag{5.36}$$

M and **N** are "generated" from Ψ and consist of Legendre and spherical Bessel functions. Writing in these function components, the θ and ϕ components of the scattered electric field are

$$\begin{aligned}E_{s\theta} &= \frac{E_o \cos\phi}{\rho} \sum_{n=1}^{\infty} i^n \frac{2n+1}{n(n+1)}\left(ia_n\left[\rho h_n(\rho)\right]' \pi_n - b_n\left[\rho h_n(\rho)\right]\tau_n\right)\\ E_{s\phi} &= \frac{E_o \cos\phi}{\rho} \sum_{n=1}^{\infty} i^n \frac{2n+1}{n(n+1)}\left(ia_n\left[\rho h_n(\rho)\right]' \tau_n - b_n\left[\rho h_n(\rho)\right]\pi_n\right)\end{aligned} \tag{5.37}$$

E_o = incident electric field intensity
$\rho = kr$
h_n = spherical Hankel function
τ_n = recurrence relation given in Equation 5.39
π_n = recurrent relation given in Equation 5.39
a_n, b_n = coefficients given in Equation 5.40
[ ]' = derivative of the function inside the bracket with respect to ρ

Furthermore, in the far field ($kr >> 1$), the Hankel functions, $h_n(kr)$, are approximated as an exponential function of ikr, which results in a simpler form for the scattered electric field.

$$
\begin{aligned}
E_{s\theta} &= \frac{E_o e^{ikr}}{-ikr}\cos\phi\sum_{n=1}^{\infty}\frac{2n+1}{n(n+1)}\left(a_n\pi_n+b_n\tau_n\right)\\
E_{s\phi} &= \frac{E_o e^{ikr}}{ikr}\cos\phi\sum_{n=1}^{\infty}\frac{2n+1}{n(n+1)}\left(a_n\tau_n+b_n\pi_n\right)
\end{aligned}
\tag{5.38a}
$$

For compactness, the foregoing equations are sometimes written in terms of the scattering amplitudes $S_1(\theta)$ and $S_2(\theta)$:

$$
\begin{aligned}
E_{s\theta} &= \frac{E_o e^{ikr}}{-ikr}\cos\phi S_1(\theta)\\
E_{s\phi} &= \frac{e^{ikr}}{ikr}\sin\phi S_2(\theta)\\
S_1(\theta) &= \sum_{n=1}^{\infty}\frac{2n+1}{n(n+1)}\left(a_n\pi_n+b_n\tau_n\right)\\
S_2(\theta) &= \sum_{n=1}^{\infty}\frac{2n+1}{n(n+1)}\left(a_n\tau_n+b_n\pi_n\right)
\end{aligned}
\tag{5.38b}
$$

Here, π_n and τ_n are recurrence relations, given by

$$
\begin{aligned}
\pi_n &= \frac{2n-1}{n-1}\pi_{n-1}\cos\theta-\frac{n}{n-1}\pi_{n-2}\\
\tau_n &= n\pi_n\cos\theta-(n+1)\pi_{n-1}\\
\pi_0 &= 0;\pi_1=1;\pi_2=3\cos\theta\\
\tau_0 &= 0;\tau_1=\cos\theta;\tau_2=3\cos(2\theta)
\end{aligned}
\tag{5.39}
$$

a_n and b_n are the coefficients in the expansion:

$$
\begin{aligned}
a_n &= \frac{m^2 j_n(mx)\left[xj_n(x)\right]'-j_n(x)\left[mj_n(mx)\right]'}{m^2 j_n(mx)\left[xh_n(x)\right]'-h_n(x)\left[mxj_n(mx)\right]'}\\
b_n &= \frac{j_n(mx)\left[xj_n(x)\right]'-j_n(x)\left[mj_n(mx)\right]'}{j_n(mx)\left[xh_n(x)\right]'-h_n(x)\left[mxj_n(mx)\right]'}
\end{aligned}
\tag{5.40}
$$

m = relative refractive index = $\mathbf{n_1}/\mathbf{n}$,

where $\mathbf{n_1}$ and $\mathbf{n}$ are the complex refractive indices of the particle and medium, respectively,

$x = ka$ = size parameter,

where $k = 2\pi/\lambda$ and a = particle radius

$j_n(z)$ = Bessel function == $\sqrt{\frac{\pi}{2z}}J_{n+0.5}(z)$,
where $J_{n+0.5}$ is the Bessel function of the first kind
$h_n(z)$ = spherical Hankel function = $j_n(z) + iy_n(z)$,
where $y_n(z) = \sqrt{\frac{\pi}{2z}}Y_{n+0.5}(z)$, and $Y_{n+0.5}$ is the Bessel function of the second kind
[ ]' = derivative with respect to the variable inside the bracket

Now we are in a position to calculate numerically the light-scattering and extinction properties of spherical particles of arbitrary $\mathbf{n}_1$ and x. Taking $\theta = \phi = 0$ in Equation 5.38 should provide us with the electric field in the forward direction after absorption and scattering by the particle. Thus, this electric field corresponds to the extinction cross section after using $\pi_n(1) = \tau_n(1) = n(n+1)/2$ in Equation 5.39 and dividing by E_o^2.

$$C_{\text{ext}} = \frac{2\pi}{k^2}\sum_{n=1}^{\infty}(2n+1)\text{Re}(a_n + b_n) \tag{5.41}$$

The scattering cross section is obtained by integrating the scattered electric field over the spherical region surrounding the particle.

$$C_{\text{sca}} = \frac{2\pi}{k^2}\sum_{n=1}^{\infty}(2n+1)\text{Re}\left(|a_n|^2 + |b_n|^2\right) \tag{5.42}$$

The MATLAB program shown in Figure 5.5 computes the extinction and scattering efficiencies, which are obtained by dividing C_{ext} and C_{sca} (Equations 5.41 and 5.42) by the particle cross-sectional area, πa^2. Two examples of the extinction and scattering efficiencies are shown in Figure 5.6a and b for a range of particle size parameter, x, for complex refractive indices of $2.5 + 0.01i$ (weakly absorbing particle (a)) and $5.0 + 0.4i$ (highly absorbing particle (b)).

5.3.1 Applications and Assessments

The Mie-scattering theory, as described, provides a way to exactly calculate the scattering and extinction properties of spheres if their optical properties are known, i.e., the complex refractive index. As noted earlier, Mie scattering is used in laser Doppler velocimetry and particle image velocimetry. For other measurements, such as Rayleigh and Raman scattering, Mie scattering is a source of interference that can overwhelm the signal strength. The particle size dependence of the Mie-scattering intensity may suggest that particle sizing measurements can be made via Mie scattering. However, there are experimental difficulties with this. A typical laser output beam has a nonuniform intensity distribution across its width, and thus the resulting signal will vary as a function of the particle size and the exact path that the particle takes through the probe beam. Secondly, the particle may clip the edges of the probe beam, or more than one particle may be present in the probe volume. These issues lead to alternate methods for particle sizing, such as phase Doppler interferometry, discussed in Chapter 6.

```
function coeff = Mie_abcd(m)

x0=0.0;
dx=0.025*2;

% i-loop for x variation
for ix=1:200
  xa(ix)=ix*dx;
  x=xa(ix)

nmax=round(2+x+4*x^(1/3))

z=m*x;
m2=m*m;

bjx(1)=sin(x)/x;
bjz(1)=sin(z)/z;
yx(1)=-cos(x)/x;
hx(1)=bjx(1)+i*yx(1);

for n=2:nmax;
nu=(n-0.5);
sqx=sqrt(0.5*pi/x);
sqz=sqrt(0.5*pi/z);
bjx(n)=besselj(nu,x)*sqx;
bjz(n)=besselj(nu,z)*sqz;
yx(n)=bessely(nu,x)*sqx;
hx(n)=bjx(n)+i*yx(n);
end

for n=2:nmax;
dbjx(n)=x*bjx(n-1)-(n-1)*bjx(n);
dbjz(n)=z*bjz(n-1)-(n-1)*bjz(n);
dhx(n)=x*hx(n-1)-(n-1)*hx(n);
display(m);
a(n-1)=(m2*bjz(n)*dbjx(n)-bjx(n)*dbjz(n))/(m2*bjz(n)*dhx(n)-hx(n)*dbjz(n));
b(n-1)=(bjz(n)*dbjx(n)-bjx(n)*dbjz(n))/(bjz(n)*dhx(n)-hx(n)*dbjz(n));
c(n-1)=(bjx(n)*dhx(n)-hx(n)*dbjx(n))/(bjz(n)*dhx(n)-hx(n)*dbjz(n));
d(n-1)=(m*bjx(n)*dhx(n)-m*hx(n)*dbjx(n))/(m2*bjz(n)*dhx(n)-hx(n)*dbjz(n));

end

%Calculate Qsca and Qext for x
Qs(ix)=0.0;
Qe(ix)=0.0;
for n=1:nmax-1
  Qsn=2/(xa(ix)*xa(ix))*(2*n+1)*(a(n)*conj(a(n))+b(n)*conj(b(n)));
  Qs(ix)=Qs(ix)+Qsn;
  Qen=2/(xa(ix)*xa(ix))*(2*n+1)*(real(a(n)+b(n)));
  Qe(ix)=Qe(ix)+Qen;

end

%i-loop end
end
plot(xa,Qs,'k-',xa,Qe,'k')
```

FIGURE 5.5 A MATLAB program to calculate the Mie scattering and extinction efficiencies.

Applications of Mie scattering in thermal and fluid measurements are ubiquitous, though not always quantitative or straightforward. The Mie-scattering theory discussed here applies to any electromagnetic wave scattering from particles. Microwave backscattering from raindrops is commonly used to infer rainfall amounts, and the Doppler shift of the wavelength can provide additional data on rainfall movement. The time delay of the backscatter signal depends on the distance to the scattering

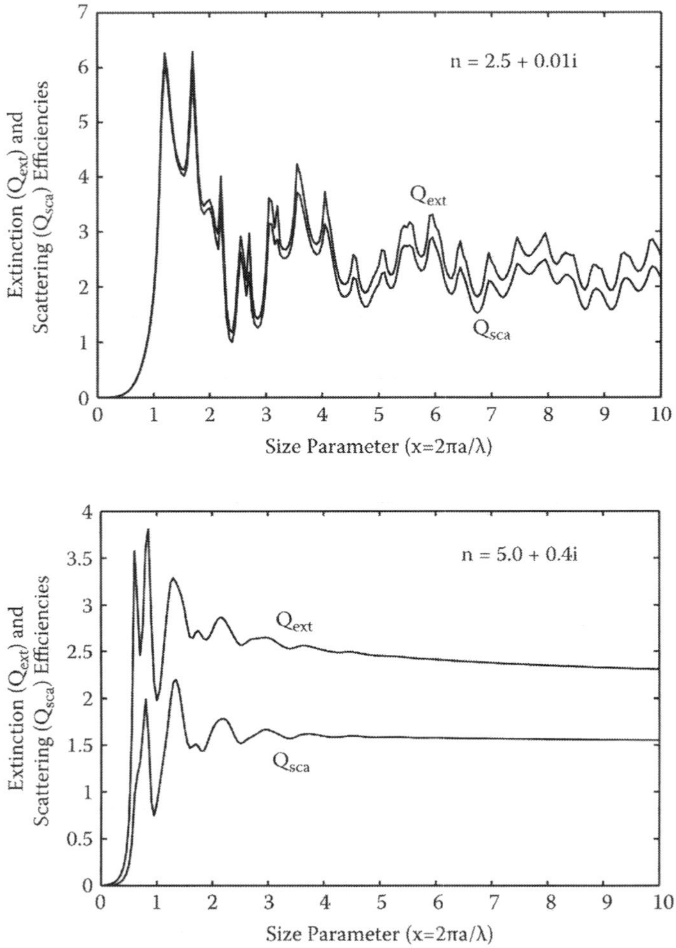

FIGURE 5.6 Mie scattering and extinction efficiencies computed using the MATLAB program shown in Figure 5.5.

particles, allowing the signal to be scanned over time to observe the spatial distribution of rainfall. Although it is reasonable to assume that raindrops are close to spherical and that there are viable models for raindrop size distributions, this type of Mie-scattering measurement is typically used only to provide approximate or relative rainfall amounts.

Similarly, aerosol measurements can, in principle, be made using Mie scattering, whether at atmospheric or micro-device scales. However, the following difficulties are commonly encountered:

1. The data for optical constants of the scattering particles, i.e., $\mathbf{n}(\lambda) = n(\lambda) + ik(\lambda)$, may not be available.
2. The particles may not be spherical.

3. The aerosol may contain a mixture of different particles whose exact composition is unknown.
4. The particle number density may be high, so that secondary scattering cannot be ignored.

For particles with unknown scattering and extinction characteristics, these optical properties can, in principle, be measured using a setup similar to the one shown in Figure 5.4. The scattering angles can be varied in both azimuthal and circumferential directions, and the photodiode detector placed in line with the laser measures light extinction. To quantify the incident light, a beam splitter is placed upstream to monitor the laser intensity.

For some simple geometrical shapes, the Mie theory can be extended to ellipsoids. For other geometries, computational methods are available to calculate optical properties, such as the discrete dipole approximation (Draine, 2003) and the methods described in Mishchenko et al. (2000). Another approach is to approximate the optical properties with those of a spheroid, since a large number of particles with random orientation exhibit optical properties similar to appropriately dimensioned spheroids. For non-geometrical shapes, such as atmospheric aerosol particles, characterization of the optical properties is an ongoing research topic.

The third and fourth difficulties just noted are more complex. Measurements only provide the scattering signal with the phase function if angular measurement data are available. Recovery of the aerosol composition represents an inversion problem, typically with an insufficient number of measured variables compared to the number of unknown variables.

Thus, in summary, Mie scattering is useful for generating relatively strong signals from particles whose characteristics are theoretically well described under ideal conditions. However, extracting quantitative data concerning the scattering particles requires further work, as discussed earlier.

5.4 RAMAN SCATTERING

Both Rayleigh and Mie scattering, discussed in the previous sections, are referred to as *elastic* scattering processes, meaning that the scattering signal has the same wavelength as the incident light. On the other hand, Raman scattering is an *inelastic* process, in the sense that Raman signals are wavelength-shifted from the incident wavelength by amounts that correspond to the energy levels of the target molecules. Figure 5.7 shows the principle of Raman scattering. The incident light is modulated by the internal oscillation of the molecule, such as the vibration of the nuclei in a diatomic molecule, as shown in Figure 5.7. Thus, Raman scattering is a spectroscopic technique in which the internal energy levels and therefore the structure of the molecules are determined through the spectral content of the scattered light. To illustrate this, let us rewrite Equation 5.2, except that we now consider the change in the polarizability of the molecule due to vibrational oscillations. We expand the polarizability as $\alpha = \alpha_o + \left(\frac{\partial \alpha}{\partial q}\right)_o q$, where q is a spatial variable, such as the internuclear distance.

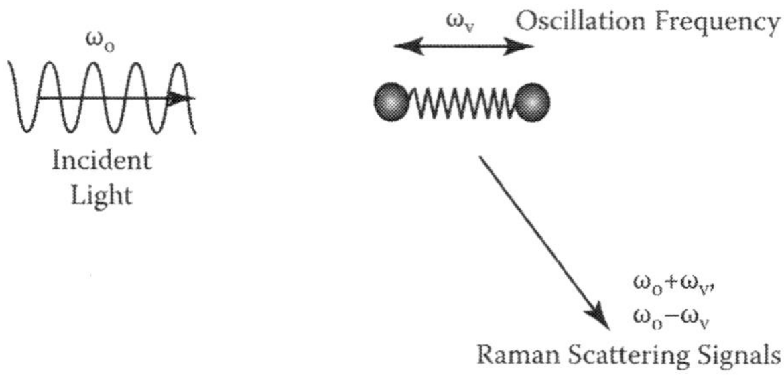

FIGURE 5.7 A model for the Raman scattering process.

If the internuclear distance changes due to molecular vibration, then the electric charge distribution also changes, leading to a change in polarizability. The foregoing expression for polarizability accounts for the change in polarizability as a function of the internuclear distances. Equation 5.2 can now be written as

$$\mathbf{p} = \alpha \mathbf{E} = \left\{ \alpha_o + \left(\frac{\partial \alpha}{\partial q} \right) q \right\} \mathbf{E} \tag{5.43}$$

Let us consider the temporal oscillation components of the electric field and the molecular vibration by writing them as

$$\begin{aligned} \mathbf{E} &= \mathbf{E_o} \cos \omega_o t \\ q &= q_o \cos \omega_v t \end{aligned} \tag{5.44}$$

The frequencies of the incident light and the molecular vibration are, respectively, ω_o and ω_v. Substituting into Equation 5.43, we get

$$\begin{aligned} \mathbf{p} &= \alpha_o \mathbf{E_o} \cos \omega_o t + \left(\frac{\partial \alpha}{\partial q} \right) q_o \cos \omega_v t \mathbf{E_o} \cos \omega_o t \\ &= \alpha_o \mathbf{E_o} \cos \omega_o t + \left(\frac{\partial \alpha}{\partial q} \right) \frac{q_o \mathbf{E_o}}{2} \left[\cos(\omega_o - \omega_v) t + \cos(\omega_o + \omega_v) t \right] \end{aligned} \tag{5.45}$$

In the second line, we have used the trigonometric relation $\cos A \cos B = 1/2\{\cos(A - B) + \cos(A + B)\}$. As discussed in the section on Rayleigh scattering, the scattering signal is proportional, among other factors, to the acceleration of the polarizability vector, **p**. Differentiating Equation 5.45 twice with respect to time introduces multiplicative terms, but the frequency components in the cosine functions remain the same. Thus, the coupling of the incident light with vibrational oscillations produces light signals at two additional frequencies—one down-shifted by the vibrational frequency and the other up-shifted. The down-shifted frequency signal due to molecular oscillation is called *anti-Stokes* Raman, and the up-shifted frequency is *Stokes*

Raman. The first term in Equation 5.45, with the original frequency ω_o corresponds to the Rayleigh signal.

The direct generation of Raman scattering signals is called *spontaneous Raman scattering*, and the experimental setup to achieve scattering measurements is based on the same setup used for Rayleigh scattering shown in Figure 5.3. To improve signal collection and separation, a dual-collimation optical setup is commonly used for spontaneous Raman measurements (Figure 5.8). Large-aperture lenses collect the Raman signal and refocus it through a variable aperture, which serves to block spurious reflections from the background. The signal is then re-collimated and sent through a combination of filters, and finally focused onto the entrance slit of a spectrometer for separation of the spectral (wavelength) components. The spectrum can be imaged using intensified cameras, such as cooled intensified charge-coupled devices (ICCDs). An example of a spontaneous Raman spectrum is shown in Figure 5.9, where the peaks for major species such as nitrogen, oxygen, and hydrocarbons (H/C) are prominently displayed. Using high-intensity lasers (Nd:YAG at 500 mJ/pulse) with cooled ICCD detectors, signals with good signal-to-noise ratios can be obtained, provided there is no interference from particle Mie scattering. It can also be seen in Figure 5.9 that the laser line leaks through the optical train even with aggressive filtering using narrowband filters and in the absence of any Mie-scattering interference.

Depending on the molecular structure, or more specifically the types of molecular bonds that exist, there are distinct vibrational modes and corresponding vibrational frequencies. Raman scattering can, in principle, be used to probe these vibrational structures and thereby determine the target molecule's identity and concentration.

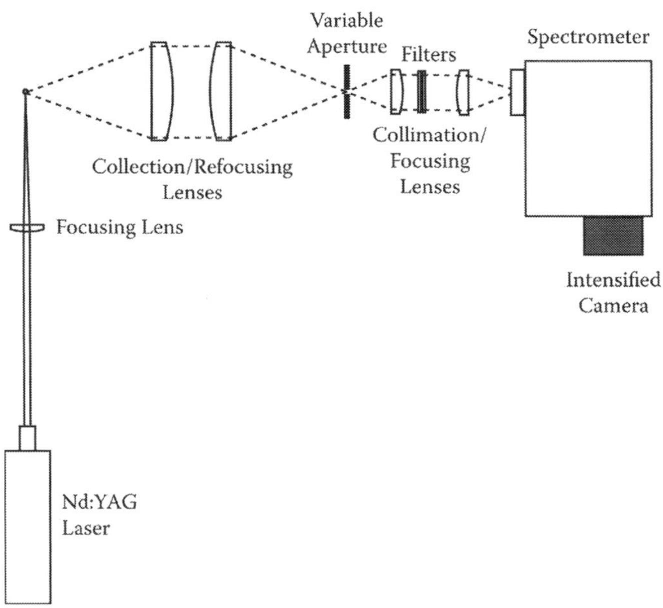

FIGURE 5.8 A complete optical setup for spontaneous Raman scattering.

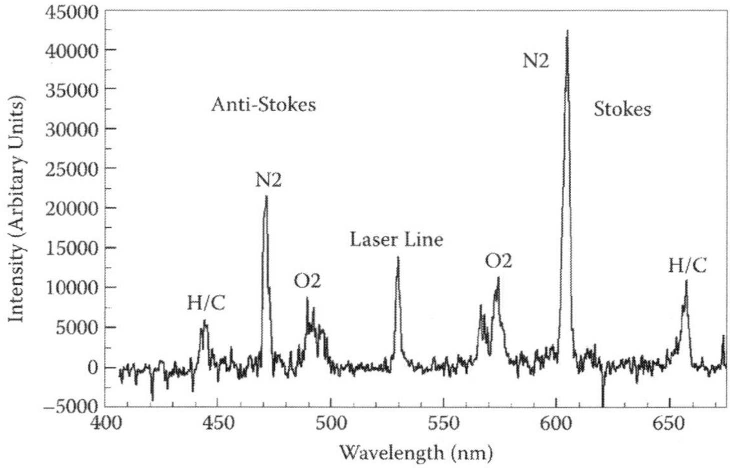

FIGURE 5.9 A spontaneous Raman spectrum showing the Stokes and anti-Stokes peaks from major species.

TABLE 5.1
Raman Frequency and Cross Sections for Some Gas Species

Species	Vibrational Frequency [cm^{-1}]	Raman Cross Sections × 10^{-30} cm^2/sr		
		$\lambda_o = 337$ nm	488	532
n_2	2330.7	3.5	0.68	0.46
h_2	4160.2	8.7	1.32	0.943
O_2	1556	4.6	0.72	0.65
NO	1877	1.5	0.15	0.2

Note: λ_o = incident light wavelength.

However, for most molecules under typical experimental conditions, the Raman cross section is very small; that is, the spontaneous Raman scattering intensity is usually several orders of magnitude lower than that of Rayleigh and Mie scattering. Table 5.1 shows the Raman cross section for some gas species, along with their primary vibrational frequencies. Therefore, more elaborate techniques, such as coherent anti-Stokes Raman spectroscopy (CARS) or surface-enhanced Raman spectroscopy, are applied to enhance the signal strength under test conditions with significant particle interference.

Example 5.2

As shown in Equation 5.45, the vibrational frequencies listed in Table 5.1 modulate the incident laser frequency to produce two "off-resonant" frequencies, $\omega_o + \omega_v$ and $\omega_o - \omega_v$, where ω_o and ω_v are the laser and vibrational frequencies, respectively. For the second-harmonic output of an Nd:YAG laser, the output wavelength is $\lambda_o = 532$ nm, which corresponds to a wavenumber ($\omega_o = 1/\lambda_o$) of 18,796.99 cm^{-1}. Note that wavenumber is used interchangeably with frequency in spectroscopy (see Equation 5.58). The Stokes and anti-Stokes Raman components are then $\omega_o + \omega_v$ and $\omega_o - \omega_v$, and for nitrogen and oxygen, these correspond to signals at the following wavelengths.

$$N_2: \text{ Stokes, } \omega_o + \omega_v = 21127.69 \text{ cm}^{-1} \rightarrow \lambda_S = 473.31 \text{ nm,}$$

$$\text{anti-Stokes, } \omega_o - \omega_v = 16466.29 \text{ cm}^{-1} \rightarrow \lambda_{AS} = 607.30 \text{ nm}$$

$$O_2: \text{ Stokes, } \omega_0 + \omega_v = 20352.99 \text{ cm}^{-1} \rightarrow \lambda_\varsigma = 491.33 \text{ nm,}$$

$$\text{anti-Stokes, } \omega_o - \omega_v = 17240.99 \text{ cm}^{-1} \rightarrow \lambda_{AS} = 580.01 \text{ nm}$$

The Raman signal can be greatly enhanced using a higher-order optical process called *CARS*. As seen in Equation 5.43, spontaneous Raman scattering arises from the polarizability that is linearly "forced" by the incident electric field. In general, however, the polarizability is also influenced by higher-order, nonlinear multiples of the electric field.

$$\mathbf{p} = \alpha^{(1)}\mathbf{E}_1 + \alpha^{(2)}\mathbf{E}_1\mathbf{E}_2 + \alpha^{(3)}\mathbf{E}_1\mathbf{E}_2\mathbf{E}_3 \tag{5.46}$$

Linear polarizability leads to Rayleigh and spontaneous Raman scattering. The second-order term is useful for describing frequency doubling in some optical devices using crystal materials (e.g., birefringent crystals). In isotropic media such as gases or liquids, the third-order term gives rise to a wave-mixing effect. If we again represent the electric fields as sinusoidal functions, as in Equation 5.44, products of the cosine functions lead to sum and difference frequency terms. The product of two cosine functions produces a pair of sum and difference frequencies, so the product of three cosine functions, as in the third-order term in Equation 5.46, leads to $\omega_1 + \omega_2 + \omega_3$, $\omega_1 + \omega_2 - \omega_3$, $\omega_1 - \omega_2 + \omega_3$, $\omega_1 - \omega_2 - \omega_3$. In the CARS setup, the first and second frequencies are kept the same ($\omega_1 = \omega_2 = \omega_p$), whereas the third frequency is set to $\omega_3 = \omega_p - \omega_v$, where ω_v is the vibrational frequency of the molecule. This results in a frequency component $\omega_1 + \omega_2 - \omega_3 = 2\omega_p - (\omega_p - \omega_v) = \omega_p + w_v$, which is the CARS signal frequency. Knowing ω_p (the "pump" frequency), we can determine the vibrational frequency of the molecule (ω_v), from the CARS frequency, $\omega_{CARS} = \omega_p + \omega_v$.

There is a theoretical basis to compute the intensity of the CARS signal by using the polarizability (Equation 5.46) in the electromagnetic wave equation (Equation C.19). The derivations can be found in Eckbreth (1996) and Tolles et al. (1977).

$$I_{CARS} \sim I_p^2 I_3 \left|\alpha^{(3)}\right|^2 l^2 \left(\frac{\sin \frac{l\Delta k}{2}}{\frac{l\Delta k}{2}} \right)^2 \tag{5.47}$$

I_p = intensity of the pump beam
I_3 = intensity of the Stokes beam with frequency $\omega_3 = \omega_p - \omega_v$
$\alpha^{(3)}$ = third-order polarizability (CARS susceptibility)
l = length of the probe volume
Δk = the phase difference $= 2\vec{k}_p - \vec{k}_3 - \vec{k}_{CARS}$

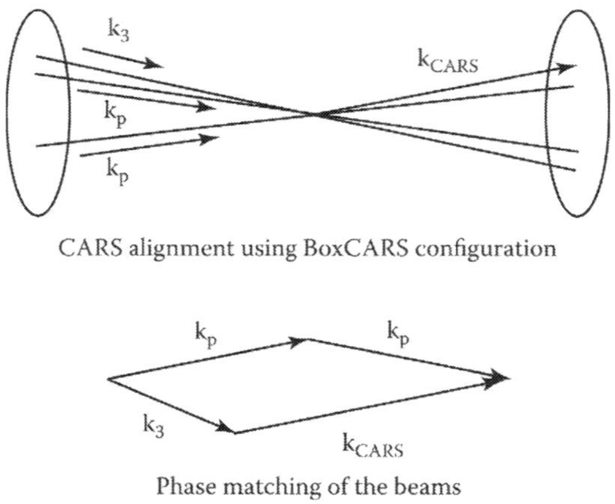

FIGURE 5.10 A "BoxCARS" phase matching used for coherent anti-Stokes Raman spectroscopy.

Examination of Equation 5.47 illustrates the strong dependence of the CARS signal intensity on the probe and Stokes beams, and therefore the signal strength can be increased by using high-intensity lasers. The phase difference term, Δk, involves wave vectors (k_p, k_3, and k_{CARS}), which are vectors in the direction of beam propagation with a magnitude of $k=2\pi/\lambda$. The terms in parentheses reach a maximum of 1 when $\Delta k=0$. This optimization is called *phase matching* and involves aligning the orientation of the pump and Stokes beams so that the phase difference becomes zero. Geometrically, there are several methods of achieving phase matching, and one of the most common methods is called *BoxCARS*, as shown in Figure 5.10. The two pump beams are sent at opposing angles, and the angle of the Stokes beam then determines the optical angle for observation of the CARS signal. In practice, a broadband laser beam (typically from a dye laser source) is used for the Stokes beam to overlap several of the vibrational frequencies in the compositional molecules, so that multiple CARS signals from different molecules can be observed as a spectrum, much like spontaneous Raman scattering. However, the nonlinear CARS susceptibility term makes the interpretation of the spectrum somewhat more difficult, requiring calibration of the signal against species concentration.

As noted earlier, Raman scattering originates from the modulation of the light frequency by internal molecular motion, such as the vibration of nuclei. Because the molecular vibrational frequencies are unique for a given molecular motion, Raman scattering allows us to identify the molecules in a gas mixture and quantify their amounts (after calibration). More detailed information is contained in the spectrum, as the molecular energy levels are reflected during spectroscopic probing.

5.5 MOLECULAR ENERGY AND SPECTROSCOPY

A molecule contains the following major energy components: kinetic energy due to its random motion, electronic energy, vibrational energy, and rotational energy. The kinetic energy of the random motion is observable through gas temperature and pressure. From statistical thermodynamics, the mean velocity of the molecule's random motion is given as

$$\bar{V}_{\text{random}} = \frac{2}{\sqrt{\pi}}\left(\frac{2kT}{m}\right)^{1/2} \tag{5.48}$$

m = molecular mass
k = Boltzmann's constant ($k = 1.3806505 \times 10^{-23}$ J/K).

This random motion results in broadening of the spectral signal peaks and also determines the frequency and magnitude of intermolecular collisions, thereby affecting spectroscopic measurements. However, for a discussion of basic spectroscopy, we will consider only the internal energy components of the molecule, i.e.,

$$E_{\text{molecule}} = E_{\text{electonic}} + E_{\text{vibrational}} + E_{\text{rotational}} = E_e + E_v + E_r \tag{5.49}$$

The electronic energy, $E_{\text{electronic}}$ or E_e, is due to the kinetic energy of the electrons orbiting the molecular nuclei. From Appendix D on Quantum Mechanics and Atomic/Molecular Structure, the electrons occupy orbitals with different energy levels. For simple diatomic molecules, these orbitals are written as $1s\sigma$, $2s\sigma$, $2p\sigma$, $2p\pi$, $3s\sigma$, $3p\sigma$, $3p\pi$, $3d\sigma$, $3d\pi$, and so on. The number in front of the orbital designation is the principal quantum number of the constituent molecular orbital, which is related to the average distance of the electron from the nucleus. The radial distance is larger for larger principal quantum numbers and is therefore associated with higher energy levels. The alphabetical letter after the number designation denotes the type of constituent atomic orbital, i.e., whether s-, p-, or other orbitals have been combined to form the molecule. Finally, the Greek letter designates the type of the combined molecular orbital as seen from the axis connecting the nuclei.

For example, for a hydroxyl molecule with the molecular formula OH, the ground-state molecular electronic configuration is

$$(1s\sigma)^2(2s\sigma)^2(2p\sigma)^2(2p\pi)^3 \tag{5.50}$$

where the numerical superscript indicates the number of electrons in that orbital. The ground state is the molecular electronic configuration with the lowest electronic energy and is therefore the most stable state. When one of the electrons moves "up" to a higher-energy orbital, the molecule is said to be *excited*. For the same OH molecule, the first excited state has the following configuration:

$$(1s\sigma)^2(2s\sigma)^2(2p\sigma)^1(2p\pi)^4 \tag{5.51}$$

One of the electrons from the $2p\sigma$ orbital has been moved up to the $2p\pi$ shell. The exact sequence of energy levels differs for different molecules, depending on the constituent atoms. For large molecules involving multiple atomic nuclei, the electronic configurations can be quite complex and may have several overlapping energy states.

When a photon is incident upon an atom or molecule, the photon energy can be absorbed if that energy corresponds to the difference in the energy states of the atom or molecule. For electronic energy states described earlier, the photon energy can send the electron to the higher energy state only if the photon energy is exactly equal to the energy difference between the two electronic states. Actually, that energy transition also has to be allowed under quantum-mechanical transition rules, and for electronic energy transitions, the transition rule is that the electron can go up (or down) to the next higher (or lower) energy state or remain within that energy state. That is, electrons cannot jump two or more levels in an electronic energy level in a single transition. During downward energy transitions, the electron returns to the ground-state orbital from the excited state, and the excess energy during this transition is given off in the form of a photon with the corresponding energy. The photon energy is given by

$$\Delta E = h\nu[J] \tag{5.52}$$

h = Planck's constant (6.63×10^{-34} J s)
ν = the light frequency of the photon

The second energy term in the molecular energy in Equation 5.49 is vibrational energy. For large molecules, there are many bonds between the constituent atoms, and there may be multiple vibrational modes, some of which overlap. For simple diatomic molecules, however, the vibrational energy can be modeled as a simple harmonic oscillator with a spring force that depends on the internuclear distance, as shown in Figure 5.11a (see Appendix D for details).

The vibrational energy levels based on the oscillator model are

$$E_v = h\nu_{\text{osc}}\left(v + \frac{1}{2}\right), \qquad v = 0, 1, 2, 3, \ldots \tag{5.53}$$

Here, h is again Planck's constant, ν_{osc} is the frequency of vibrational oscillation, and v is the vibrational quantum number. The vibrational energy of real molecules includes other vibrational modes (anharmonic vibrations) other than those represented in Equation 5.53, but they can be accounted for by adding correction terms to Equation 5.53. Similar to electronic transitions, only certain transitions are allowed from one vibrational energy state to another: transitions are allowed only for $\Delta v = \pm 1$.

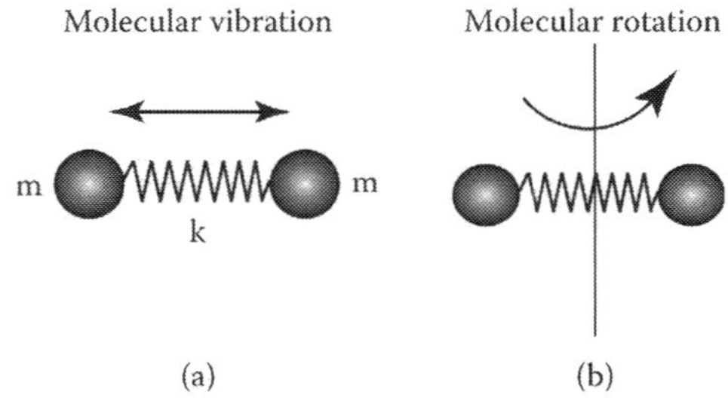

FIGURE 5.11 (a) Diatomic molecule vibration and (b) diatomic molecule rotation.

Molecular energy state can be moved to the next higher or lower vibrational energy state, but not any other.

The rotational energy of the molecule is modeled as the two nuclear masses rotating about the center of mass (Figure 5.11b).

$$E_r = \frac{1}{2} I\omega^2 \tag{5.54}$$

I = moment of inertia of the nuclear masses
ω = angular velocity of rotation

The quantum-mechanical solution for the dynamics of this rigid rotator model gives the rotational energy as follows:

$$E_r = \frac{h^2}{8\pi^2 I} J(J+1) = hcB_v J(J+1), \qquad B_v = \frac{h}{8\pi^2 cI} \tag{5.55}$$

where J is the rotational quantum number. Because the actual bond between the nuclei in a molecule is not rigid, correction terms are again used in Equation 5.55 to obtain realistic rotational energy levels. For rotational energy levels, the allowed transitions are $\Delta J = 0, \pm 1, \pm 2$.

The energy levels discussed here can be visualized in a molecular energy diagram (Figure 5.12). Within an electronic energy level, there are typically subdivisions representing vibrational energy levels and further subdivisions for rotational energy levels. The energy state of the molecule can change from one rotational energy state

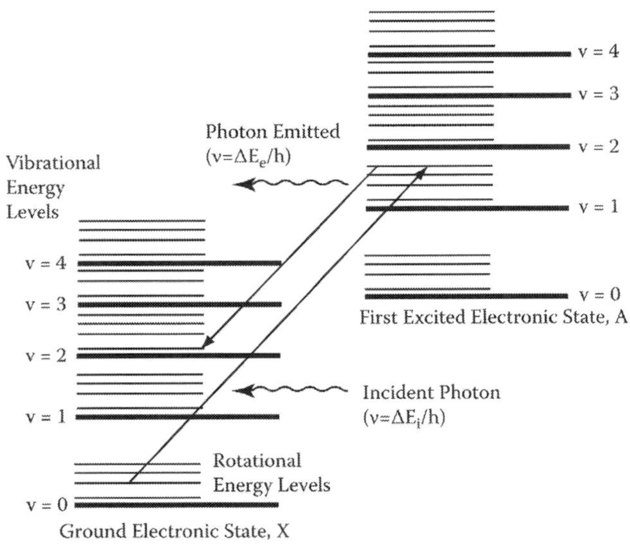

FIGURE 5.12 Molecular energy levels and transitions.

to another, from one vibrational energy state to another, or all of them combined, as long as the so-called selection rules are followed. The selection rules are summarized as follows:

$$\begin{aligned} &\text{Electronic transitions} : \Delta n = 0, \pm 1 \\ &\text{Vibrational transitions} : \Delta v = \pm 1 \\ &\text{Rotational transitions} : \Delta J = 0, \pm 1, \pm 2 \end{aligned} \tag{5.56}$$

Equation 5.52 gives the relationship between the energy difference and the photon frequency. Thus, for any energy transition following the foregoing selection rules, there is a corresponding photon frequency. If light or a stream of photons with this frequency is incident upon the molecule, the molecule can absorb exactly that energy and move to the corresponding higher energy state. Conversely, when a molecule at a higher energy state returns to a lower energy state, it conserves energy by emitting a photon with the frequency given by Equation 5.52. Therefore, by observing which light frequency components are absorbed by the molecules, we can determine their energy states. This is the basis for absorption spectroscopy. For electronic transitions induced by photons, a molecule with elevated electronic energy is said to be *excited* and is unstable. The unstable molecule tends to return to the original, stable energy state (called the *ground* state), emitting a photon in the process to conserve energy. The emitted photons bear the light frequency signature of the energy difference involved in this downward transition, as determined by Equation 5.52, and this is the basis for fluorescence spectroscopy. Thus, spectroscopy is a technique in which the relationship between photon frequency and particle energy states is used to infer the atomic or molecular internal structure.

For most molecules, various transitions involve the energy difference, ΔE, and the corresponding frequency, v, as shown in Table 5.2. In spectroscopy, wavenumber or wavelength is more commonly used as the spectral unit, so the frequency is converted to these quantities.

$$(\text{Wavelength}) = \lambda = c / v \left[\mu\text{m or nm}\right] \tag{5.57}$$

$$(\text{Wavenumber}) = \frac{1}{\lambda}\left[\text{cm}^{-1}\right] \tag{5.58}$$

TABLE 5.2
Typical Energy and Spectral Parameters for Various Transitions

Transitions	Energy Difference[J]	Frequency [s^{-1}]	Wavelength [nm]	Wavenumber [cm^{-1}]
Electronic, outer shells	$10^{-19} \sim 10^{-20}$	$10^{14} \sim 10^{15}$	$10^{2} \sim 10^{3}$ (uv ~ visible)	$10^{4} \sim 10^{5}$
Electronic, inner shells	$10^{-18} \sim 10^{-19}$	$10^{13} \sim 10^{14}$	$10 \sim 10^{2}$ (x-rays ~ uv)	$10^{5} \sim 10^{6}$
Vibrational	$10^{-20} \sim 10^{-21}$	$10^{15} \sim 10^{16}$	$10^{3} \sim 10^{4}$ (infrared)	$10^{3} \sim 10^{4}$
Rotational	$10^{-21} \sim 10^{-22}$	$10^{16} \sim 10^{17}$	$10^{4} \sim 10^{5}$ (microwave)	$10^{2} \sim 10^{3}$

As one can see in Table 5.2, electronic transitions in the outer shell are accessible using visible or ultraviolet (*UV*) light sources (typically lasers) and, also, *UV* optics are needed to observe the emission in these bands. Vibrational energy transitions are mostly in the infrared wavelength range, and the microwave is needed for probing the rotational energy structure. For a typical rotational-vibrational-electronic transition shown in Figure 5.12, we can estimate the energy of that transition using the respective energy levels as determined by Equations 5.47, 5.51, and 5.53. If we denote the quantum numbers in the upper energy state with a single prime (′) and those in the lower energy state with a double prime (″), then the energy difference between the upper and lower states is

$$\begin{aligned}\Delta E &= \left(E_e' - E_e''\right) + \left(E_v' - E_v''\right) + \left(E_r' - E_r''\right) \\ &= \left(E_e' - E_e''\right) + h\nu_{osc}\left(v' - v''\right) + hcB_v\left[J'(J'+1) - J''(J''+1)\right]\end{aligned} \tag{5.59}$$

Example 5.3

Consider a transition for a hydroxyl (OH) radical involving the following terms: $n' = 1$, $n'' = 0$, $v' = 1$, $v'' = 0$, $J'' = 5$, $J'' = 4$. In spectroscopic work, energy terms are converted to wavenumbers (Equation 5.58) by dividing by hc. The difference in the electronic energy is calculated $\left(E_e' - E_e''\right)/hc = 32{,}450\ \text{cm}^{-1}$. The vibrational and rotational energies can be computed from $\nu_{osc}/c \equiv \omega_e \approx 3739.94\,\text{cm}^{-1}$ and $B_v \approx 18.867\,\text{cm}^{-1}$. The total energy difference is $36{,}378.6\,\text{cm}^{-1}$, which corresponds to a wavelength of 274.87 nm. For an upward transition, a laser beam tuned to this wavelength is needed; conversely, for a downward transition, a signal at this wavelength will be observed. In practice, many downward transitions occur, and the spectrum is filled with large and small peaks depending on the strength of the transitions.

5.6 LASER-INDUCED FLUORESCENCE

Laser-induced fluorescence (LIF) has been discussed briefly in Chapter 3 for its use in flow visualization. LIF makes use of outer-shell electronic transitions, accompanied by rotational-vibrational transitions, and due to its high signal strength, it provides an effective means of quantitatively observing a specific target molecule. For the large organic molecules used as fluorescent dyes for flow visualization, there are many electrons, so multiple transition energies overlap, leading to the so-called broadband absorption and emission characteristics. For smaller molecules, there are distinct, individual transition lines, and the light source (laser) must be tuned to an exact wavelength (down to 1/100th of a nanometer) to induce specific energy transitions. Therefore, a typical setup for LIF experiments involves a narrowband, tunable laser. Four types of lasers can be used for LIF: dye lasers, optical parametric oscillators, excimer lasers, and tunable solid-state lasers. For broadband fluorescence involving dye molecules, exact tuning is not required. Other than the tunability of the laser, the LIF setup is quite similar to that for Rayleigh (Figure 5.3) or Raman scattering (Figure 5.9). Table 5.3 shows examples of chemical species produced in flames and the corresponding excitation wavelengths for LIF measurements.

TABLE 5.3
Examples of Chemical Species Measured with Laser-Induced Fluorescence (LIF)

Chemical Species	Excitation Wavelength [nm], Transition
OH	306.4, $A^2\Sigma$-$X^2\Pi(0,0)$
	286.4, $A^2\Sigma$-$X^2\Pi(1,0)$
CH	431.5, $A^2\Delta$-X^2n
NO	226.5, $A^2\Sigma$-$X^2\Pi_{1/2}$
C_2	516.5, $A^2\Pi_g$-$X^3\Pi_u$

The capital letters designate the electronic energy states (X represents the ground state, A the first excited state, and the Greek letters indicate the shape of the molecular orbitals), whereas the numbers in parentheses indicate the vibrational quantum numbers.

In practice, during LIF measurements, it is customary to first run excitation scans to identify and calibrate the incident laser wavelengths that lead to molecular excitation and, consequently, LIF signal, followed by fluorescence scans to identify the downward transition wavelengths. The principles of these scans are illustrated in Figure 5.13. For excitation scans, the laser wavelength is varied to scan across the possible excitation wavelengths, and the resulting fluorescence signal is collected using broadband detection. Each time the laser output wavelength matches an upward transition wavelength, i.e., the excitation wavelength, a signal will be detected. The order in which these peaks appear in the excitation spectrum depends on the relative energy difference between the upper and lower energy states—the larger the energy difference, the shorter the excitation wavelength. Conversely, for fluorescence scans, the laser is tuned to a fixed excitation wavelength, and the resulting signal is scanned over a range of wavelengths using a monochromator or spectrometer. Once the molecules are excited by the incident light, the downward transition can follow many routes depending on the transition probabilities. Based on the excitation and fluorescence scans, one can then identify strong excitation wavelengths as well as strong detection wavelength bands. As will be seen later, these scans can also be used to determine the population distributions of the vibrational and rotational states, which depend on temperature.

We can begin examining some of the signal dependencies by considering a simple model of LIF. The model includes two energy levels corresponding to the ground state and the excited state, as shown in Figure 5.14. If N_1 and N_2 are the number densities of the molecules in energy states 1 and 2, respectively, then their time rate of change is due to either upward or downward transitions. This can be expressed as differential equations using the various rates for upward (photon absorption) and downward (photon emission or collisional loss) transitions (Eckbreth, 1996).

$$\frac{dN_1}{dt} = -N_1 b_{12} + N_2\left(b_{21} + A_{21} + Q_{21}\right)$$
$$\frac{dN_2}{dt} = N_1 b_{12} - N_2\left(b_{21} + A_{21} + Q_{21}\right) \tag{5.60}$$

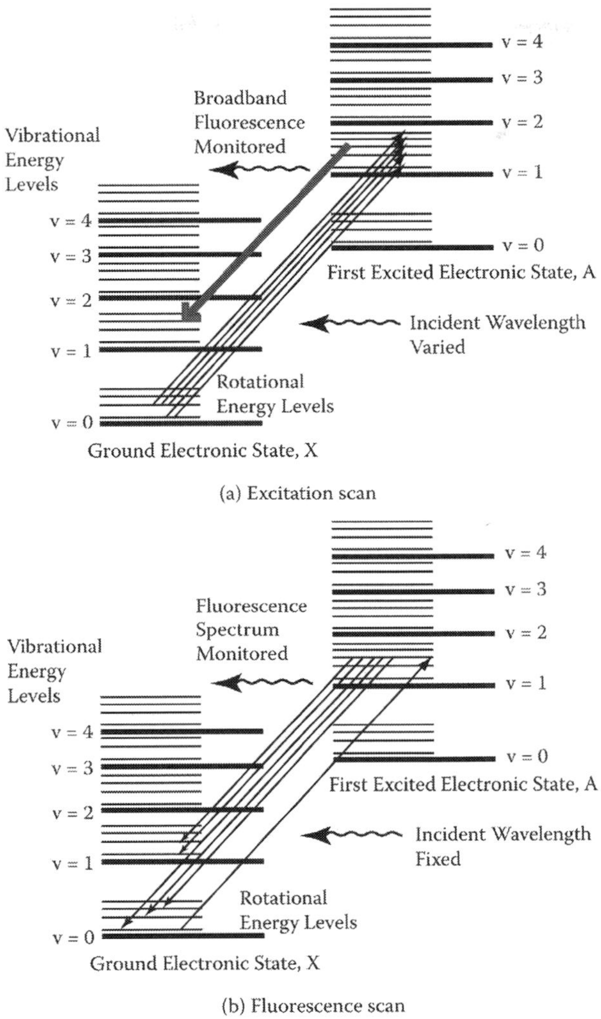

FIGURE 5.13 Excitation and fluorescence scans for laser-induced fluorescence (LIF).

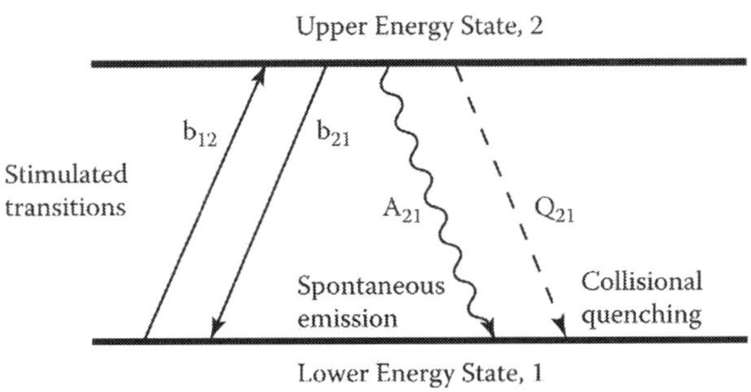

FIGURE 5.14 A two-level model of laser-induced fluorescence.

b_{12} = rate constant for the transition from 1 to 2 due to photon energy absorption
b_{21} = rate constant for the transition from 2 to 1 due to stimulated photon emission
A_{21} = rate constant for the transition from 2 to 1 due to spontaneous photon emission
Q_{21} = rate constant for the transition from 2 to 1 due to quenching (collisional energy loss)

In the foregoing equation, b_{12} is the rate constant for the transition from energy state 1 to 2, representing the quantum-mechanical probability of that transition. Not all transitions are equally probable; those with higher probabilities exhibit stronger peaks in the spectrum. The rate constants indicate the fraction of molecules that actually undergo a given transition upon excitation. We distinguish between stimulated and spontaneous emissions, with the former representing a resonant behavior in the presence of photons of the transition energy. For example, in laser cavities, excited atoms or molecules resonate with nearby photons, following the energy transitions to emit photons of exactly the same wavelength; this process is called *stimulated* emission. Because stimulated processes are induced by resonant photons, their rate constants depend on the photon flux, or the incident laser energy, I_λ, as follows:

$$b_{12\,\text{or}\,21} = \frac{B_{12\,\text{or}\,21} I_\lambda}{C_\lambda} \tag{5.61}$$

$B_{12\text{ or }21}$ = transition probability
I_λ = incident spectral intensity $\left[\dfrac{W}{\text{cm}^2\ \text{sr}\ \mu\text{m}}\right]$
C_λ = conversion factor having the same unit as I_λ

Spontaneous emission differs in that the relaxation of the molecular energy to lower states occurs naturally, without influence from surrounding photons. Quenching refers to the process in which molecules lose their excited-state energy through molecular collisions.

Returning to Equation 5.60, initially the molecules are in the ground state, so $N_1 = N_1^o$ (the initial molecular number density), and $N_2 = 0$ at $t = 0$. The molecules can transition from state 1 to 2 upon excitation, but the overall number is conserved, which leads to $N_1 + N_2 = N_1^o$, or $N_1 = N_1^o - N_2$ at all times. We can substitute this expression into Equation 5.60 to get

$$\begin{aligned} \frac{dN_2}{dt} &= b_{12}\left(N_1^o - N_2\right) - N_2\left(b_{21} + A_{21} + Q_{21}\right) = b_{12}N_1^o - rN_2 \\ r &= b_{12} + b_{21} + A_{21} + Q_{21} \end{aligned} \tag{5.62}$$

The preceding rate equation for the excited-state molecular number density has the following solution:

$$N_2(t) = \frac{b_{12}N_1^o}{r}\left(1 - e^{-rt}\right) \tag{5.63}$$

Immediately upon laser fluence ($t << 1/r$), N_2 starts to increase linearly, and after some time ($t > 1/r$) it reaches a steady-state value of $b_{12}N_1^o / r$. The transient period ($t << 1/r$) is typically on the picosecond level, so most measurements are made in the steady-state time scales.

Now, as noted previously, the fluorescence signal depends on the spontaneous transition probability, A_{21}, and the number density N_2. Therefore, the fluorescence signal intensity, F_λ, can be written using the steady-state solution of Equation 5.63 as:

$$F_\lambda \sim N_2 A_{21} = \frac{b_{12} N_1^o A_{21}}{r} \tag{5.64}$$

It is useful to define a threshold spectral intensity for the incident laser beam, I_λ^{sat}, called the *saturation intensity*, based on the ratio of the spontaneous and quenching effects to the stimulated effects:

$$I_\lambda^{\text{sat}} = \frac{A_{21} + Q_{21}}{B_{12} + B_{21}} C_\lambda \tag{5.65}$$

C_λ = conversion factor having the same unit as I_λ

Using Equations 5.61 and 5.65 in Equation 5.64, we get, after a little algebraic manipulation,

$$F_\lambda \sim N_1^o \frac{B_{12} A_{21}}{\left(B_{12} + B_{21}\right)\left(1 + \dfrac{I_\lambda^{\text{sat}}}{I_\lambda}\right)} \tag{5.66}$$

Based on Equation 5.66 for the fluorescence signal intensity, we can consider two different regimes for LIF. First, if the incident laser spectral energy is very small relative to I_λ^{sat}, i.e., $I_\lambda \ll I_\lambda^{\text{sat}}$, then $1 + I_\lambda^{\text{sat}}/I_\lambda \approx I_\lambda^{\text{sat}}/I_\lambda$, and then Equation 5.66 becomes

$$F_\lambda \sim N_1^o \frac{B_{12} A_{21}}{A_{21} + Q_{21}} \frac{I_\lambda}{C_\lambda} \quad \left(I_\lambda \ll I_\lambda^{\text{sat}}\right) \tag{5.67}$$

This is called the *linear regime* because F_λ depends linearly on I_λ. The difficulty with quantitative measurements in the linear regime is the presence of Q_{21}, the quenching rate. The quenching effect is due to energy relaxation with various collision partners, the number concentrations of which are usually unknown. For this reason, it is preferable to use the other limiting condition for quantitative measurements: the *saturation regime*.

For large laser fluence, $I_\lambda \gg I_\lambda^{\text{sat}}$, the ratio $I_\lambda^{\text{sat}}/I_\lambda$ can be dropped from Equation 5.66 altogether, leading to

$$F_\lambda \sim N_1^o \frac{B_{12} A_{21}}{B_{12} + B_{21}} \quad \left(I_\lambda \gg I_\lambda^{\text{sat}}\right) \tag{5.68}$$

The preceding equation shows that, in addition to the quenching effect being absent in the fluorescence signal, F_λ is also independent of I_λ, provided that the condition for

TABLE 5.4
Saturation Laser Spectral Irradiances for Some Combustion Radicals

Species	$I_\lambda^{sat}\ [\mathrm{W/cm^2cm^{-1}}]$
OH	1.3×10^6
CH	3.0×10^5
NO	1.2×10^7
CN	3.0×10^5
C_2	6.0×10^7
NH	2.7×10^6

Source: Adapted from Eckbreth, A. C. *Laser Diagnostics for Combustion Temperature and Species*, 2nd ed., Gordon and Breach, New York, 1996.

saturation is satisfied $I_\lambda \gg I_\lambda^{sat}$. This is quite convenient because there is no need to maintain or monitor the incident laser energy. Table 5.4 lists the saturation spectral irradiances for radical species (those species that are produced and rapidly destroyed through chemical reactions) in flames (Eckbreth, 1996). I_λ^{sat} is given in units of watts per cm^2 (cross-sectional area of the focused beam) per cm^{-1} (unit wavenumber interval).

Example 5.4

For Nd:YAG lasers, a typical laser pulse is approximately 10 ns in duration with a 0.5 cm^{-1} linewidth (the spectral width in the wavenumber unit). Assuming that the pulse duration and linewidth are maintained during laser spectral tuning through a dye laser and frequency doubler, to reach the saturation irradiance of 1.3×10^6 [W/cm^{2}cm^{-1}] for OH, the final laser energy needs to be only 0.051 mJ or 51 μJ. This energy level is much lower than the available laser output from most Nd:YAG-based spectral tuning systems. For planar LIF imaging, the beam is spread out in space, so the spectral irradiance will be much smaller.

Now we consider the temperature effect on LIF measurements. In Equations 5.56–5.58, the LIF signal depends on N_1^o, which is the initial number density of the molecules in a particular energy state that the incident laser is tuned to bring about an upward transition. That energy state involves electronic, vibrational, and rotational quantum numbers, and as will be shown later, the fraction of molecules in a particular rotational energy state has a sensitive dependence on temperature. That is, there are many possible energy states, and the temperature will affect the distribution of rotational energy among the molecules. In fact, the fraction of molecules, f_J, having a particular rotational energy or, equivalently, rotational quantum number, is given by the Boltzmann distribution.

$$f_J = \frac{N_J}{N} = \frac{hcB_v}{kT}(2J+1)\exp\left[-\frac{B_vJ(J+1)hc}{kT}\right] \tag{5.69}$$

N_J = number density of molecules with the rotational quantum number J
N = total number density of the molecules
h = Planck's constant = $6.62606876 \times 10^{-34}$ Js
c = speed of light
B_v = a constant representing the rotational energy characteristics for a given molecule
k = Boltzmann's constant = $1.3806503 \times 10^{-23}$ J/K
T = temperature [K]

For species concentration measurements in nonisothermal flows, we want $N_1^o = f_i N$ to be as independent of temperature as possible. To find the rotational quantum number least dependent on temperature, we can take Equation 5.71 and differentiate it with respect to temperature.

$$\frac{\partial f_J}{\partial T} = 0 \rightarrow J^2 + J - \left(\frac{k}{hcB_v}\right)T = 0 \tag{5.70}$$

For example, for OH in flames, the use of the $Q_1(8)$ line for excitation is associated with only a 10% change in the lower energy state number density for the temperature range of 1000–2600 K. The notation Q 1(8) means that we are using the rotational quantum number J = 8 in the lower energy state for excitation with ΔJ = 0 in going from the lower to the upper energy state. The subscript "1" denotes the vibrational quantum number, also of the lower energy state.

5.7 THERMOMETRY BASED ON LIF

Temperature measurements (thermometry) can, in principle, be made using LIF. They are based on the known distribution of rotational energy in molecules as a function of temperature, i.e., the Boltzmann distribution in Equation 5.69. In the linear regime, the fluorescence signal is given by Equation 5.67 (rewritten after normalizing by I_λ/C_λ):

$$\frac{F_\lambda}{(I_\lambda / C_\lambda)} \equiv F_J^{'} \sim N_1^o B_{12} \frac{A_{21}}{A_{21} + Q_{21}} = N_1^o B_{12} \eta_J \tag{5.71}$$

$$\eta_J \equiv \frac{A_{21}}{A_{21} + Q_{21}} = \text{fluorescence quantum yield}$$

Equation 5.69 tells us that, by using a laser tuned to a specific rotational transition wavelength and measuring $F_J^{'}$, we get a signal that is proportional to N_1^o for that rotational level, J, at a given temperature T. To differentiate between energy states with different J, we write $N_1^o = \left(N_1^o\right)_J = f_J N$ (Equation 5.69) and substitute it into Equation 5.71.

$$F_J^{'} \sim \left(N_1^o\right)_J B_{12}\eta_J = \left(f_J N\right) B_{12} = \frac{NhcB_v}{kT} B_{12}\eta_J (2J+1)\exp\left[-B_v J(J+1)\frac{hc}{kT}\right] \tag{5.72}$$

Now, if we perform the excitation scan in the transition wavelength range, from large to small wavelengths, we will be exciting one transition after another, as shown in

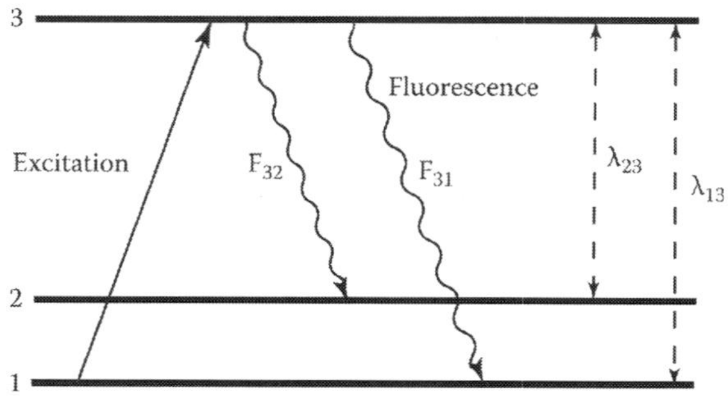

FIGURE 5.15 A two-line LIF for temperature measurements.

Figure 5.15. In doing so, we will be scanning various transitions starting from lower rotational energy states in the upper level to higher ones, because the photon energy increases with decreasing wavelength (increasing frequency). That is, we will be scanning across the rotational quantum number, *J*, in increasing order. The order of the rotational energy levels, as denoted by the rotational quantum number *J*, depends on the particular molecule, and thus data on the rotational energy transitions must be available. Taking the logarithm of both sides of Equation 5.70, we get

$$\ln\left(\frac{F_J^{'}}{\eta_J B_{12}(2J+1)}\right) = -B_v J(J+1)\frac{hc}{kT} + \ln\left(\frac{NhcB_v}{kT}\right) \tag{5.73}$$

Plotting the left-hand side as a function of $hcB_vJ(J+1)$ yields a straight line with a negative slope 1/kT. The second term on the right-hand side is treated as a constant.

For steady-state processes, where there is the luxury of time to run excitation scans, the foregoing method can provide reasonably accurate temperature measurements. For dynamical processes, a "two-line" LIF thermometry can be used, which is an abbreviated version of the full excitation scan. Instead of scanning across a range of *J* values, two *J* values are selected and excited, and the resulting fluorescence signals are compared to determine the slope 1/kT. This process is illustrated in Figure 5.15.

5.8 VIBRATIONAL AND ROTATIONAL SPECTROSCOPY

In Raman scattering (Section 5.4), we saw how molecular vibration can modulate the incident laser energy and therefore its wavelength to produce signals that yield information concerning the molecular vibrational energy levels. We also saw in Sections 5.6 and 5.7 that these energy levels depend not only on the molecular structure but also on the distribution of energy levels as dictated by the Boltzmann distribution. Probing the vibrational and rotational energy structures of molecules using Raman scattering, LIF, or infrared absorption is called vibrational/rotational spectroscopy.

The distribution of energy among molecules, such as in Equation 5.69, is predicted by the Boltzmann distribution, which gives the general distribution of energy among particles (including molecules).

$$\frac{N_j}{N} = \frac{g_j e^{-E_j/kT}}{\sum_{j=1}^{\infty} g_j e^{-E_j/kT}} \tag{5.74}$$

N_j = number density of particles in the j-th energy state
N = total number density
E_j = energy of the state "j"
g_j = degeneracy of the state, i.e., the number of states with the same energy
k = Boltzmann's constant = $1.3806503 \times 10^{-23}$ J/K
T = temperature

Equations 5.51 and 5.53, respectively, gave the vibrational and rotational energy of the molecules. First, by substituting the vibrational energy $E_v = h\nu_{osc}\left(v + \frac{1}{2}\right)E_j$ into Equation 5.72, we obtain

$$\frac{N_v}{N} = \frac{\exp\left(-\frac{\left(v+\frac{1}{2}\right)h\nu_{osc}}{kT}\right)}{\sum_{v=0}^{\infty}\exp\left(-\frac{\left(v+\frac{1}{2}\right)h\nu_{osc}}{kT}\right)} \tag{5.75}$$

The denominator in Equation 5.75 is called the *partition function*, and it is an infinite series that converges to $\frac{1}{1-\exp\left(-\frac{h\nu_{OSC}}{kT}\right)}$. Equation 5.75 can then be written as

$$\frac{N_v}{N} = \exp\left(-\frac{\left(v+\frac{1}{2}\right)h\nu_{osc}}{kT}\right)\left\{1-\exp\left(-\frac{h\nu_{osc}}{kT}\right)\right\} \tag{5.76}$$

Calculations using Equation 5.76 show that the variation of the vibrational population for most molecules is relatively small as a function of temperature. At low-to-intermediate temperatures, the molecules occupy only the lowest vibrational quantum numbers, and the degree to which higher vibrational quantum numbers are occupied varies relatively slowly with increasing temperature. Thus, the vibrational energy distribution for most molecules is only weakly sensitive to temperature changes. This is advantageous when measuring species concentrations, for example, because temperature variations have little effect on the measurements. On the other hand, if one is trying to measure temperature using vibrational spectra, more sensitive variations, as found in rotational energy distributions, are preferred, as described next. An alternate method of extracting temperature data from vibrational spectra is by comparing the anti-Stokes and Stokes peaks, as shown in Figure 5.9. For nitrogen molecules, for example, the ratio of the anti-Stokes to Stokes peaks is nearly linear above 1000 K, allowing for a high-temperature probe using this approach. The drawback, however, is that below 750 K, the anti-Stokes signal is very weak, making this approach of little use at or below that temperature.

For rotational energy, $E_r = hcB_vJ(J+1)$ (from Equation 5.55), the degeneracy g_j is $2J+1$, and there is also the so-called nuclear-spin degeneracy, g_{NS}. The rotational partition function (the denominator in Equation 5.74) converges to kT/hcB_v, and we have

$$\frac{N_J}{N} = \frac{(2J+1)\, g_{NS} \exp\left(-\dfrac{hcB_v J(J+1)}{kT} \right)}{\dfrac{kT}{hcB_v}} \tag{5.77}$$

Figure 5.17 depicts the changes in the rotational population as a function of temperature. At 300 K, the population density peaks at a low rotational quantum number ($J = 3$). Even with a modest increase in temperature, higher rotational energy levels are occupied, and this change becomes more prominent as the temperature increases further. Thus, the change in the rotational energy distribution is quite sensitive to temperature, making it possible to perform temperature measurements based on rotational spectra. The rotational energy distribution in Figure 5.16 was generated using Equation 5.77, and the measured rotational spectra can be fit to those computed (as in Figure 5.16) or compared with calibrated profiles at different temperatures.

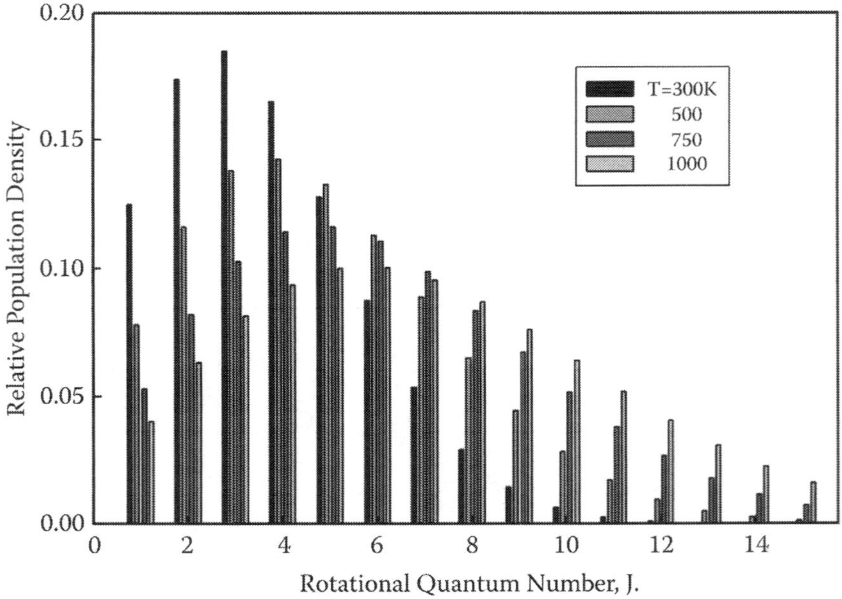

FIGURE 5.16 An example of the change in the rotational population with increasing temperatures.

5.9 FOURIER-TRANSFORM INFRARED SPECTROSCOPY

Fourier-transform infrared spectroscopy (FTIR) is a widely used method in vibrational spectroscopy. We can examine typical vibrational transition frequencies using Equation 5.59. In spectroscopy, it is customary to use "term values" or the wavenumber, expressed in units of cm^{-1}; thus, we convert Equation 5.59 into term values by dividing by hc.

$$\begin{aligned}\frac{\Delta E}{hc} \equiv \nu &= \frac{\left(E_e^{'} - E_e^{''}\right)}{hc} + \frac{\left(E_v^{'} - E_v^{''}\right)}{hc} + \frac{\left(E_r^{'} - E_r^{''}\right)}{hc} \\ &= \frac{\left(E_e^{'} - E_e^{''}\right)}{hc} + \omega_e\left(v' - v''\right) + B_v\left[J'(J'+1) - J''(J''+1)\right]\end{aligned} \tag{5.78}$$

Vibrational transitions typically occur in combination with rotational transitions, so we will use the latter two terms in Equation 5.78. Using OH as an example, the energy constants are $\omega_e = 3739.94\,\text{cm}^{-1}$ and $B_v = 18.867\,\text{cm}^{-1}$. OH radicals are found in flames and in atmospheric ozone reactions ($H + O_3 \rightarrow OH + O_2$). Some correction terms to these energy constants account for non-rigid rotation and ro-vibrational interactions, but for our purposes, the preceding numbers suffice to illustrate the point. Considering the magnitudes of ω_e and B_v, vibrational energy is much higher than rotational energy. Taking $\Delta v = +1$ and $J'' = 4 \rightarrow J' = 5$, we get from Equation 5.78, $\nu = 3828.61\,\text{cm}^{-1}$. This term value (wavenumber) corresponds to a wavelength of 2.545 μm. Thus, the ro-vibrational transition involves photon energy corresponding to the infrared wavelength range. This is typical for most molecules, making FTIR a technique applied to probe the vibrational structure of molecules in gas, liquid, or solid phases. Table 5.5 shows some typical vibrational wavenumbers associated with various molecules.

TABLE 5.5
Vibrational Absorption Bands for Some Molecules

Molecule Name	Molecular Formula	Absorption Bands, Wavelength [μm]	Absorption Bands, Wavenumber [cm^{-1}]
Acetylene	C_2H_2	2.95 ~ 3.14, 6.97 ~ 8.16, 11.76 ~ 16.66	600 ~ 850, 1225 ~ 1435, 3180 ~ 3390
Carbon dioxide	CO_2	2.69, 2.76, 4.25, 14, 15	667, 714, 2353, 3623, 3717
Carbon monoxide	CO	2.3, 4.7	2128, 4348
Ethylene	C_2H_4	10.52	950
Methane	CH_4	3.4, 7.4, 7.58, 7.87	1270, 1319, 1351, 2941
Nitric monoxide	NO	5.24 ~ 5.42	1842 ~ 1909
Nitric dioxide	NO_2	5.7 ~ 5.75, 6.14 ~ 6.27, 7.88 ~ 7.98	1252 ~ 1268, 1595 ~ 1628, 1740 ~ 1755
Nitrous oxide	N_2O	7.83, 16.97	589, 1277
Ozone	O_3	9, 9.59, 14.2	704, 1042, 1111
Propane	C_3H_8	3.1 ~ 3.5, 7.09 ~ 8.47	1180 ~ 1410, 2850 ~ 3220
Sulfur dioxide	SO_2	7.34, 8.68, 19.3	518, 1151, 1362
Sulfur trioxide	SO_3	7.51, 9.35, 15.33, 18.79	532, 652, 1069, 1330
Water vapor	H_2O	1.87, 2.69, 3.2, 6.3	1587, 3125, 3704, 5348

The unique aspect of FTIR is that it scans a wide infrared wavelength range using a fast, nondispersive method. Figure 5.17 shows the optical arrangement used in FTIR spectrometers. The light source is typically a heated cavity that emits radiation close to blackbody radiation. Radiative heat transfer calculations show that if the depth-to-diameter ratio of a cavity is larger than 10, the effective emissivity of the radiation from the cavity is greater than 0.99. An emissivity close to one means that the source emits nearly like a blackbody, for which the radiative emission is well characterized by Planck's law (Appendix C). As seen in Figure 5.17, the continuous-spectrum radiation from the infrared light source is split and reflected off a fixed and a moving mirror so that the two beams are mixed or interfered with one another (optical heterodyning). The moving mirror introduces a time-dependent phase difference and mixes it with the original, unshifted beam. The component of the radiation with a certain wavelength, λ, will constructively interfere if the path length difference caused by the moving mirror is 0, 2λ, 4λ, and so on. Conversely, if the path length difference is an odd multiple of the wavelength, λ, 3λ, 5λ …, there will be destructive interference.

It is convenient to observe the effect of the phase lag on the beam by using the "Euler" notation for the electromagnetic wave (as in Equation C.20b). The beam reflected from the fixed mirror can be written as

$$B_1(\lambda) = B(\lambda)\exp i\left(\frac{2\pi x}{\lambda} - ct\right) \tag{5.79a}$$

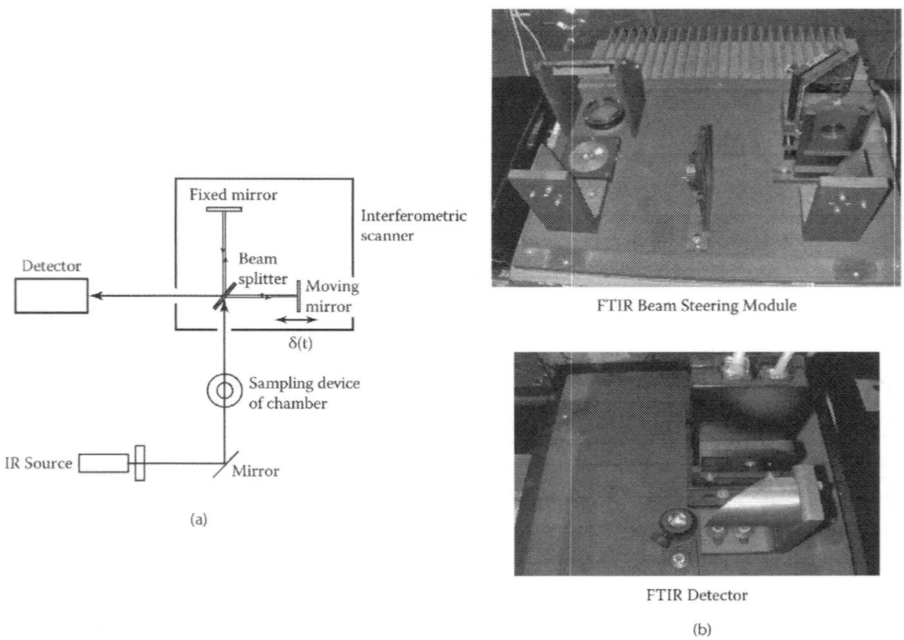

FIGURE 5.17 The optical configuration of FTIR.

and the beam from the moving mirror has a time-dependent phase shift added to the above equation.

$$B_2(\lambda) = B(\lambda)\exp i\left(\frac{2\pi x}{\lambda} - ct + \frac{2\pi\delta(t)}{\lambda}\right) = B(\lambda)\exp i\left(\frac{2\pi x}{\lambda} - ct\right)\exp i\left(\frac{2\pi\delta(t)}{\lambda}\right) \tag{5.79b}$$

The electromagnetic wave equation is linear in the sense that the two beams can be added together, and the x- and t-dependent terms are no longer needed because the detection will average over the time scale of the measurements, and the detector location, x, is fixed. Thus, the mixed beam can be expressed as

$$B_T(\lambda) = B(\lambda)\left[1 + e^{\frac{2\pi\delta(t)}{\lambda}}\right] \tag{5.80}$$

It is the time-dependent term that is of use in FTIR. If we consider a mirror scan that encompasses a wavelength range for $B(\lambda)$, which includes continuous-wavelength components, then the observed signal can be expressed as an integral.

$$F(t) = \int_0^T B(\lambda)\exp i\left(\frac{2\pi\delta(t)}{\lambda}\right)dt = \frac{1}{2}\int_{-T}^{T} B(\lambda)\exp i\left(\frac{2\pi\delta(t)}{\lambda}\right)dt \tag{5.81}$$

A Fourier transform (Equation 1.28) of the preceding relationship then gives the original spectrum, $B(\lambda)$, which is what we seek.

$$B(\lambda) = \frac{1}{T}\int_{-T}^{T} F(t)\exp\left(\frac{2\pi\delta(t)}{\lambda}\right)dt \tag{5.82}$$

The mirror position, $\delta(t)$, is controlled by an onboard motion controller and is thus a known function of time. If the signal is detected in synchronization with $\delta(t)$, then $F[\delta(t)]$ is available from the temporally resolved measurement. Thus, $F(x)$ can be digitally integrated using the Fourier-transform formula of Equation 5.82 to obtain the spectra. If $B(\lambda)$ contains any loss of wavelength components due to absorption by the molecules, it will show up through Equation 5.82, and this is the operating principle of FTIR measurements.

Figure 5.18 shows a typical FTIR spectrum obtained from an ethylene-air flame. The FTIR spectrum is converted into the absorption spectrum, as in Figure 5.18, from the raw spectrum, which is essentially a transmission spectrum (the detector shows which components have been transmitted through the medium). Due to the soot particles in the flame, there is a background pedestal originating from the soot radiation. Soot particles, when heated in the flame, radiate like a blackbody. Prominent infrared absorption bands of carbon dioxide and water vapor are observed in Figure 5.18. Carbon dioxide is a greenhouse gas, absorbing much of the infrared radiation outbound by the Earth's surface. This propensity for infrared absorption is evident in the FTIR spectrum near carbon dioxide bands centered at 667, 714, and 2353 cm^{-1}. It can also be pointed out that the FTIR spectrum contains multiple bands

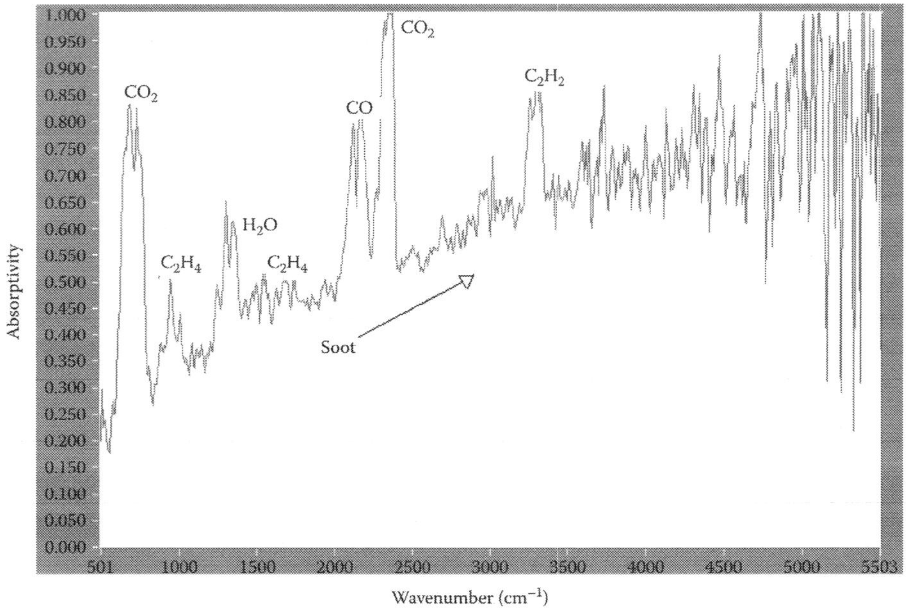

FIGURE 5.18 A typical FTIR spectrum.

from the same molecules due to the number of different vibrational energy modes in the molecule. Vibrational spectroscopy, such as FTIR, identifies vibrational modes such as C=O (double bond) stretching, C–O (single bond) bending, C–H stretching, etc., and it is the spectroscopist's function to relate these bonds to the specific molecules. In complex spectra, this does introduce ambiguities as to the origin of the observed absorption peaks. FTIR thus does not specifically identify the molecules but only the vibrational modes of the molecules.

The spectra in Figure 5.18 were obtained through direct absorption sampling, i.e., the modulated IR beam from the FTIR spectrometer was sent directly through the volume containing the target molecules. Obviously, direct sampling is the simplest way to use FTIR, as it only requires a clear (IR-transmitting) path across the sample. However, for gas samples, for which direct sampling is mostly applicable, the amount of absorbed radiation may be very small, leading to weak FTIR signatures. As long as absorption can be made to occur through the target sample, there are several other methods through which FTIR can be deployed to enhance absorption levels. Some of these FTIR sampling methods are shown in Figure 5.19. To increase the total absorption, a long-path or a multipass gas cell may be used (Figure 5.19b). In Appendix C, it is shown that the absorption increases exponentially as a function of the path length. For liquids, on the other hand, the absorptivity may be broadband and also very high, and efforts may be needed to fabricate a sampling cell with a thin sampling volume. In both cases, IR-transmitting windows are needed, and some of the common materials for IR measurements are shown in Table 5.6. For liquid and solid powders, reflectance sampling can be used as shown in Figure 5.19c–e. The attenuated total reflectance (ATR) uses multiple surface reflections from the sample surface by using

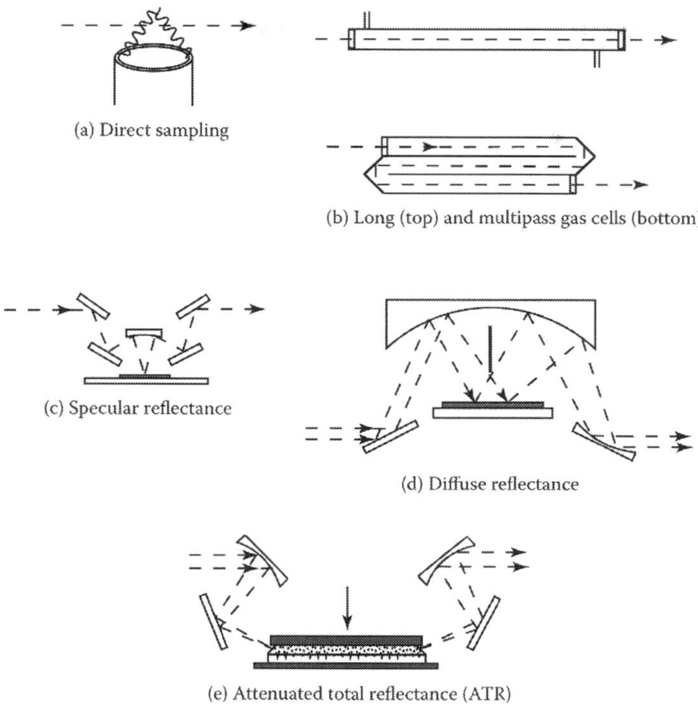

FIGURE 5.19 FTIR sampling methods.

TABLE 5.6
Some IR-Transmitting Materials and Their Cutoff Limits

Material	Cutoff Wavenumber [cm^{-1}]	Cutoff Wavelength [μm]
Glass	3000	3.3
Quartz	2500	4.0
Lithium fluoride	1500	6.6
Calcium fluoride	1200	8.3
Barium fluoride	850	11.7
Irtran-2	750	13.4
Sodium chloride	600	16.5
Potassium bromide	350	28.6
Silver chloride	350	28.6
KRS-5	250	40.0
Cesium bromide	250	40.0
Cesium iodide	200	50.0
Polyethylene	100 (does not transmit above 650 cm^{-1})	100.0 (does not transmit below 15.4 μm)

Note: The cutoff limit is the wavenumber or wavelength beyond which no significant IR is transmitted.

a crystal element pressed against the sample (Figure 5.19e). Most of these sampling devices can be fabricated in the laboratory, except perhaps the ATR sampler, which requires a precise entry angle into the crystal to ensure proper reflectance paths.

FTIR is a prominent example of absorption spectroscopy, and it has the capability to scan over a wide wavelength range in the infrared. If the target molecule has a specific absorption wavelength band, then diode or other laser sources with light output in that wavelength band can be used for monochromatic absorption detection of that target molecule. For example, the carbon–hydrogen (C–H) stretching mode of vibration has an absorption wavenumber of 2950 cm^{-1}, corresponding to a wavelength of 3.4 μm, and He–Ne laser light output at 3.392 μm coincides with this absorption band. In Table 5.5, the C–H absorption bands can be found in methane and propane. This absorption band has been used to measure fuel concentrations in engines. Such an approach can be considered an abbreviated version of absorption spectroscopy that allows for deployment in practical systems, such as engines, using a set of compact and low-cost components.

5.10 OTHER DIAGNOSTIC METHODS

5.10.1 Tunable Lasers

Advances in lasers and detector technology have enabled other methods for thermal and flow diagnostics. One example is absorption spectroscopy using tunable-diode lasers for measurements of temperature and species. There are also other tunable lasers, such as excimer lasers, dye lasers, and optical parametric oscillators, which allow the user to scan or select a specific narrowband output wavelength. Excimer lasers operate through recombination of positive gas ions with halides or through reactions of excited gas atoms with halogens. Typical gas–halide combinations and their output wavelengths are shown in Table 5.7. As indicated in the table, excimer lasers emit ultraviolet light and are useful for LIF measurements.

Similar to other lasers, the gas–halide combination is first excited to elevated electronic energy states through "pumping," which in excimer lasers can be achieved by high-energy electron beams, electron beam discharge, or microwave excitation. The excited excimers emit photons during return to the ground state, and precise tuning of the output beam is enabled through a prism or grating. Since a single pass through the prism or grating may not generate high output power, narrowband tunable oscillators are built into the laser for multiple-pass amplification.

TABLE 5.7
Excimer Lasers

Gas–Halide/Halogen Combination	Output Wavelength [nm]	Electronic Transition
ArF	193	B → X
KrF	248	B → X
XeCl	308	B → X
XeF	351	B → X

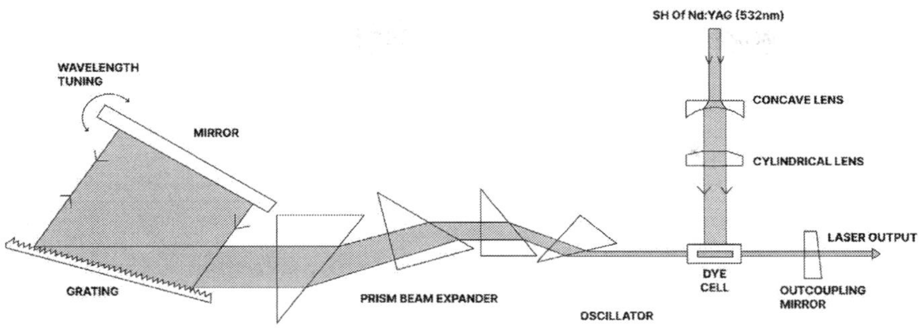

FIGURE 5.20 Tuning of dye laser output.

Dye lasers use liquid dyes pumped by a pulsed laser (e.g., frequency-doubled Nd:YAG output at 532 nm). Narrowband tuning is again accomplished using a grating, typically as shown in Figure 5.20.

Light generation in tunable-diode lasers originates from p-n junction(s), where electrical current applied at the junction elevates the electron energy from the valence band to a higher state (the so-called conduction band). Decay of electron energy back to the valence band (recombination) results in photoemission due to conservation of energy. The frequency (or wavelength) of the emitted photons depends on the band gap or the type (material) of the p-n junction. Common materials for diode lasers are: GaAs (gallium arsenide), AlGaAs (aluminum gallium arsenide), InGaP (indium gallium phosphide), GaN (gallium nitride), InGaAs (indium gallium arsenide), GaInNAs (indium gallium arsenide nitride), InP (indium phosphide), and GaInP (gallium indium phosphide). Most of the materials above generate light in the near-infrared. However, GaInP-based diode lasers emit in the red (used in laser pointers), GaN in blue, and PbSe (lead selenide) in the mid-infrared. These solid-state lasers are commercially available for a wavelength range from 190 to 4000 nm. Precise and narrow wavelength output is again achieved using a diffraction grating or, to some extent, by varying the junction temperature.

5.10.2 Tunable-Diode Laser Absorption Spectroscopy

Tunable-diode lasers are useful for detecting temperature and gas species, even in harsh environments. The fundamental principle of operation for tunable-diode laser absorption spectroscopy (TDLAS) is Beer's law of light absorption for a mixture of gas species.

$$I_\lambda(\lambda)/I_{\lambda o} = \exp\left\{-\sum_i \left(n_i L \sum_j S_{i,j} G_{i,j}\right)\right\} \tag{5.83}$$

I_λ = measured intensity at wavelength λ
$I_{\lambda o}$ = laser intensity at wavelength λ
n_i = number density of species i
L = path length
$S_{i,j}$ = temperature-dependent absorption line strength of line j, species i
$G_{i,j}$ = line broadening function for line j, species i

TABLE 5.8
Detectable Gases Using TDLAS

Gas	Near-Infrared Absorption Wavelength [nm]	Mid-Infrared Absorption Wavelength [nm]
Water (H_2O)	1390	5940
Carbon dioxide (CO_2)	1960	4230
Carbon monoxide (CO)	1570	4600
Nitric oxide (NO)	1800	5250
Nitric dioxide (NO_2)	680	6140
Sulfur dioxide (SO_2)	—	7280
Methane (CH_4)	1650	3260
Acetylene (C_2H_4)	1520	7400
Hydrogen sulfide (H_2S)	1570	—
Ozone (O_3)	—	9500
Ammonia (NH_2)	1500	10,300
Oxygen (O_2)	760	—

The amount of absorption depends on the concentration of the absorbing gas species at that wavelength. By tuning the laser output to different absorption lines, a range of gas species can be detected. A set of gases measurable through tunable-diode laser absorption spectroscopy is shown in Table 5.8. Detection limits depend on the absorption line strength, laser bandwidth, and path length, but typically range from 2 to ~12 ppb for the gases listed in Table 5.8. Temperature measurement is enabled by comparing the intensity of two or more absorption lines that depend on temperature.

5.10.3 Tomographic Reconstruction of Two-Dimensional Concentration Fields

For line-of-sight absorption measurements (e.g., tunable-diode laser absorption spectroscopy), only the integrated signal along the path length is available. However, localized species concentration can be obtained for axisymmetric flows using tomographic reconstruction. Multiple absorption line signals at several different angles are needed for accurate mapping of two-dimensional fields, as shown in Figure 5.21. Abel inversion is one of the methods that can be used. The local scalar quantity, f(r), can be found using the inversion equation:

$$f(r) = -\frac{1}{\pi}\int_{r}^{R}\frac{A'_{\lambda o}(x)}{\sqrt{x^2-r^2}}dx \tag{5.84}$$

$A'_{\lambda o}$ ($= dA_{\lambda o}/dx$) is found from the signal (Figure 5.22).

$$A_{\lambda o}(x) = \frac{I_{\lambda}(x)}{I_{\lambda o}(x)} \tag{5.85}$$

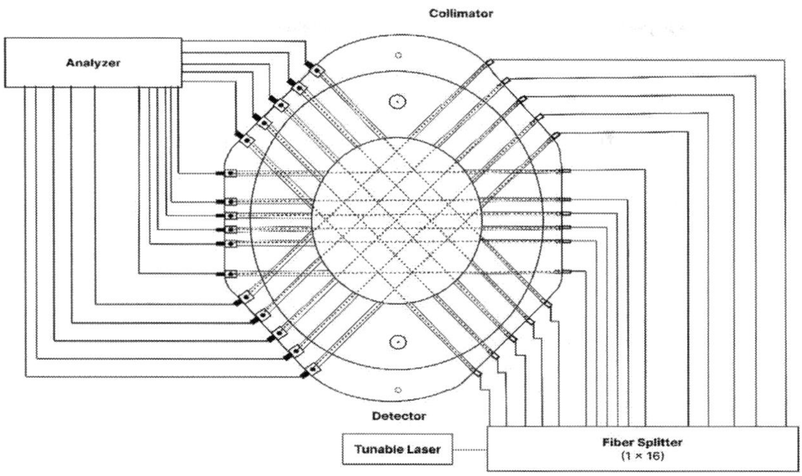

FIGURE 5.21 Experimental setup for tomography based on absorption measurements.

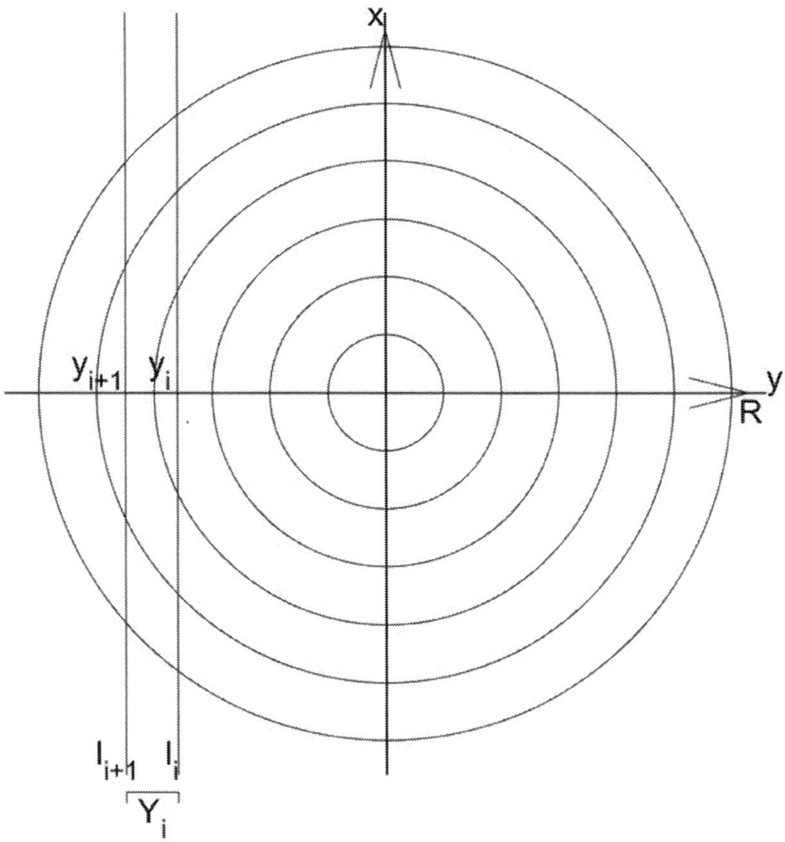

FIGURE 5.22 Tomographic inversion.

5.10.4 Laser-Induced Breakdown Spectroscopy (LIBS)

When a sufficiently intense laser beam is focused onto gas or solid, it will break down the constituent molecules to atomic components. The breakdown process produces a spark plasma (Figure 5.23), and the excited atoms and ions will emit radiation that can be analyzed to determine the components of the sample. The breakdown phenomenon involves first a multi-photon (from the laser) ionization step that produces large number of seed electrons. Then, the energetic electrons knock out yet more electrons from other molecules to cause the breakdown. The laser power flux required for breakdown of air is in the order of 10^{10} W/cm^2 (which translates to 10^{29} photons/cm^2). Using Nd:YAG laser, for example, at 100 mJ per pulse of 10 ns pulse duration will produce a laser fluence of approximately 10^{10} W/cm^2 when focused to a 0.1 mm-diameter spot. The breakdown plasma reaches temperatures in excess of 10^5 K, and pressure of several atm, the latter resulting in a popping sound as a shock wave emanates from the focal point. This process can be estimated using the excited species number density and ensuing emission.

$$N_j / N_{jo} = \exp\left(-\frac{\Delta E}{kT}\right) \tag{5.86}$$

N_j = number of atoms in the excited state
N_{jo} = number of atoms in the lower state
ΔE = band gap (energy difference between the lower and excited states)
k = Boltzmann constant (1.38066 x 10^{-23} J/K)
T = temperature

For negligible self-absorption of radiation, the subsequent emission is given by

$$I_{e,j} = A_{ij} h \nu_{ij} N_j \tag{5.87}$$

$I_{e,j}$ = emitted intensity for species j, at light frequency ν_{ij}
A_{ij} = transition probability of spontaneous emission

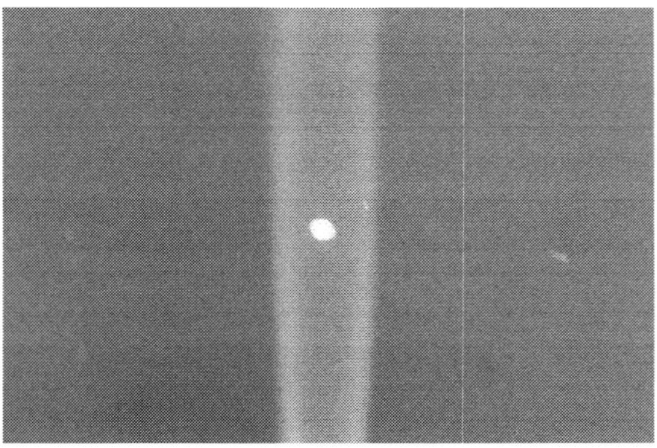

FIGURE 5.23 A laser spark in a flame environment. Emission from the spark contains spectral information on atomic composition.

h = Planck's constant (6.624 x 10^{-34} J s)
ν_{ij} = light frequency from ($\Delta E = h\nu_{ij}$)

From Equation 5.87, we see that the concept of LIBS is based on the emitted intensity from atomic species j at a specific wavelength (or frequency, ν_{ij}), which depends on the number concentration N_j. However, other factors affect the direct quantitative correlation between the measured emitted intensity and atomic species concentration. These include nonuniform plasma, plasma temperature, and matrix effects, the latter referring to recombination and interactions with other atomic species in the plasma. Detection of elemental species and their ratios can nonetheless be used to infer the chemical composition and, to some extent, the gas temperature. One method for temperature measurement using LIBS is based on the assumption that nitrogen signal intensity is inversely proportional to temperature under constant pressure. Alternatively, the fact that the breakdown threshold is proportional to the pressure can be used, which leads to $I_{\text{threshold}} = \alpha T^{\beta}$ again under constant pressure (Kiefer et al., 2012).

5.10.5 Magnetic Resonance Imaging (MRI)

Magnetic resonance imaging is an established technology in medical diagnostics. However, it has uses in thermal-fluid measurements as a non-intrusive tomographic method without optical access. MRI works through a phenomenon called nuclear magnetic resonance. Each atom in a molecule is composed of a nucleus and electrons. If there is an odd number of protons in the nucleus, then there is a net "spin" of the protons that acts as an oscillating electric dipole. Following Ampere's law (Equation C.7 in the Appendix C), this time-dependent motion of the electric dipole leads to a magnetic field. Ampere's law states that the curl of the magnetic field is equal to the current density and time derivative of the electric field.

$$\nabla \times \vec{H} = \vec{J} + \frac{\partial \vec{D}}{\partial t} \tag{C.7}$$

$$\vec{H} = \text{magnetic field intensity } [\text{A/m}]$$

$$\vec{J} = \text{current density } \left[\text{A/m}^2\right]$$

Under a strong magnetic field, the magnetic dipole of the nucleus in specific atoms (in the molecule) aligns with the magnetic field. Typical MRI scanners using superconducting magnets generate a magnetic field at the 1.5 T (Tesla) level. In comparison, the Earth's magnetic field is 50 μT, while a small household magnet has about 5 mT strength. In medical MRI imaging, H atoms in water molecules, with a strongly discernible magnetic dipoles (due to their abundance in the body), are used. The aligned magnetic dipoles are then "illuminated" with radio waves, which, upon "de-illumination (radio waves shut off)", return to their original state.

The transition from the resonant (with radio wave illumination) to the original state (de-illuminated) emits radio waves that can be scanned at multiple angles as a function of time. The 360° scans as a function of time are deconvoluted (similar to the Abel inversion discussed above) to generate cross-sectional images. The same method can be used for cross-sectional images of the flow composition, and velocities (Iraqi et al., 2025). The latter technique is called phase-contrast MRI velocimetry, which makes use of the fact that the radio waves generated contain a small but finite Doppler shift due to the motion of the nucleus. The flow motion involves the velocity of the constituent molecules (atoms → nucleus). An example of phase-contrast MRI applied to a mixing flow in a solid channel is shown in Figure 5.24. MRI can also be used for thermometry through a process called "chemical shift,"

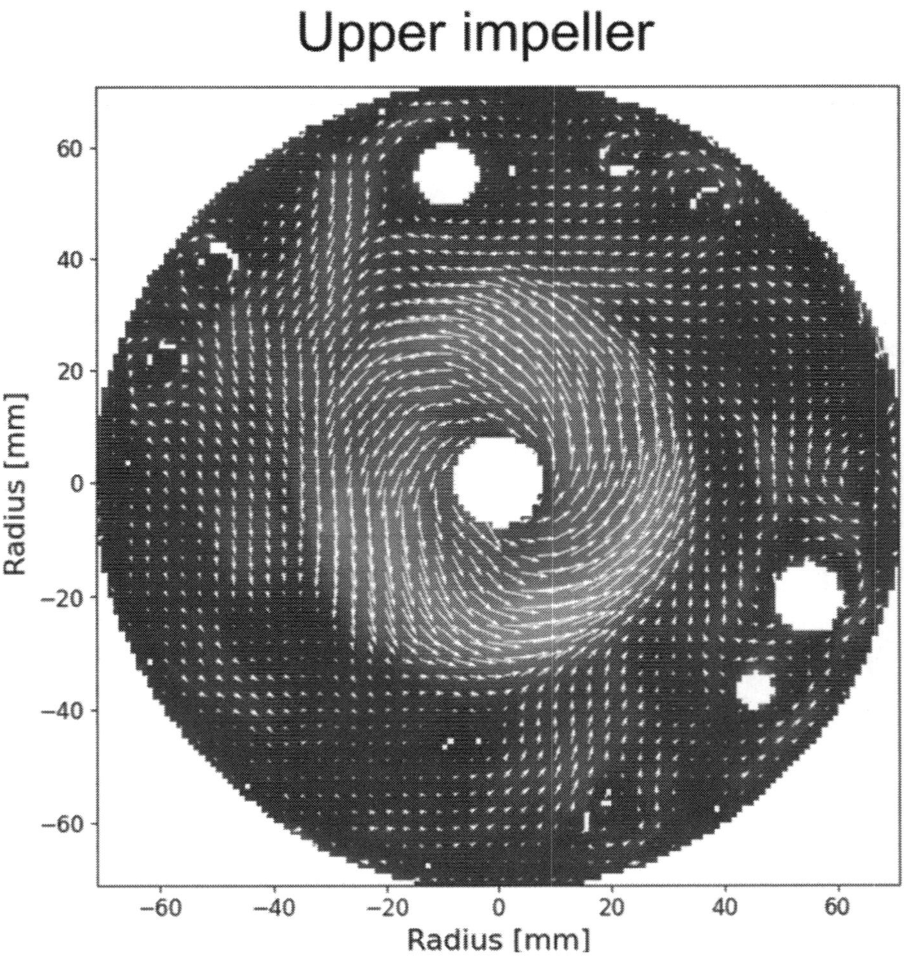

FIGURE 5.24 A phase-shift MIR image of the flow velocity vectors within a solid (no optical access) impeller casing (Lenczyk et al., 2025). Image is courtesy of Hamburg University of Technology (Prof. Alexander Penn).

which refers to the shielding or de-shielding of the target protons by electron clouds. The electron clouds are "shifted" depending on the bonding or arrangement of atoms in a molecule. This bonding, and thus the degree of de-shielding, has a small but measurable dependence on temperature, which can be calibrated for MRI thermometry (Serial et al., 2023).

PROBLEMS

5.1. Visible light has wavelength components from 400 to 760 nm. We can see on a daily basis that the cloud particles act as Mie scatterers, while atmospheric gas molecules act as Rayleigh particles. Using the criterion for the applicability of the Rayleigh theory (Equation 5.1), what must be the size ranges of the cloud and atmospheric gas molecules?

5.2. a. Plot the longitudinal component of the Rayleigh scattered signal for $\alpha = 0.1$ for a fixed r of 1 m, as a function of the wavelength, λ.
b. Plot the transverse component of the Rayleigh scattered signal for $\alpha = 0.1$, $r = 1$ m, and $\lambda = 532$ nm, as a function of the scattering angle, θ.

5.3. For a particle diameter of 100 nm with a refractive index of $\mathbf{n} = 1.2 + 2.75i$, incident laser intensity of 100 mJ at 532 nm, estimate the Rayleigh scattering signal intensity at a distance of 0.5 m.

5.4. For isotropic scattering of the particle in Problem 5.3, what is the (a) scattering cross section; (b) differential scattering cross section, and (c) scattering efficiency?

5.5. For the Rayleigh scattering measurements described in Problem 5.3, now we have a polydisperse particle following a number distribution of $N(D) = 29{,}685\exp(-735D)$. D is in nm. Find the Rayleigh scattering coefficient, K_{sca}, for particle diameter range from 0.1 nm to 500 nm.

5.6. Show that the soot particle volume fraction, mean diameter, and the number density are related to the scattering, extinction, and refractive properties by Equations 5.27–5.29.

$$f_v = \frac{\lambda K_{ext}}{6\pi \left| \mathrm{Im}\left\{ \frac{\mathbf{n}^2 - 1}{\mathbf{n}^2 + 2} \right\} \right|} \tag{5.27}$$

$$D_{63} = \lambda \left(\frac{4}{\pi^2} \frac{\left| \mathrm{Im}\left\{ \frac{\mathbf{n}^2 - 1}{\mathbf{n}^2 + 2} \right\} \right|}{\left| \frac{\mathbf{n}^2 - 1}{\mathbf{n}^2 + 2} \right|} \frac{K_{sca}}{K_{ext}} \right)^{1/3} \tag{5.28}$$

$$N = \frac{12 f_v}{\pi D_{63}^3} \tag{5.29}$$

5.7. Using MATLAB or equivalent, write a computer program to calculate and plot the Mie extinction and scattering efficiencies as a function of the particle size parameter, x, for complex refractive index of and $1.5+0.0i$ (weakly absorbing particle) and $1.2+2.75i$ (highly absorbing particle).

5.8. For an incident laser wavelength of 532 nm, find the anti-Stokes and Stokes wavelengths of the Raman scattering signal if the molecule has a vibration frequency of 2330.7 cm^{-1}.

5.9. For an incident laser wavelength of 532 nm, estimate the Raman scattering signal intensity for nitrogen, hydrogen, and oxygen molecules, using the Raman cross sections listed in Table 5.1. Compare with the signal intensities at 488 and 337 nm.

5.10. In the BOXCAR setup shown in Figure 5.10, the wavevectors are vectors with orientation along the beam direction and vector magnitude (length) given by $2\pi/\lambda$. If the pump beam wavelength is 532 nm at an angle of 10 degrees from horizontal and the vibrational frequency 2330.7 cm^{-1} with the Stokes beam (k_3) at an angle of 15 degrees from horizontal, what should be the detector angle with respect to the horizontal? Use law of sines: $\sin\alpha/a=\sin\beta/b=\sin\gamma/c$.

5.11. Calculate the energy associated with light signals with wavelengths of 225 nm, 280 nm, 532 nm, and 2.2 μm.

5.12. Determine the wavelength of the light signal corresponding to vibrational transitions of $\Delta v=\pm 1$, for a molecule with a vibrational frequency of $v_{osc}=3557.5\ cm^{-1}$.

5.13. Determine the wavelength of the light signal corresponding to rotational transitions of $\Delta J=\pm 1, \pm 2$, for $B_v=122.7\ cm^{-1}$.

5.14. For hydroxyl (OH) molecules undergoing the following transition, determine the corresponding light wavelength, $n'=1$, $n''=0$, $v'=1$, $v''=0$, $J'=4$, $J''=6$. For OH, $(E_e^{'}-E_e^{''})/hc=32{,}450\ cm^{-1}$ $v_{osc}/c\equiv(\omega_e\approx 3739.94\ cm^{-1}$ and $B_v\approx 18.867\ cm^{-1}$.

5.15. Plot the rotational energy distribution (Equation 5.71) for a molecule with $B_v=18.867\ cm^{-1}$ as a function of the rotational quantum number, J, at temperatures of 300, 500, 750, 1000, 1250, 1500, and 2000 K.

5.16. Plot the vibrational energy distribution (Equation 5.78) for a molecule with $v_{osc}=2557.0\ cm^{-1}$ as a function of the vibrational quantum number, v, at temperatures of 300, 500, 750, 1000, 1250, 1500, and 2000 K.

5.17. In a Fourier-transform infrared spectrometer, the mirror position $\delta(t)$ in μm is given by $\delta(t)=0.1(t+20)$ for $t=-20$ to 0 ms and $\delta(t)=2-0.1t$ for $t=0$ to 20 ms. For an observed signal $F(t)=1.25\exp i(-0.25t^2)\cos(2.5t)$, numerically integrate Equation 5.84 to find $B(\lambda)$.

$$B(\lambda)=\frac{1}{T}\int_{-T}^{T}F(t)\exp i\left(\frac{2\pi\delta(t)}{\lambda}\right)dt \tag{5.84}$$

BIBLIOGRAPHY

Bohren, C.F. and Huffman, D.R., *Absorption and Light Scattering by Small Particles,* Wiley, New York, 1983.

Draine, B.T. and Flatau, P.J., Users guide for the discrete dipole approximation code DDSCAT 6.0, https://arxiv.org/abs/astro-ph/0309069, 2003.

Eckbreth, A.C. *Laser Diagnostics for Combustion Temperature and Species*, 2nd ed., Gordon and Breach, New York, 1996.

Iraqi, Lenczyk, Penn and Schlüter, *Establishing Jet Loop Reactors as Scalable Bioelectrochemical Reactor Systems for Anodic and Cathodic Production Processes,* ICMF 2025, Toulouse, France.

Kiefer, J., Troeger, J.W., Li, Z., Seeger, T., Alden, M., and Leipertz, A., Laser-induced breakdown flame thermometry, *Combustion and Flame*, 159, pp. 3576–3582, 2012.

Lenczyk, Benders and Penn, *Advancements in the Application of a 3T MRI System for Analysing the Hydrodynamics in Stirred Tank Reactors*, NMRPM 2025, Kaiserslautern, Germany

Mishchenko, M. I., Hovenier, J. W., and Travis L. D. (Eds.), *Light Scattering by Non-Spherical Particles: Theory, Measurements and Applications*, Academic Press, Boston, 2000.

Rabenstein, F. and Leipertz, A., Two-dimensional temperature determination in the exhaust region of a laminar flat-flame burner with linear Raman scattering, *Applied Optics*, 36(27), p. 6989, 1997.

Santoro, R. J., Semerjian, H. G., and Dobbins, R. A., Soot particle measurements in diffusion flames, *Combustion and Flame,* 51, p. 203, 1983.

Serial, M.R., Benders, S., and Rotzetter, P., Temperature distribution in a gas-solid fixed bed probed by rapid MRI, *Chemical Engineering Science*, 269, p. 118457, 2023.

Tolles, W.M., Nibler, J.W., McDonald, J.R., and Harvey, A.B., A review of the theory and application of coherent anti-Stokes Raman spectroscopy, *Applied Spectroscopy*, 31(4), pp. 253–171, 1977.

6 Particle Sizing and Two-Phase Flow Measurements

6.1 INTRODUCTION

Flows laden with particles are frequently found in thermal and flow processes. Fuel sprays in automotive and gas-turbine engines, evaporators, cooling towers, pulverized-coal/air flow in power plants, and flow with internal boiling (gas bubbles) are some examples of the so-called particle-laden, two-phase flows. Many industrial processes involve mixing liquid/solid, gas/solid, or liquid/gas combinations of materials. Also, atmospheric processes such as aerosols, raindrops, and cloud particles have a profound effect on local and global meteorology. For these two-phase flow processes, it is of interest to determine the particle size, velocity, and, in some cases, chemical composition so that one can assess the effects of the particles on the overall thermal and flow properties.

There is a class of particle-sizing methods based on different operating principles. These include simple sieve analysis and image analyses (optical microscopy, scanning electron microscopy, holography). There are standard sieves with openings down to 0.020 mm, as specified by ISO (International Standards Organization) and ASTM (American Society for Testing and Materials). However, these require a large number of particles samples to be extracted from the flow for analysis. Optical microscopy can be used for *in situ* (Latin meaning "in place," i.e., without being removed from the flow) measurements, and depending on the magnification and image resolution, can measure particle sizes down to about 1 μm. For example, shadowgraph images using high magnification can be digitally processed to provide particle size data. Electron microscopy can resolve particle sizes down to the nanometer range, but it again requires the particles to be extracted, sampled, and prepared for either SEM (scanning electron microscopy) or TEM (transmission electron microscopy) analysis. The image data can be digitally processed to provide particle size data after calibration of the image size dependence on the particle diameter.

Acoustic measurements, similar in some regards to optical measurements, are applied either through transmission, pulse echo, or interferometry of ultrasonic waves to infer particle size distribution and flux. If the particle density and acoustical properties are known, then ultrasonic attenuation or scattering can be used to determine the particle density and flux. This is similar to optical methods using electromagnetic wave absorption and scattering, except that the acoustical wave equation is used.

In this chapter, we will mostly discuss methods of *in situ* particle size and flux measurements, such as laser diffraction, light scattering, and phase Doppler

DOI: 10.1201/9781003611837-6

interferometry. Before discussing these and other methods, let us begin by exploring some of the parameters that are used to characterize a large group of particles.

6.2 PARAMETERS TO CHARACTERIZE GROUPS OF PARTICLES

The measured particle size can be grouped according to the size range. An example of particle sizes grouped into different bins is shown in Table 6.1. The data in Table 6.1 can be plotted as shown in Figure 6.1, and this representation is called the particle size distribution. The bin width, ΔD_i, varies in Table 6.1 to illustrate some of the statistical definitions, but for some measurements, the bin size is kept constant. With the data in Table 6.1 or the particle size distribution, we can compute various statistical parameters, such as the mean, variance, skewness, and kurtosis.

The arithmetic mean diameter is the average particle diameter over the entire size distribution.

$$\langle D \rangle = \frac{\sum_{i=1}^{N} n_i D_i \Delta D_i}{\sum_{i=1}^{N} n_i \Delta D_i} = \sum_{i=1}^{N} p_i D_i \Delta D_i \tag{6.1}$$

A more useful parameter is the generalized mean, the average diameter over the entire size distribution with different "moments" of particle size.

$$\langle D_{pq} \rangle \left[\frac{\sum_{i=1}^{N} n_i D_i^p \Delta D_i}{\sum_{i=1}^{N} n_i D_i^q \Delta D_i} \right]^{\frac{1}{p-q}} \tag{6.2}$$

TABLE 6.1
An Example of a Particle Size Grouped into Different Size Bins

Particle Size Range [μm]	Representative Particle Size, D_i [μm]	ΔD_i [μm]	n_i, Number of Particles	$p_i \Delta D_i = \frac{n_i \Delta D_i}{\Sigma n_i \Delta D_i}$, Size Distribution Function
0.0–10.0	5.0	10.0	25	0.03151
10.0–20.0	15.0	10.0	127	0.16005
20.0–30.0	25.0	10.0	275	0.34657
30.0–40.0	35.0	10.0	155	0.19534
40.0–50.0	45.0	10.0	75	0.09452
50.0–60.0	55.0	10.0	54	0.06805
60.0–70.0	65.0	10.0	32	0.00403
70.0–85.0	77.5	15.0	15	0.02836
85.0–100.0	92.5	15.0	12	0.02268
100.0–120.0	110.0	20.0	5	0.01260
			N = Total number of particles = 7935	$\Sigma p_i \Delta D_i = 1$

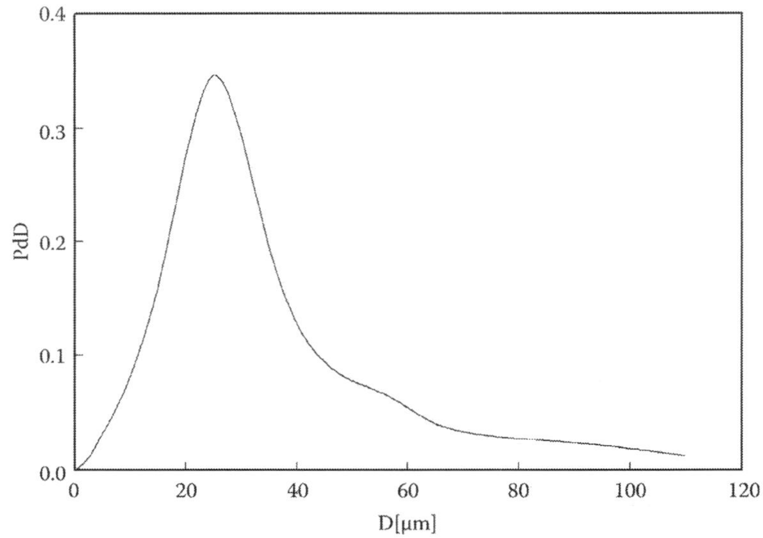

FIGURE 6.1 An example of particle size distribution.

The arithmetic mean diameter corresponds to $<D_{10}>$. The volume-to-surface mean, $<D_{32}>$, is called the Sauter mean diameter and represents the mean diameter that has the same volume-to-surface ratio as the entire group of particles. The *median* represents the diameter that divides the particle size distribution into two equal halves, each having the same number of particles. The *mode* is the most frequent diameter, i.e., the highest point in the particle size distribution, whereas the *variance* is a measure of the width of the size distribution, as shown in Equation 6.3.

$$\sigma^2 = \frac{\sum_{i=1}^{N} n_i \left(D_i - \langle D \rangle\right)^2}{N-1} \tag{6.3}$$

The standard deviation is an alternative measure of the spread of the particle size distribution function and is equal to the square root of the variance.

$$\sigma = \sqrt{\frac{\sum_{i=1}^{N} n_i \left(D_i - \langle D \rangle\right)^2}{N-1}} \tag{6.4}$$

Skewness is the degree of distortion from a symmetrical distribution. For a perfectly symmetrical distribution, skewness is zero.

$$g = \frac{N}{(N-1)(N-2)} \sum_{i=1}^{N} n_i \left(\frac{D_i - \langle D \rangle}{\sigma} \right)^3 \approx \frac{\sum_{i=1}^{N} n_i \left(D_i - \langle D \rangle\right)^3}{\sigma^3 N} \tag{6.5}$$

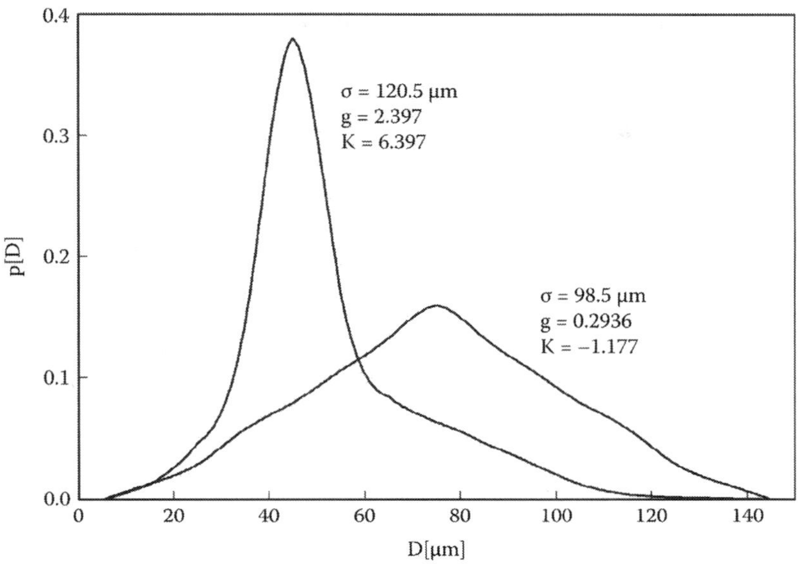

FIGURE 6.2 Example of different kurtosis for particle size distribution.

Finally, the kurtosis is a measure of the smoothness of the size distribution.

$$K = \frac{N(N+1)}{(N-1)(N-2)(N-3)} \sum_{i=1}^{N} n_i \left(\frac{D_i - \langle D \rangle}{\sigma} \right)^4 - \frac{3(N-1)^2}{(N-2)(N-3)}$$
$$\approx \frac{\sum_{i=1}^{N} n_i \left(D_i - \langle D \rangle\right)^4}{\sigma^4 N} - 3 \tag{6.6}$$

A size distribution with a sharp peak has a positive kurtosis, whereas a "flat" distribution has a negative kurtosis, as illustrated in Figure 6.2. The flat and nearly symmetric distribution in Figure 6.2 has negative kurtosis and small skewness.

Example 6.1

For the particle size distribution shown in Table 6.1, we can construct a table for some intermediate parameters needed to obtain the particle size distribution function and the arithmetic mean, D_{10}. From Equations 6.1 or 6.2, the sum of the right-most column in the table below gives the arithmetic mean diameter (34.36 μm). Although this example was for nonuniform particle size bins (ΔD_i), it is much more convenient to use uniform size bins. Such statistical data can then be easily processed using embedded functions in Excel or MATLAB to obtain higher-order moments, e.g., skewness and kurtosis.

Particle Size Range [μm]	Representative Particle Size, D_i [μm]	$n_i\Delta D_i$ [μm]	$p_i\Delta D_i = \frac{n_i\Delta D_i}{\Sigma n_i\Delta D_i}$, Size Distribution Function	$p_iD_i\Delta D_i$
0.0–10.0	5.0	250	0.0315	0.1575
10.0–20.0	15.0	1270	0.1600	2.401
20.0–30.0	25.0	2750	0.3466	8.664
30.0–40.0	35.0	1550	0.1953	6.837
40.0–50.0	45.0	750	0.0945	4.253
50.0–60.0	55.0	540	0.0681	3.743
60.0–70.0	65.0	320	0.0403	2.621
70.0–85.0	77.5	225	0.0284	2.198
85.0–100.0	92.5	180	0.0227	2.098
100.0–120.0	110.0	100	0.0126	1.386
		$\Sigma\, n_i\Delta D_i = 7935$	$\Sigma\, p_i\Delta D_i = 1$	$\Sigma\, p_iD_i\Delta D_i = \langle D\rangle = 34.36\,\mu m$

In Table 6.1, $p_i\Delta D_i$ is referred to as the histogram, and $p_i(D_i)$ as the *size distribution function*. Typically, one would like to fit the measured particle size data to a size distribution function so that the entire group of particles can be characterized by a small number of parameters, such as the mean and the variance. Table 6.2 shows some commonly used size distribution functions, and these functions are illustrated in Figure 6.3. It can be seen in Figure 6.3 that the normal distribution is simply a representation of random samples centered at the mean. The other functions replicate the size distributions found in sprays. The log-normal function skews the distribution toward small diameters, as is found in many sprays. The Rosin-Rammler function has the sharpest peak at a small diameter, and the parameters n and b are adjusted to fit the drop size distribution data. The Marshall-Palmer distribution originated in describing raindrop size distributions.

6.3 LIGHT SCATTERING AND EXTINCTION MEASUREMENTS OF PARTICLE SIZE

Soot particle size measurements using light scattering and extinction measurements were discussed in Chapter 5. Using the Rayleigh scattering and extinction coefficients, $\langle D_{63}\rangle$ can be determined along with the soot volume fraction, f_v, which is related to $\langle D_{30}\rangle$ (Santoro et al., 1982).

$$K_{sca} = \frac{N}{4}\left(\frac{\pi}{\lambda}\right)^4 \left|\frac{\mathbf{n}^2-1}{\mathbf{n}^2+2}\right|^2 \int_0^\infty P(D)\,D^6\,dD \tag{5.24}$$

$$K_{ext} = \frac{\pi^2 N}{\lambda}\left|\mathrm{Im}\left\{\frac{\mathbf{n}^2-1}{\mathbf{n}^2+2}\right\}\right| \int_0^\infty P(D)\,D^3\,dD \tag{5.25}$$

TABLE 6.2
Some Commonly Used Size Distribution Functions

Distribution Function	Functional Form	Mean	Mode
Normal (Gaussian)	$p(D)=\frac{1}{\sqrt{2\pi}\sigma}\exp\left(-\frac{\left(D-\langle D\rangle\right)^2}{2\sigma^2}\right)$	$<D>$	$<D>$
Log-normal	$p(D)=\frac{1}{\sqrt{2\pi}D\sigma\left(\ln(D)\right)}\exp\left(-\frac{\left(\ln(D)-\langle\ln(D)\rangle\right)^2}{2[\sigma\left(\ln(D)\right]^2}\right)$ $<\ln(D)>$ and $\sigma(\ln(D))$ are the mean and the standard deviation of the $\ln(D)$, respectively	$<D>$	$\frac{<D>^4}{(\sqrt{2\pi}D\sigma\left(\ln(D)\right)}$
Rosin-Rammler-Sperling-Bennet distribution	$p(D)=nbD^{n-1}\exp(-bD^n)$, n and b are adjustable parameters	$\frac{\Gamma\left(\frac{1}{n}+1\right)}{\sqrt[n]{b}}$, Γ is the gamma function	$n\sqrt{\frac{n-1}{bn}}$
β-Distribution	$p(D)=\frac{x^m\left(\langle D\rangle-D\right)^n}{\beta(m+1,n+1)}$, β is the beta function, and m and n are adjustable parameters	$\frac{(m+1)\langle D\rangle}{(n+m+2)}$	$\frac{m\langle D\rangle}{m+n}$
Marshall-Palmer	$p(D)=\lambda e^{-\lambda D}$, is the inverse of the mean diameter	$1/\lambda$	

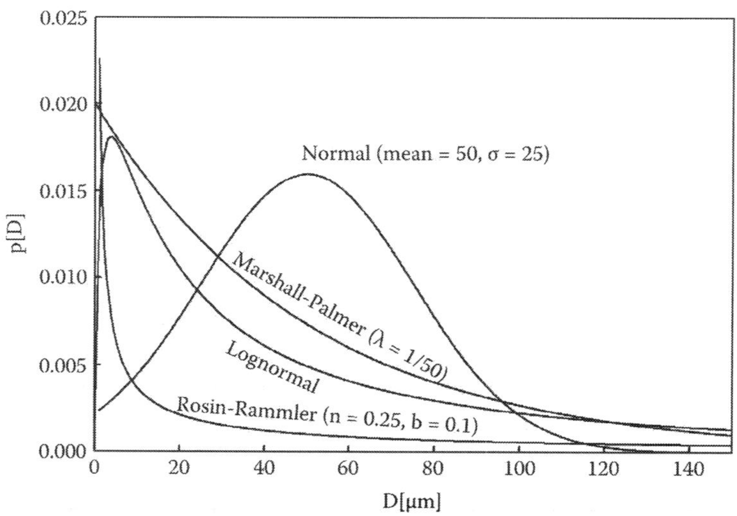

FIGURE 6.3 Some examples of size distribution functions.

$$D_{63} = \lambda \left(\frac{4}{\pi^2} \frac{\left| \text{Im}\left\{ \frac{\mathbf{n}^2 - 1}{\mathbf{n}^2 + 2} \right\} \right|}{\left| \frac{\mathbf{n}^2 - 1}{\mathbf{n}^2 + 2} \right|} \frac{K_{sca}}{K_{ext}} \right)^{1/3} \tag{5.28}$$

$$f_v = \frac{\lambda K_{ext}}{6\pi \left| \text{Im}\left\{ \frac{\mathbf{n}^2 - 1}{\mathbf{n}^2 + 2} \right\} \right|} \tag{5.27}$$

Similar measurements of light extinction can be extended to Mie-scattering particles, except that the Mie extinction cross section (efficiency), C_{ext} (or Q_{ext}) needs to be used. From Beer's law, the extinction coefficient (K_{ext}) can be found by measuring I_λ, $I_{\lambda o}$, and specifying the path length, L. Then, K_{ext} can be related to the particle size as follows:

$$K_{ext} = \frac{1}{L} \ln\left(\frac{I_{\lambda o}}{I_\lambda} \right) = N \int_{D=0}^{\infty} P(D) C_{ext}(D) dD = \frac{\pi}{4} N \int_{D=0}^{\infty} P(D) Q_{ext}(D) D^2 dD \tag{6.7}$$

For Mie scattering and absorption, the extinction coefficient is, in general, nonmonotonic, so that an assumed form of the $P(D)$ is needed in Equation 6.7. By fitting the parameters in $P(D)$, such as those in Table 6.2, with the observed K_{ext} at two or more wavelengths, the function $P(D)$ is estimated. Thus, particle size and volume fraction measurements using Rayleigh scattering may be useful in some laboratory experiments. However, there are some fundamental difficulties in general applications. First, the extinction measurement is a line-of-sight measurement (it measures everything in the beam path), so there is poor spatial resolution. For axisymmetric, steady flows, this may be circumvented by taking many cuts across the flow and using tomographic inversion to recover the pointwise data. Second, only the higher moment of the mean diameter (D_{63}) is obtained. Other limitations include the fact that the general shape of the particle size distribution needs to be known *a priori*, and that the scattering cross section is usually available only for spherical particles or particles of simple geometry. Any detailed measurement of nonspherical particles is a challenging topic in thermal and flow diagnostics, with few established techniques.

We may also consider Mie-scattering measurements from particles. From Equation 5.38b, the Mie-scattering intensity for spherical particles can be calculated:

$$\frac{I_{\lambda,s}}{I_{\lambda o}} = \frac{|S_1(\theta)| + |S_2(\theta)|}{2\pi k^2 r^2} \tag{6.8}$$

However, the functions S_1 and S_2 are typically very sensitive to variations in particle size, D, and are also nonmonotonic, making it difficult, if not intractable, to retrieve the particle size distribution. Mie scattering does produce a diffraction effect in the forward scattering direction, $\theta = 0$, which can be used for particle size characterization, as will be discussed in Section 6.5.

6.4 LASER-INDUCED INCANDESCENCE MEASUREMENTS OF PARTICLE VOLUME FRACTION AND SIZE

As noted earlier, *in situ* measurement of particle characteristics is a challenging area in thermal and flow diagnostics. A method called laser-induced incandescence (LII) is a useful technique for determining soot particle volume fraction. In contrast to the light scattering/extinction method described in the previous section, LII provides spatially resolved measurements of the volume fraction and can also be used for imaging the soot field. Although the following analysis of this method is for carbon particles found in soot, a similar approach is feasible for other materials with appropriate optical and thermal properties.

LII is based on thermal heating of soot particles by focused, incident laser light to the point of vaporizing the carbonaceous particles in the soot. Upon heating, the elevated temperature of the soot particles results in blackbody radiation that is approximately proportional to the soot volume fraction. Let us begin by examining the thermal energy balance of particles heated by laser irradiation.

$$\frac{4}{3}\pi r^3 \rho_s c_s \frac{dT}{dt} = \alpha q_i\left(\pi r^2\right) - h\left(4\pi r^2\right)\left(T - T_\infty\right) + \Delta H \frac{dM}{dt} - \varepsilon\sigma\left(4\pi r^2\right)\left(T^4 - T_\infty^4\right) \tag{6.9}$$

r = particle radius
ρ_s = particle density
c_s = particle specific heat
T = particle temperature
α = absorptivity
q_i = incident radiative power flux from the laser
h = heat transfer coefficient
T_∞ = ambient temperature

ΔH = heat of vaporization
M = mass of the particle
ε = emissivity
σ = Stefan-Boltzmann constant for blackbody emissive flux

The left-hand side is the time rate of change of the particle energy, and the right-hand side includes the rate of laser energy absorbed by the particle, convective heat loss, heat loss due to vaporization, and radiative exchange of energy. At high laser power, the soot undergoes steady-state vaporization, and the foregoing energy balance is simplified by using $dM/dt = 4\pi^2\rho_s(dr/dt)$ (Melton, 1984),

$$\alpha q_i + 4\rho_s \Delta H \frac{dr}{dt} \approx 0 \tag{6.10}$$

The continuity relation (mass balance) at the surface is

$$-\rho_s \frac{dr}{dt} = \rho_v U_v \tag{6.11}$$

where U_v is the vapor velocity, which can be equated with the thermal velocity, $(RT/2M_v)^{1/2}$, M is the molecular weight of the carbon vapor, and R is the gas constant. Then, we have

$$\alpha q_i = 4\Delta H \rho_v U_v \tag{6.12}$$

The right-hand side, for carbon particles, is a known function of temperature, including the thermal velocity, U_v.

$$4\Delta H \rho_v U_v = 1.22 \times 10^8 \frac{1}{\sqrt{T}} \exp\left(23.9\left(1 - \frac{3915}{T}\right)\right) \tag{6.13}$$

After using the fact that the square-root term is a slowly varying function of temperature at $T \approx 4000$ K (carbon vaporization temperature), we obtain an expression for temperature as a function of αq_i. For small carbon particles, α can be approximated as r/r_c (r_c = threshold particle radius <100 nm for carbon particles). Using these facts and Equation 6.13 in Equation 6.12, we can solve for the soot particle temperature, T.

$$\frac{1}{T} \approx \frac{1}{3915}\left[1 - \frac{1}{23.9}\log\left(\frac{r}{r_c} q_i\right)\right] \tag{6.14}$$

The LII radiation can be determined using Planck's law (Equation C.38), modified by the carbon emissivity, ε.

$$I_{LII} = \varepsilon\left(4\pi r^2\right)\frac{2C_1}{\lambda_{LII}^5\left(e^{-C_2/\lambda_{LII}T} - 1\right)} = \frac{r}{r_c}\left(4\pi r^2\right)\frac{2C_1}{\lambda_{LII}^5\left(e^{-C_2/\lambda_{LII}T} - 1\right)} \tag{6.15}$$

We have again used $\varepsilon = \alpha = r/r_c$, and λ_{LII} is the wavelength at which the LII signal is observed, in microns.

Substituting Equation 6.14 into Equation 6.15, and writing only the terms that contain r, we have (we obtain a factor of $r^{0.154/\lambda m}$ from the exponential term in Equation 6.15).

$$I_{LII} \sim r^{3+0.154/\lambda_{LII}} \quad [\lambda_{LII} \text{ in } \mu\text{m}] \tag{6.16}$$

For a group of soot particles with a size distribution function, $p(r)$, the total LII signal is

$$I_{LII} \sim N \int_{r=0}^{\infty} p(r) r^{3+0.154/\lambda_{LII}} dr \tag{6.17}$$

N = number density of soot particles

Equation 6.17 shows that the LII signal is approximately proportional to the mean r^3 or the volume of the soot particles, and this serves as the basis for measuring the soot volume fraction using LII. For λ_{LII} from 0.4 to 0.7 µm, the exponent varies from 3.38 to 3.22. In actual measurements, the excitation is made with either the fundamental Nd:YAG laser output at 1064 nm or the frequency-doubled 532 nm, and the LII signal is observed away from the excitation or other interference wavelengths. Usually, a laser fluence of 2×10^8 W/cm^2 is sufficient to induce soot incandescence (Quay et al., 1994).

We can see from Equation 6.9 that the particle temperature will start to decrease at the end of the laser pulse at a rate that depends on the particle size. Time-resolved LII has been used to infer the particle size by monitoring the rate of decrease of the LII signal at temporal resolutions of 50 ns or less. Using the particle properties, Equation 6.9 is numerically solved (without the αq_i (πr^2) term because the cooling occurs following the laser pulse) for representative particle sizes, and the observed temperature decay is compared with the computed temperature history to infer the particle size. In LII measurements with pulsed lasers, a signal processor called a boxcar averager is usually used, which generates gate windows of specified time width and time delay with respect to the laser pulse. Thus, by moving this time delay in increments of 50 ns or less, a precise time history of the LII signals can be obtained (Quay et al., 1994).

6.5 LASER DIFFRACTION

Laser diffraction involves interrogation of the Mie-scattering signal in the forward direction ($\theta << 1$). For $\theta << 1$, Equation 6.8 reduces to

$$I(\theta) = \frac{I_o}{k^2 r^2} \alpha^4 \left[\frac{J_1(\alpha \sin\theta)}{\alpha \sin\theta} \right]^2 \tag{6.18}$$

$I(\theta)$ = scattered light intensity at an angle of θ
I_o = incident light intensity
$k = 2\pi/\lambda$

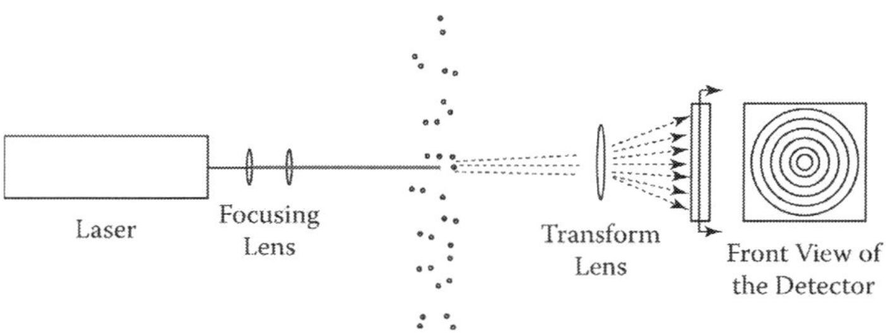

FIGURE 6.4 A schematic of the setup used for laser diffraction particle size measurements.

r = distance to the detector
$\alpha = \pi D/\lambda$ = dimensionless size parameter
J_1 = Bessel function of the first kind

Equation 6.18 can be integrated from some angle range, θ_1 to θ_2, for a cloud of particles with a uniform diameter D and number density N, which yields

$$\Delta I = C_1 N D^2 \left[J_0^2(\alpha \sin \theta_1) + J_1^2(\alpha \sin \theta_1) - J_0^2(\alpha \sin \theta_2) - J_1^2(\alpha \sin \theta_2) \right] \quad (6.19)$$

Here, the constant C_1 absorbs all the constant terms in Equation 6.18, such as the incident light intensity, wavelength, distance to the detector, etc., which are fixed in the experiment.

For a more general cloud of particles with a size distribution function, $p(D)$, Equation 6.19 becomes

$$\Delta I = C_1 N \sum_{i=1}^{M} p(D_i) D_i^2 \left[J_0^2(\alpha_i \sin \theta_1) + J_1^2(\alpha_i \sin \theta_1) - J_0^2(\alpha_i \sin \theta_2) - J_1^2(\alpha_i \sin \theta_2) \right] \quad (6.20)$$

The laser diffraction method works by taking measurements of ΔI between successive angles (from θ_i to θ_{i+1}) in a concentric set of ring-shaped detectors and inverting the signal to obtain $p(D)$. The inversion requires many precise measurements of ΔI; alternatively, a form for $p(D)$ is assumed, and then the measured ΔI's are used to find the best-fit parameters for the size distribution function.

Figure 6.4 shows the setup used for laser diffraction particle size measurements.

6.6 PHASE DOPPLER ANALYSIS

For a spherical particle with a diameter significantly greater than the laser wavelength, the laser light will be refracted by a different amount that depends on the particle diameter and the refractive index, as shown in Figure 6.5. The optical paths are evidently different in Figure 6.5, in that both the beam entry and exit points differ. This causes a path length difference and, therefore, a phase difference, depending on the particle size. For spherical particles, this phase difference is a monotonic function of the particle diameter. This can be used as a basis for spherical particle size

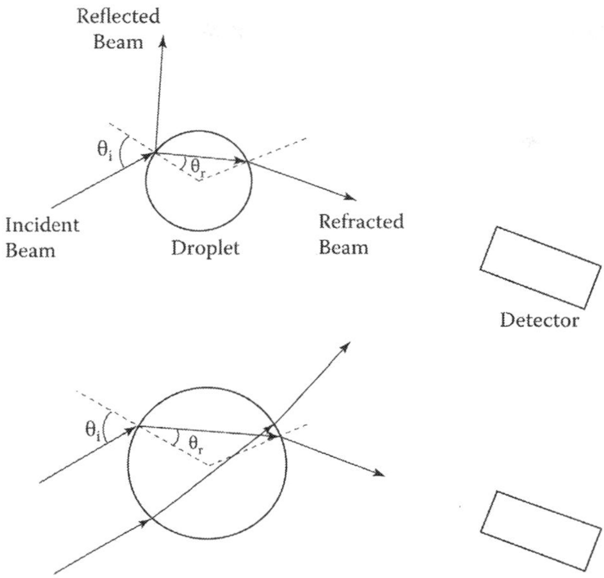

FIGURE 6.5 Light refraction from a spherical particle.

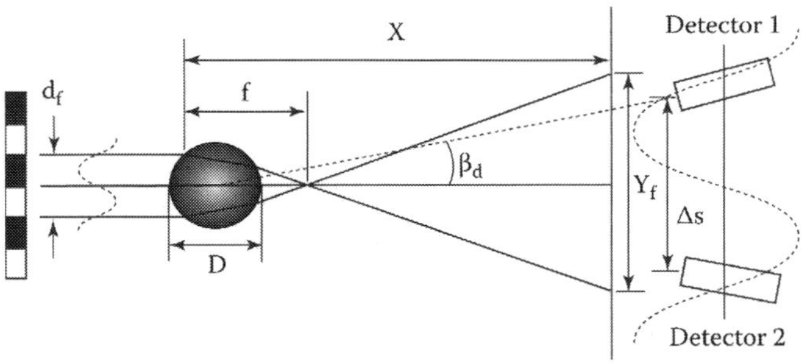

FIGURE 6.6 A simple model of phase Doppler anemometry.

measurements. If this method is used within the laser Doppler velocimetry (LDV) setup, both particle velocity and size data can be extracted. This approach is called phase Doppler anemometry (PDA).

Before considering the formal model, let us consider a simpler interpretation of the PDA signal by looking at the image of the fringe pattern formed at the intersection of two beams, as shown in Figure 6.6. The particle is crossing a fringe with a spatial wavelength of $d_f = \lambda_L/2\sin(\theta/2)$, as was used in Equation 4.14 in Chapter 4 for LDV. A spherical particle will refract the beams and project an image of the fringe on the detector plane. From Figure 6.6, we can deduce that the spatial wavelength of this projected image is related to the actual fringe spacing by

$$\frac{d_f}{f} \approx \frac{Y_f}{X - f} \tag{6.21}$$

In PDA measurements, the distance from the particle to the detector, X, is much larger than the focal length of the spherical droplet, f, which is given by

$$f = \frac{m}{m-1}\frac{D}{4} \tag{6.22}$$

D = particle diameter
m = ratio of the refractive index of the particle to that of the surrounding medium = n_p/n

So, we have

$$Y_f \approx \frac{X}{f} d_f = \frac{4Xd_f}{D}\frac{m-1}{m} \tag{6.23}$$

If the particle was stationary, then this would be the spatial wavelength of the fringe projected on the detector plane. However, as discussed in Chapter 4 for the LDV system, the particle moves across the fringe at a speed of v_{py} and produces a signal of frequency $f_D = v_{py}/d_f$. In other words, the projected waveform in Figure 6.6 also has a *temporal* frequency of f_D.

Now, if we have two detectors separated by a half-angle, β_d, resulting in a spatial separation of $\Delta s = X\sin\beta_d$, then the phase difference between the waveforms observed at the two detectors is

$$\Delta\Phi = \text{phase difference} = 2\pi\Delta s / Y_f \tag{6.24}$$

For example, if Δs is $Y_f/2$, then the phase difference would be π. Using Equation 6.23 in 6.24, we get

$$\Delta\Phi = 2\pi\frac{\Delta s}{Y_f} = \frac{2\pi X \sin\beta_d}{Y_f} = \frac{2\pi D}{\lambda_L}\frac{m}{m-1}\sin\frac{\theta}{2}\sin\beta_d \tag{6.25}$$

D = particle diameter
λ_L = laser wavelength
m = ratio of the refractive index of the particle to that of the surrounding medium = n_p/n
$\theta/2$ = beam crossing half-angle
β_d = half-angle for the detector

Note that this interpretative model contains several approximations, such as using the spherical focal length from the entry point of the beam, and in calculations of some distance terms. Thus, the foregoing model works only for small angles (and zero elevation angle, as discussed in the following text). Nonetheless, it serves as a useful intuitive model of the phase difference observed in PDA, given as a function of the diameter and the refractive index ratio.

A geometrical model for the PDA signal is shown in Figure 6.7. We can consider the two beams intersecting to create the fringe pattern in the LDV setup, and a finite-size particle passing through a beam intersection, as shown in Figure 6.7. The two beams enter the particle and are refracted by Snell's law (Appendix B)

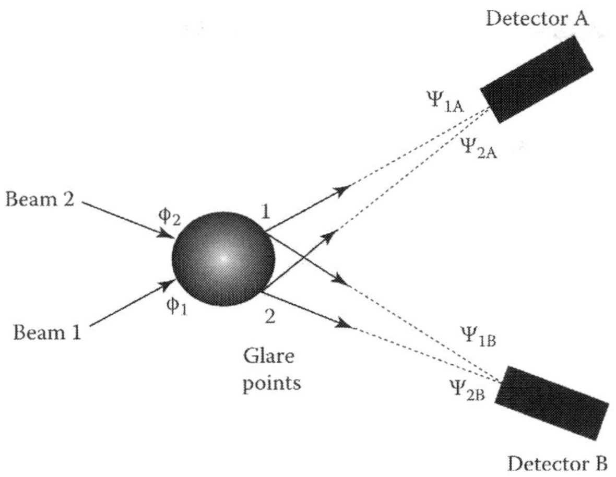

FIGURE 6.7 The phase differences that arise in phase Doppler anemometry.

due to the refractive index difference. At the exit points (also called glare points), the beams are again refracted for the same reason. In general, the two beams may arrive at the particle at different phases, and they will also take different paths across the particle, both of which result in phase differences at the entry and glare points. Moreover, from the glare points to the two detectors (as used in PDA), there will arise a phase difference. If we denote the phase difference at the entry points as ϕ, and the phase difference to the detectors ψ, we can write the electric field of the scattered signal beams emanating from glare points 1 and 2 as

$$E_1 \sim E_{1o}\left[\sin\left(2\pi f_D t+\phi_1+\psi_{1d}\right)\right] \tag{6.26a}$$

$$E_2 \sim E_{2o}\left[\sin\left(2\pi f_D t+\phi_2+\psi_{2d}\right)\right] \tag{6.26b}$$

E_{o1}, E_{o2} = intensities of the incident laser beam
f_D = Doppler frequency shift due to the particle motion relative to the fringes (Equation 4.14)
$\phi_{1,2}$ = phase of incident laser beams 1 and 2 at the entry point
$\psi_{1d,2d}$ phase of the scattered light from the exit points for beams 1 and 2 to the detector

The crossing of the two beams will result in the fringe pattern, and thus the signal will have a temporal frequency, f_D. The wavelength terms are omitted in Equation 6.26 because only the phase terms differ between the two beams. From glare points 1 and 2, the phases will be different depending on whether the beam is headed toward detector A or B, so we write ψ_{1d} or ψ_{1d}, where $d = A$ or B.

At the detector (A or B), the two signals emanating from glare points 1 and 2 are superposed. For $E_{o1} = E_{o2} = E_o$, we have

$$E_1 + E_2 \sim E_o\left[\sin\left(2\pi f_D t + \phi_1 + \psi_{1d}\right) + \sin\left(2\pi f_D t + \phi_1 + \psi_{1d}\right)\right] \tag{6.27}$$

Again, the signal intensity is proportional to the square of the electric field.

$$I_d \sim \left(E_1 + E_2\right)^2 \sim E_o^2\left[\sin\left(2\pi f_D t + \phi_1 + \psi_{1d}\right) + \sin\left(2\pi f_D t + \phi_2 + \psi_{2d}\right)\right]^2 \tag{6.28}$$

The squared bracket contains a cross product of the sine functions, $2\sin(2\pi f_D t + \phi_1 + \psi_{1d})$ $\sin(2\pi f_D t + \phi_2 + \psi_{2d})$, which produces the phase difference and sum terms through a trigonometric identity, $2\sin A \sin B = \cos(A - B) - \cos(A + B)$. It is the phase difference term that is of use for the phase Doppler method.

$$\Phi_A = \phi_1 + \psi_{1A} - \phi_2 - \psi_{2A} = \text{phase term at the detector } A \tag{6.29a}$$

$$\Phi_B = \phi_1 + \psi_{1B} - \phi_2 - \psi_{2B} = \text{phase term at the detector } B \tag{6.29b}$$

By comparing the foregoing phase terms, we can detect the phase difference appearing in the two Doppler bursts as recorded by two detectors separated by an angle, as shown in Figure 6.8.

$$\Delta\Phi = \psi_{1A} - \psi_{2A} - \psi_{1B} + \psi_{2B} \tag{6.30}$$

For spherical particles, the foregoing phase difference (between the phase measured at detector A and B) is a monotonic function of the particle diameter and the refractive index.

If two detectors are employed to detect the phase difference, then the signals at the two detectors can be written using $d = 1, 2$ in the subscript for ψ_{1d}, ψ_{1d}.

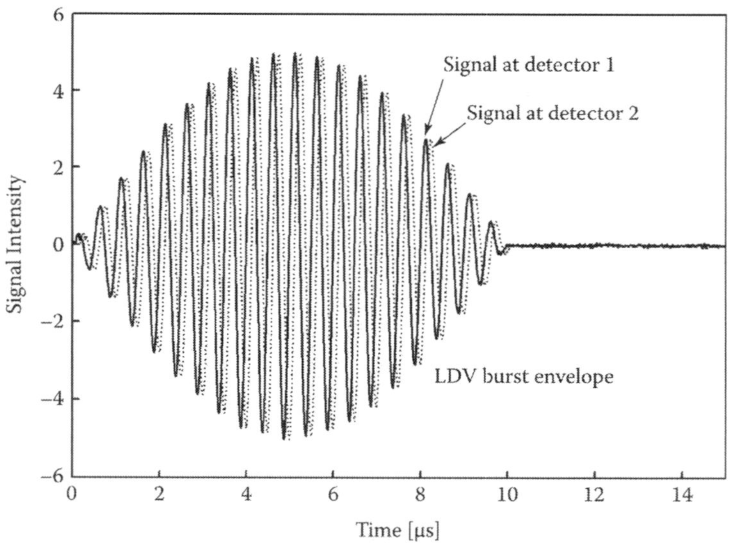

FIGURE 6.8 Typical phase Doppler signals.

$$I_1(t) \sim I_o\left[\cos\left(2\pi f_D t\right)-\left(\phi_1-\phi_2+\psi_{11}-\psi_{21}\right)\right]$$
$$I_2(t) \sim I_o\left[\cos\left(2\pi f_D t\right)-\left(\phi_1-\phi_2+\psi_{12}-\psi_{22}\right)\right] \tag{6.31}$$

We can write the difference in the phase terms as

$$\Delta\Phi=\left(\phi_1-\phi_2+\psi_{11}-\psi_{21}\right)-\left(\phi_1-\phi_2+\psi_{12}-\psi_{22}\right)=\psi_{11}-\psi_{21}-\psi_{12}+\psi_{22} \tag{6.32}$$

The calculations to relate the phase difference to the particle diameter involve geometrical optics (essentially Snell's law and trigonometry) and are somewhat lengthy, but we can at least see that the path difference will lead to the phase difference in Figure 6.7. For detectors located on the same plane as the beam plane and at offset angles of $\beta_d = \pm\theta/2$ (Figure 6.9), the phase difference is

$$\Delta\Phi=\frac{4\pi D}{\lambda}\left[m-1-\sqrt{1+m^2-2m\cos\frac{\theta}{2}}\right] \tag{6.33}$$

In general, the phase difference can be calculated for arbitrary detector angles, α_d and β_d, as shown in Figure 6.9 (Albrecht, 2002). In Figure 6.9, the angle α_d is the "elevation" angle of the detectors measured from the beam plane (the x–z plane in Figure 6.9), and the angle β_d is the half-angle between the detectors measured in the plane formed by the origin and the two detectors.

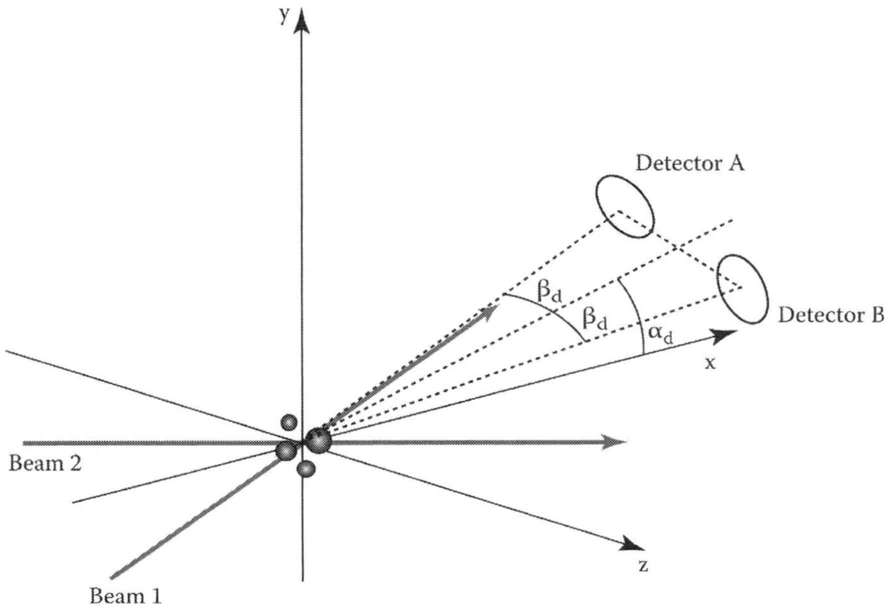

FIGURE 6.9 Detector angles for phase Doppler anemometry.

$$\Delta\Phi = \frac{4\pi D}{\lambda}\left[\sqrt{1+m^2-m\sqrt{2\left(1+\sin\beta_d\sin\frac{\theta}{2}+\cos\beta_d\cos\alpha_d\cos\frac{\theta}{2}\right)}}\right.$$
$$\left.-\sqrt{1+m^2-m\sqrt{2\left(1-\sin\beta_d\sin\frac{\theta}{2}+\cos\beta_d\cos\alpha_d\cos\frac{\theta}{2}\right)}}\right] \tag{6.34}$$

It can be verified that Equation 6.34 reduces to Equation 6.33 when $\alpha_d=0$ and $\beta_d=\theta/2$. In the phase Doppler configuration shown in Figure 6.9, the two detectors yield Doppler bursts with a relative shift (see Figure 6.8 as an example) given by Equation 6.34. The maximum droplet size that can be measured without ambiguity corresponds to that giving a phase difference of 2π from Equation 6.25. To increase the drop size range, the angle β_d can be decreased, or a three-detector combination is often used. Except for the use of multiple detectors to measure the phase, the setup is identical to the LDV shown in Chapter 4. As shown in Figure 6.8, the Doppler burst frequency is used to determine the particle velocity, whereas the phase difference is related to the particle size, giving simultaneous particle size and velocity data. One of the attractive features of PDA is that it provides simultaneous droplet size and velocity data from the respective phase and frequency of the signals.

6.7 PARTICLE MASS FLUX MEASUREMENTS

Thus far, we have considered experimental methods to determine particle size. In many applications, however, we would also like to obtain the particle mass flux data. Although PDA, in principle, gives simultaneous particle size and velocity information, data dropout due to nonspherical particles or unresolved phase shifts associated with noisy signals results in frequent miscounts of the total particle flux. For that reason, the phase Doppler anemometer is not associated with reliable particle flux measurements. Accurate particle mass flux measurements are challenging, and few methods are widely applicable. Here, we will discuss some of the commonly used methods for measuring particle flow rate, with the attendant caution that each method has limitations and, in any event, needs to be adapted for specific applications.

6.7.1 Fiberoptic Correlation Method

Fiberoptic measurements make use of Beer's law, $I_\lambda=I_{\lambda o}\exp(-K_{\text{ext}}L)$, along with Equation 6.7, to relate light transmission to the particle number density and size. In general, particle size distribution and number density are difficult to quantify based on a single measurement of transmittance. Thus, calibration is made between the measured transmittance and the particle number density and mean size, as obtained by an independent method (e.g., direct sampling). Then, the transmittance furnishes the mass density of the particles in the probe paths. For particle mass flux measurements, the particle mean velocity is also required. To infer particle velocity, a set of two probes is used for cross-correlation of the signals, as shown in Figure 6.10. For compactness and easy installation, fiberoptic probes are advantageous; however, this

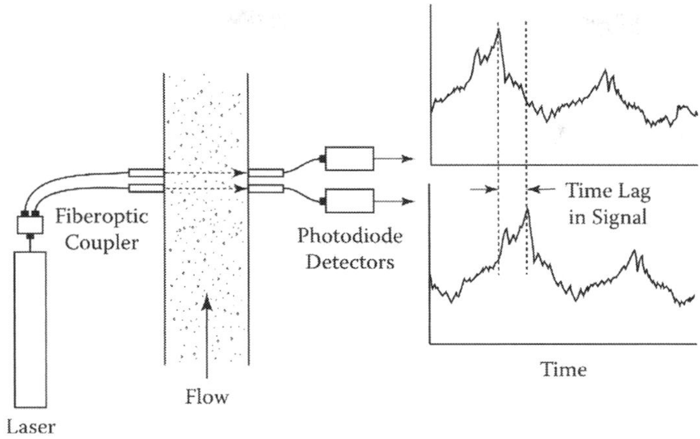

FIGURE 6.10 Fiberoptic cross-correlation method for particle flow-rate measurement.

method can also be used with conventional optical components. The cross-correlation between the two signals is defined as

$$R(\tau) = \lim_{T\to\infty} \frac{1}{T} \int_{t=0}^{T} I_1(t) I_2(t+\tau)\,dt \tag{6.35}$$

T is taken as a sufficiently large sampling time of the signals, and τ is an arbitrary time delay between the two signals, I_1 and I_2. In the calculation of the cross-correlation, $R(\tau)$, τ is varied until the maximum $R(\tau)$ is found. The value of τ, then, represents the transit time of the signal observed by probe 1 to reach probe 2. Thus, the mean speed of the particle motion is the distance between the probes divided by τ,

$$V_p = \frac{L}{\tau} \tag{6.36}$$

The average mass flux of the particles is then taken as the product of the mean speed and the particle mass concentration obtained from the transmittance measurement.

$$(\text{Mass Flux}) = (\text{Mass Concentration}) \times (\text{Mean Speed}) = C_p V_p \left[\text{kg}/\left(\text{m}^2\text{s}\right)\right] \tag{6.37}$$

6.7.2 Acoustic Measurements

Acoustic transmission (typically ultrasonic) through particle-containing media is also used for mass flux measurements. The ultrasonic sound level can be described in several different ways. From acoustics, one way to measure the sound level is by using sound intensity. Sound intensity is defined as follows:

$$\text{Sound intensity} = I = (\text{pressure}) \times (\text{air velocity}) = pu \tag{6.38}$$

Because for acoustic wave propagation (governed by the wave equation), the pressure and air velocity are related by $p/u = \rho c$ (density times sound speed), the sound intensity averaged over time may be written as

$$I = \langle p^2 \rangle / \rho c \tag{6.39}$$

where the parenthesis indicates averaging over time.

A more convenient way to characterize the sound level is to use the mean-squared pressure. This is called the sound pressure level in decibels [dB]. The sound pressure is defined as

$$L_p = 10\log_{10}\left[p^2\right]/p_{\text{ref}}^2 = 10\log_{10}\left\langle \text{p}^2 \right\rangle - 10\log_{10} p_{\text{ref}}^2 \left[\text{dB}\right] \tag{6.40}$$

This gives the familiar decibel level for sound, where p_{ref} is the lowest sound pressure that the human ear can detect, $p_{\text{ref}} = 2 \times 10^{-5}$ Pa. The pressure can be measured using pressure transducers with a good frequency response (high-quality microphones serve as ideal pressure transducers in the sonic range of 20 Hz to 20 kHz).

Because the sound level is in logarithmic form, any attenuation (or gain) factor can simply be subtracted (or added) from the source sound level. For sonic flowmeters, the detectable signal for the particle-laden flow can be written as

$$L_s = SL - TL + PG + DG - NL \tag{6.41}$$

L_s = detectable flow signal
SL = source sound level
TL = transmission loss
PG = processing gain = electronic amplification factor
DG = directivity gain = acoustic gain through waveguides
NL = noise level

Similar to optical transmission measurements, transmission loss due to the particle mass concentration in the flow needs to be calibrated. Sound attenuation depends on particle loading and sound frequency. A calibration equation for the data can be substituted into the TL term in Equation 6.41.

$$TL = \alpha\left(C_p, f\right)r + 20 \log r \tag{6.42}$$

$\alpha(C_p,f)$ = attenuation factor at a given particle loading (mass concentration, C_p) and frequency (f)
r = distance from the sound source to the detector
20log r = beam spreading loss (the loss is proportional to r^2)

The foregoing calibrated transmission loss term gives the particle loading, with other terms in Equation 5.41 being constant or known. We again need the speed of the particles to estimate the mass flux. The cross-correlation, the Doppler effect (as used in optical measurements), or a transit-time analysis can be used for the

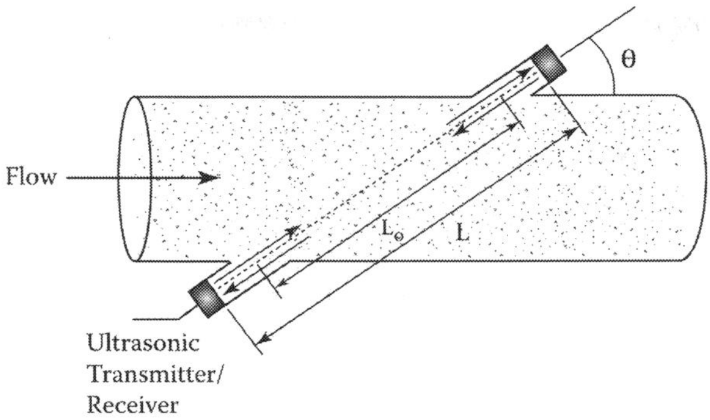

FIGURE 6.11 Counter-propagation configuration for measurement of convection speed.

mean particle velocity. The transit-time analysis makes use of the fact that sound propagation is affected by the speed of sound *and* the convection speed of the fluid. A counter-propagation configuration (Figure 6.11) is used to determine the fluid convection speed. The transit times for the ultrasonic wave to traverse a distance L_o in the upstream and downstream directions are given by the referenced work (Soo, 1999).

$$t_u - t_m = \frac{L_o}{C - V\cos\theta} \tag{6.43a}$$

$$t_d - t_m = \frac{L_o}{C + V\cos\theta} \tag{6.43b}$$

C = speed of sound
V = convection speed of the flow
θ = inclination angle of the probes
t_m = transit time in the dead space $(L - L_o)$

Equations 6.43a and 6.43b can be combined, after dropping the term involving V^2/C^2, to give

$$V = \frac{C^2 (t_u - t_d)}{2L_o \cos\theta} \tag{6.44}$$

Pulses of ultrasonic waves can be sent in opposite directions, as shown in Figure 6.11, and by measuring the transit time for the pulses to be received at the opposite detectors, the convection speed can be retrieved. Its accuracy depends on the ability to measure the time-difference term in Equation 6.35, and typically nanosecond resolutions are required to achieve high accuracy.

6.7.3 Capacitance Probes

Figure 6.12 shows a capacitance probe installed in a tube with an internal solid/gas flow. The principle of the capacitance probe is identical to that of a capacitor. The capacitance in the volume between two conducting plates is given by

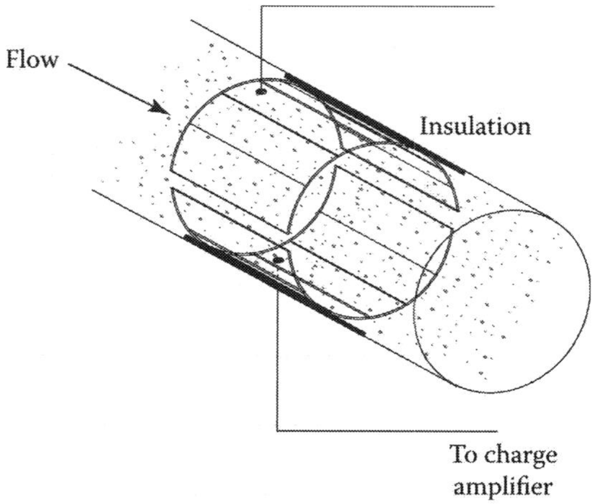

FIGURE 6.12 Capacitance probe used for particle concentration measurement.

$$C = \frac{\varepsilon_o \varepsilon A}{d} \tag{6.45}$$

ε_o = dielectric constant in vacuum = 8.85 pF/m
ε = dielectric constant of the actual medium between the two plates
A = area covered by the plates
d = the distance between the plates

Because the solid particles typically have a higher or at least different dielectric constant than gas, the capacitance is going to vary as a function of the solid particle concentration. This variation can be incorporated into Equation 6.45.

$$C = \varepsilon_o \left\{ \varepsilon_s \phi + \varepsilon_g (1 - \phi) \right\} \frac{A}{d} \tag{6.46}$$

ε_o = dielectric constant in vacuum = 8.85 pF/m
ε_s = dielectric constant of the solid particle
ε_g = dielectric constant of the gas
ϕ = volume fraction of the solid particles

If the capacitance at zero solid particle volume fraction is first measured ($C_g = C$ @ $\phi = 0$), then the measured capacitance can be related to an arbitrary volume fraction, ϕ, as follows (Soo, 1999).

$$\phi = \frac{(C - C_g)/C_g}{(\varepsilon_s - \varepsilon_g)/\varepsilon_g} \tag{6.47}$$

Once the particle volume fraction is determined, then the cross-correlation or transit-time method is again applied to determine the average speed of the particles.

6.8 ISOKINETIC SAMPLING

Isokinetic sampling is a direct probe-based method to extract particles from the flow for further analysis. Although probe-based measurement methods are discussed in Chapter 7 of this book, a brief discussion is presented here because direct sampling is useful in two-phase flow measurements as a way to calibrate or verify other methods discussed earlier. *Isokinetic* means the particles are drawn into the probe at the same velocity as they were traveling in the undisturbed flow. To achieve isokinetic sampling, the flow velocity at the point of extraction needs to be known through an alternative method (e.g., LDV) and appropriate back pressure must be applied to draw the sample into the flow at the same velocity. Figure 6.13 shows a typical arrangement for isokinetic sampling. The velocity at the sampling probe is monitored using a differential pressure transducer, as shown in Figure 6.13. To keep the sampling velocity constant, an adjustable three-way valve is used to "mix" the high pressure and low pressure in response to the differential pressure reading. The sampled particles are analyzed using different methods, such as cyclone separation or cascade impaction.

Once the particles are sampled, there are several ways that the size distribution and other parameters can be obtained, including optical microscopy and other methods described in this chapter. A classical device that is still used for the analysis of sampled particle size is the cascade impaction method, shown in Figure 6.14. The sampled particles are sent through a series of sampling plates. The geometry of the flow requires the particles to navigate the turns, and if the particle inertia is too large, the particle will impact and be collected on the sampling plate. The smaller particles move around the flow obstruction through the cascade until their inertia becomes so small that they follow the flow onto the sampling plate. The sampling plate is coated with adhesive material for effective collection and is later removed for optical or gravimetric (weighing of the collected particles) methods.

The Stokes number is used to estimate the size of the particles that will be collected at a given stage. It can be defined as the ratio of the particle transit distance to

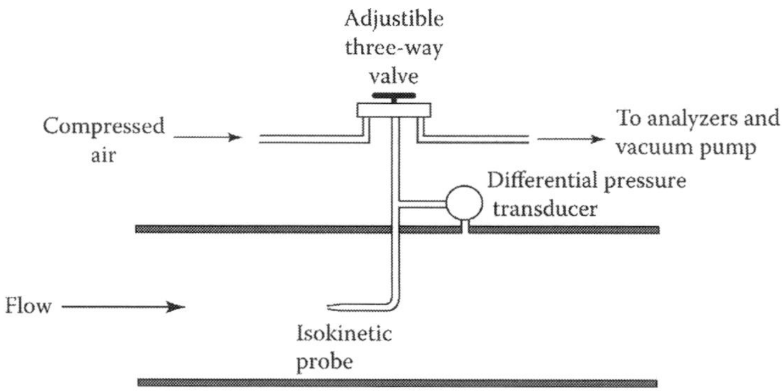

FIGURE 6.13 A typical arrangement for isokinetic sampling of suspended particles.

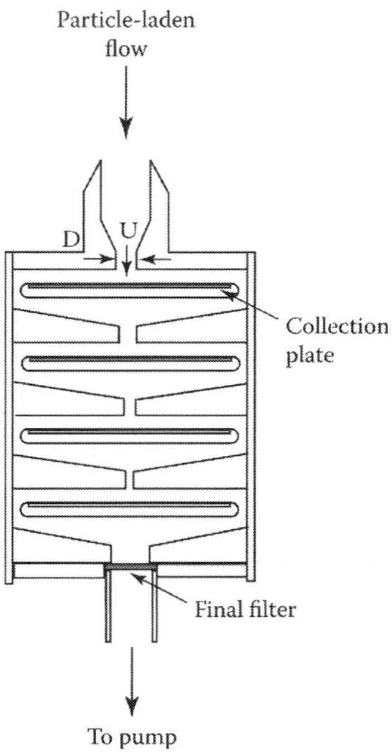

FIGURE 6.14 Cascade impaction analyzer.

the characteristic length scale of the flow. If we take the latter to be half of the nozzle diameter, $D/2$, we have

$$St = \frac{\delta}{D/2} = \frac{2\delta}{D} = \frac{2U\tau_S}{D} \tag{6.48}$$

δ = particle transit distance
D = diameter of the nozzle
U = velocity at the nozzle
τ_S = Stokes relaxation time = time to reach the terminal velocity

Now, the Stokes relaxation time, or the time to reach the terminal velocity, can be written as U_t/g, where U_t is the terminal velocity and g is the gravitational acceleration. The terminal velocity is found from the steady-state force balance (drag force equals the weight of the particle) on the particle as follows:

$$C_D \frac{\pi d_p^2}{4} \frac{\rho U_t^2}{2} = \frac{\pi}{6} \rho_p g d_p^3 \tag{6.49}$$

C_D = the drag coefficient
d_p = particle diameter
ρ, ρ_p = gas and particle density

We neglect the contribution of the buoyancy force in the foregoing force balance. Using $C_D = 24/\text{Re}$ ($\text{Re} = \rho U_t d_p/\mu$) for the particles of interest,

$$U_t = \frac{\rho_p g d_p^2}{18\mu} \tag{6.50}$$

μ = fluid viscosity

Using Equation 6.50 to write $\tau_S = U_t/g$, and substituting it in Equation 6.49, we get

$$St = \frac{C\rho_p d_p^2 U}{9\mu D} \tag{6.51}$$

The foregoing Stokes number can be modified for submicron particles or low-pressure conditions using the Cunningham slip correction factor, C.

$$C = 1 + \frac{0.163}{pd_p} + \frac{0.0549}{pd_p}\exp\left(-6.66pd_p\right) \tag{6.52}$$

p = static pressure in atm
d_p = particle diameter in μm

For $p = 1$ atm and $d_p > 1$ μm, C is close to 1. If we take $St = 1$, i.e., the particle will reach the collection plate located $D/2$ from the nozzle, the characteristic particle size is

$$d_p = \sqrt{\frac{9\mu D}{C\rho_p U}} \tag{6.53}$$

Thus, by making the nozzle diameter (D) smaller for the consecutive cascade stages (which will also increase U for constant flow-rate sampling), smaller and smaller particles are collected until the remaining fine particles are captured by the final filter in Figure 6.14.

6.9 IMAGING METHODS

Imaging methods have existed for quite some time. However, advances in hardware/software enable more powerful applications in drop size and velocimetry. High-resolution imaging devices can focus on a small field of view to detect particles down to the 10 μm range. Typically, an in-line light source and imaging camera configuration is used to maintain constant illumination, as shown in Figure 6.15. Using calibrated conversion of imaged to actual particle size, rapid and automated hardware/software combinations are commercially available for measurements of particle size, shape parameters (sphericity, aspect ratio, compactness, etc.), velocity (using translation of the particle in sequential images), and mass flux. AI and other algorithms (discussed in Chapter 3) facilitate image analysis, leading to particle size measurements. Using imaging probes, a particle size range of 10 μm to 1 mm can be detected, for measurement volumes on the order of a few hundred mm. Modern

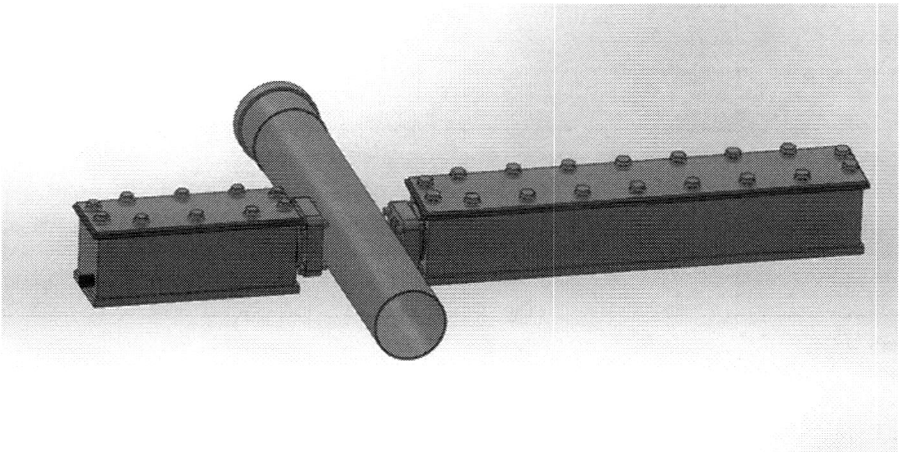

FIGURE 6.15 In-line light source/imaging device arrangement for particle sizing and velocimetry.

imaging cameras have ~10 M-pixel spatial resolution and operate at 50 ~ 100 frames per second. An LED flashlight is used for uniform backlight illumination, at short (<10 μs) pulse durations and pulse frequencies of up to 10 kHz.

Imaging hardware and software can also be developed in the laboratory. The hardware consists of a fast (short-pulse) and uniform light source, and a high-resolution (spatial and temporal) imaging device. The analysis algorithm typically consists of binarization, intensity thresholding, and edge detection (discussed in Ch. 3). Object detection is facilitated using AI functions (e.g., in Matlab®). Various particle shape parameters can be determined using a standard formula.

Area-equivalent diameter: $d_A = \sqrt{\dfrac{4A}{\pi}}$
A = area
Circularity (← Sphericity): $C_A = 4\pi \dfrac{A}{P^2}$
P = perimeter
Aspect ratio : $\text{AR} = \max(y)/\max(x)$
$\max(x)$ = maximum length in x-direction
$\max(y)$ = maximum length in y-direction
Compactness : $C_P = \dfrac{d_A}{\max(d)}$
d_A = area-equivalent diameter
$\max(d)$ = maximum diameter

PROBLEMS

6.1. Fill in the blanks in the following data table for particle size distribution.

Particle Size Range [μm]	Representative Particle Size, D_i [μm]	ΔD_i [μm]	n_i, Number of Particles	$p_i \Delta D_i = \frac{n_i \Delta D_i}{\sum n_i \Delta D_i}$, Size Distribution Function
0.0–10.0	5.0	10.0	125	
10.0–25.0	12.5	15.0	1127	
25.0–35.0	30.0		2475	
35.0–50.0			1455	
50.0–65.0			775	
65.0–80.0			540	
80.0–100.0			325	
100.0–125.0			150	
125.0–150.0			120	
150.0–200.0			55	
			N = Total number of particles =	$\Sigma p_i \Delta D_i = 1$

6.2. For the particle size distribution in Problem 6.1, find the arithmetic mean diameter (D_{10}) and the Sauter mean diameter (D_{32}) and D_{63}.

6.3. Plot the following size distributions.

a. Gaussian distribution with $<D> = 50\ \mu m$ and $\sigma = 12.75\ \mu m$.

b. Log-normal distribution with $<\ln(D)> = \ln(50)$ and $\sigma(\ln(D)) = \ln(12.75)$.

c. Rosin-Rammler-Sperling-Bennet distribution with $n = 1$, $b = 2$.

d. Marshall-Palmer distribution with $D = 50\ \mu m$.

6.4. For the particle size distribution shown in Table 6.1, determine the extinction and scattering coefficients, K_{sca} and K_{ext}, for $\lambda = 488$ nm and $\mathbf{n} = 1.75 + 2.25i$.

6.5. For the particles in Problem 6.4, determine D_{63} and the volume fraction, f_v.

6.6. For the particles in Problem 6.4, what percentage of the light will be attenuated after traveling 0.25 m through the particles?

6.7. Show that for soot particles, the laser-induced incandescence signal has a nearly cubic dependence on the particle radius (Equation 6.17) and therefore soot volume fraction.

$$I_{LII} \sim r^{3+0.154/\lambda_{LII}} \quad [\lambda_{LII} \text{ in } \mu m] \tag{6.16}$$

6.8. Plot the laser diffraction pattern as a function of small deflection angle, θ, using Equation 6.18 for $D = 10$, 50 and 100 mm. $\lambda = 512$ nm, and $r = 1.0$ m.

6.9. For the particle size distribution shown in Table 6.1, and for $\lambda = 512$ nm, and $r = 1.0$ m, write a MATLAB program to compute and plot the laser diffraction intensity as a function of θ.

6.10. Calculate the phase difference in a phase Doppler interferometer using Equation 6.25 as a function of particle diameter for $D=1$ to 100 mm, for $\lambda_L=512$ nm, $m=1.5$, laser wavelength, $\theta/2=7.5°$, and $\beta_d=7.5°$.

6.11. Compare the plot of Problem 6.10 with the plot generated from an alternate equation (Equation 6.33).

6.12. Plot the general phase difference as a function of D for $\theta/2=7.5°$, $\alpha_d=10°$ and $\beta_d=7.5°$, using Equation 6.34.

6.13. Show that Equation 6.34 reduces to Equation 6.33 when $\alpha_d=0$ and $\beta_d=\theta/2$.

6.14. For two fiberoptic signals located 20 mm apart, write a MATLAB program to calculate the cross-correlation as a function of time interval, τ, using Equation 6.35.

$$I_1 = 2.5\sin(2.5t + 2.05)\sin(37.5t + 2.75)$$

$$I_2 = 2.75\sin(2.45t)\sin(35.5t)$$

6.15. Using the cross-correlation of Problem 6.14, at what τ does the maximum occur? What is the estimated particle velocity based on the cross-correlation?

6.16. Using Equations 6.52 and 6.53, estimate the particle size that can be collected in a cascade impactor with a 2 mm nozzle in air at $p=0.5$ atm, $T=300$ K, for particle density of 5.87 g/cm^3. The nozzle velocity is approximately 0.075 m/s.

6.17. a. Plot the characteristic particle size for the above condition, assuming $C=1$, as a function of nozzle diameter for $U=1.25/D^2$.

b. Write a MATLAB program to iterate for C and the characteristic particle size, using the same conditions at $p=1$, 0.5, and 0.1 atm.

BIBLIOGRAPHY

Albrecht, H.-E., Borys, M., Damaschke, N., and Tropea, C., *Laser Doppler and Phase Doppler Measurement Techniques*, Springer-Verlag, New York, 2002.

Allen, T., *Powder Sampling and Particle Size Determination*, Elsevier, Amsterdam, 2003.

Melton, L.A., Soot diagnostics based on laser heating, *Applied Optics*, 23(13), p. 2201, 1984.

Quay, B., Lee, T.-W., Ni, T., and Santoro R.J., Spatially-resolved measurements of soot volume fraction using laser-induced incandescence, *Combustion and Flame*, 97, pp. 384–392, 1994.

Santoro, R.J., Semerjian, H.G., and Dobbins, R.A., Soot particle measurements in diffusion flames, *Combustion and Flame*, 51, p. 203, 1982.

Soo, S. L. (Ed.), *Instrumentation for Fluid-Particle Flow*, Noyes Publications, Park Ridge, NJ, 1999.

7 Gas-Sampling Measurements

7.1 INTRODUCTION

Gas sampling is still widely used for analyzing chemical composition. As the phrase gas sampling suggests, a volume of gas is sampled (usually continuously) from the flow and sent through analyzers to discern the various gas molecules present in the sample. Gas-sampling analyzers represent a mature technology, can be calibrated using reference gases, reach sensitivities down to ppm levels, and are rugged and readily available from various vendors. There is a wide variety of analyzers for the detection of pollutants and toxic gases, fuel, and oxidizers, and for identifying a complete spectrum of molecules (gas chromatography and mass spectroscopy). In many chemically reacting flows, including engine combustion, power plants, environmental monitoring, and chemical processing, it is of interest to determine the constituents of gas going in and out of systems for efficiency, process control, and pollutant emissions. For these reasons, gas analyzers represent an important diagnostic method for direct monitoring of chemical composition or calibration of other sensor devices. On the other hand, the sampling probes do introduce fluid dynamic perturbations to the flow and also suffer from relatively poor spatial resolution. The analyzers are often bulky and heavy, and sampling requires care to preserve the content of the gas from the sampling point to the analyzer. In this chapter, we will begin by discussing the considerations for accurate physical sampling of gas using probes and go on to present various analysis methods for the detection and monitoring of gas species concentrations. Figure 7.1 shows a general schematic of a gas-sampling probe connected to a gas analyzer. A pump draws the sample from the test section through the sampling probe and the analyzer. To optimize the sampling conditions at the probe, pressures are monitored and controlled in the sampling line.

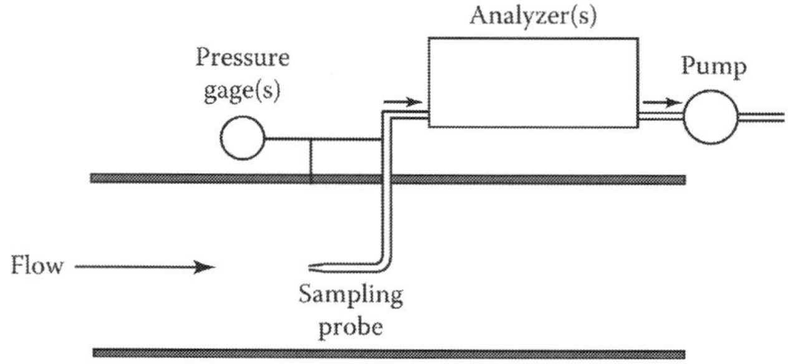

FIGURE 7.1 A gas sampling and analysis system.

DOI:10.1201/9781003611837-7

7.2 SAMPLING PROBES

Ideal sampling probes should extract the true content of the gas from the flow and deliver the sample to the analyzer in that state while minimally affecting the flow. In real measurement systems, some common errors due to gas sampling are (Turns, 1991)

1. Flow disturbance due to the physical presence of the probe
2. Alteration of the flow streamlines and concentration field due to the sink action
3. Change in the local chemical reaction due to the thermal or catalytic effect of the probe surface
4. Inadequate thermal quenching of the gas-phase reaction (i.e., chemical reactions occurring in the sampling line from the probe to the analyzer)
5. Catalytic reaction on the probe or sampling line walls
6. Condensation of gas species to liquid
7. Mixing of different gas samples in intermittent or turbulent flows (lack of temporal resolution)

Errors 1 and 2 in the above list originate from the finite dimensions of the probe and the pressure difference at the probe with respect to the flow pressure necessary to extract the gas sample, respectively. The pressure difference at the probe inlet and the ambient will alter the local streamline, and in subsonic flows, this alteration in the flow will affect the flow upstream. These two effects are, to some extent, circumvented by using isokinetic sampling (extraction of the gas sample at the same flow velocity as in the undisturbed flow) and thin-walled microprobes. More will be said about isokinetic sampling later in this section.

Error 3 is due to the probe acting as a thermal sink. To minimize Error 4 (continuing chemical reaction in the probe or sampling line), chemical reactions should be suppressed (quenched) by cooling the probe tube and the sampling line. The cooled probe surface, however, will also lower the local gas temperature in the flow, which will affect the chemical reactions at that point and, therefore, the gas sample. Some probe materials may act as a catalyst for extraneous gas-phase reactions near the probe. To minimize the catalytic effect, probes are typically made of quartz or stainless steel. The "freezing" or quenching of any further chemical reactions in the probe and in the sampling line can be achieved to some extent by using (1) cooling (e.g., a co-annular water tube around the probe), (2) aerodynamic expansion, which would result in cooling and dilution, and (3) dilution with an inert gas; or some combination of these methods can be used. Figure 7.2 shows some typical geometries for water-cooled (Figure 7.2a) and air-cooled, expansion (Figure 7.2b) probes.

Although Error 5 is minimized by using quartz or stainless steel tubes, some catalytic reactions can still occur, particularly for nitric oxides. Condensation of water or fuel vapor must also be avoided. The last error, Error 7, is associated with the lack of temporal resolution and, therefore, difficulty in resolving turbulent or intermittent effects. This is an inherent difficulty in most gas-sampling systems.

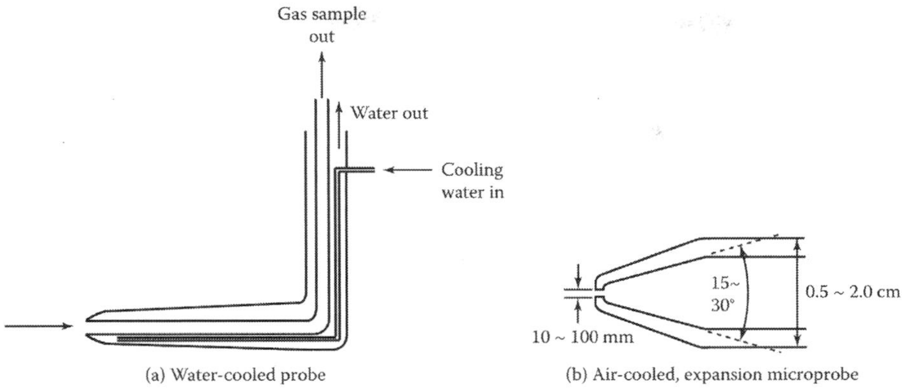

FIGURE 7.2 (a) Water-cooled probe; (b) air-cooled, expansion probe (microprobe).

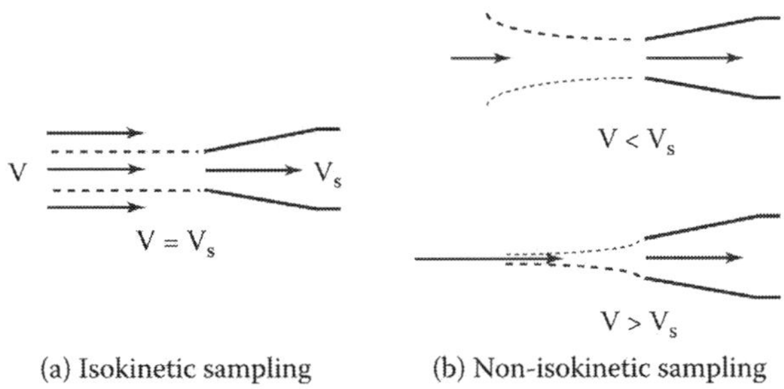

FIGURE 7.3 Isokinetic sampling.

Due to the finite volume of gas extracted and the transit time between the extracted sample and the analyzer, the sampling analysis yields spatially and temporally averaged gas composition.

To optimize gas sampling in spite of the above constraints, an *isokinetic* probe is preferred. In instances where simplicity is desired in the probe design, so-called *sonic* probes can be used. Non-isokinetic and non-sonic probes are even simpler to design and implement, but this introduces the errors described earlier. Isokinetic probes withdraw the gas at the same velocity as the flowing stream. This matching of velocity causes minimum flow disturbance and can be considered the most effective gas sampling at that point. The isokinetic probe is also useful in sampling particles from flows, as discussed in Chapter 6. The requirement for isokinetic sampling is the ability to control the back pressure (sink pressure) of the sampling line so that all the flow incident upon the probe opening enters the probe (i.e., streamlines are parallel when entering the probe), as shown in Figure 7.3. A secondary pressure probe can be used to monitor the dynamic pressure at the sampling point to adjust the back pressure to maintain the isokinetic condition. For steady-state processes, a velocity measurement method such as a hot-wire anemometer can also be used to find the correct back pressure that matches the sampling velocity to that of the flow.

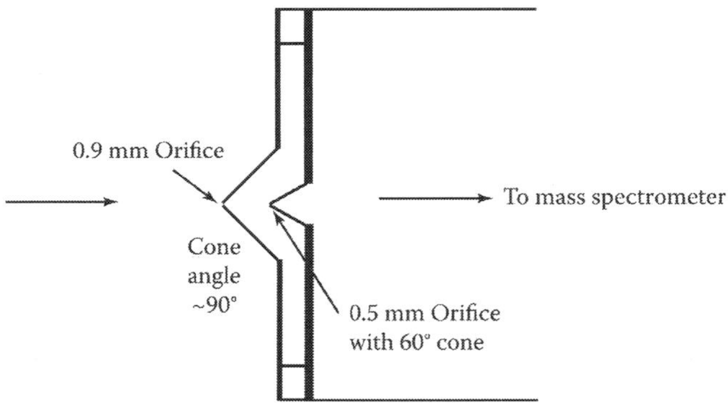

FIGURE 7.4 A sampling probe used for mass spectrometers.

A simpler way to sample the gas is to apply a constant but sufficiently low back pressure to cause the flow to choke at the small probe opening (i.e., achieve the sonic condition). This type of probe is also referred to as a *sonic probe* or a *microprobe* and is typically made of quartz (see Figure 7.2b). The sonic condition at the probe tip will ensure that the sampling flow rate is constant (as opposed to being isokinetic), essentially determined by the ratio of the pressures at the probe tip and the sink. For mass spectrometers, a special sonic probe with a large expansion angle is used, as shown in Figure 7.4. For sampling both stable and radical molecular species during mass spectrometry, a large pressure ratio up to 100:1 from ambient to sink is used. Mass spectrometry will be discussed later in this chapter.

Once the gas sample is acquired, it can be sent to various analyzers, which can detect and quantify carbon oxides (CO and CO_2), nitric oxides (NO and NO_2), unburned hydrocarbons (UHCs), oxygen (O_2), sulfur dioxide (SO_2), and a wide range of stable and radical species in general. This versatility in applying different analyzers, along with high sensitivities down to ppm levels is the strength of gas-sampling analyses, in spite of the sampling difficulties discussed earlier. In the remainder of this chapter, various analysis methods for sampled gas are discussed.

7.3 NONDISPERSIVE INFRARED (NDIR) ANALYZERS

In Chapter 5, absorption spectroscopy was discussed. Absorption spectroscopy is based on light being absorbed in the distinct energy bands of the target molecule. The spectral location (wavelength) and the amount of light absorbed provide information on the identity and concentration of the molecule. A general spectroscopic setup would involve a spectrometer or a dispersive element that decomposes the spectral content of the optical signal. Nondispersive spectroscopy is a simplified version of absorption spectroscopy, where the bulk and expense of the dispersive system are avoided by using filters to observe specific absorption bands. For these reasons, nondispersive analyzers are usually custom-made for the detection and monitoring of specific gases. Nondispersive analyzers consist of light sources, sample and reference gas cells, and optical or gas filters, along with signal conditioning devices (e.g., optical choppers). Because most gas molecules have appreciable absorption bands in the infrared,

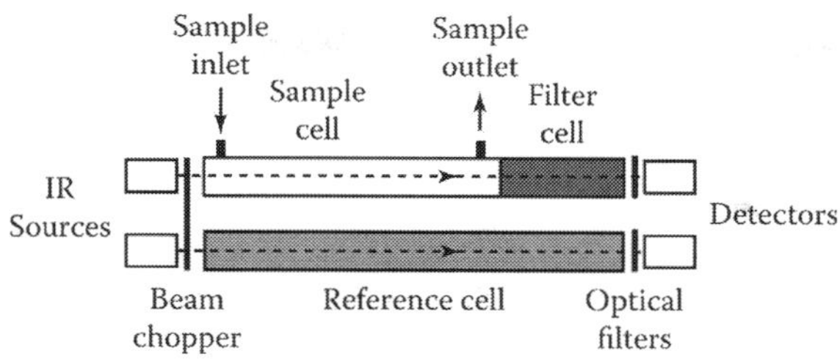

FIGURE 7.5 Nondispersive infrared analyzer.

infrared light sources are often used, and therefore the name nondispersive *infrared* (NDIR) analyzers. Figure 7.5 shows a schematic of an NDIR analyzer. Optical or other light filters are placed in the light path to eliminate interference from other molecules in adjacent infrared absorption bands. For example, carbon dioxide has a prominent absorption band centered at 4.2 μm, whereas carbon monoxide absorbs strongly at 4.6 μm. Using filtering schemes, the interference from carbon monoxide is nearly completely suppressed so that carbon dioxide can be monitored even in the presence of carbon monoxide. For example, an interference filter centered at 4.2 μm can be used in the optical filter slot to reject carbon monoxide interference, or the gas filter cell can be filled with carbon monoxide to minimize its effect in the sample gas.

The absorption characteristics of a gas are given by the absorption coefficient, the amount of light intensity absorbed per unit path length of the sample gas. Furthermore, we can separate the contribution to the absorption coefficient from the gas concentration (mole fraction) and other effects (temperature and pressure) as follows:

$$\alpha_\lambda = \alpha_\lambda(\lambda, T, p, X_i) = X_i f(\lambda, T, p) \tag{7.1}$$

λ = wavelength
T = temperature
p = pressure
X_i = mole fraction of the gas to be analyzed

The preceding equation states that the absorption coefficient is proportional to the number density of the molecules and to some function of the temperature and pressure, $f(\lambda,T,p)$, at a given wavelength. The absorption coefficient is used in Beer's law to calculate the attenuation of the source light across the sample cell.

$$I_\lambda(L) = I_\lambda(0)\exp\left[-\int_{s=0}^{L} \alpha_\lambda ds\right] \tag{7.2}$$

$I_\lambda(L \text{ or } 0)$ = light intensity at $s=L$ or $s=0$, i.e. at the ends of the sample cell.
s = path length across the sample cell (from $s=0$ to $s=L$)
L = optical path length across the sample cell

By solving for α_λ and X_i in turn using Equation 7.1, we can directly relate the amount of light attenuated to the mole fraction or concentration of the sample gas.

TABLE 7.1
Gas Species Detectible with NDIR Analyzer

Gas	Minimum Detection Level [ppm]	Maximum Detection Level [%]
Carbon monoxide, CO	50	100
Carbon dioxide, CO_2	50	100
Nitric monoxide, NO	100	100
Sulfur dioxide, SO_2	100	100
Methane, CH_4	50	100
Propane, C_3H_8	50	100
n-hexane, C_6H_{14}	50	6
Ammonia, NH_3	500	100
Acetylene, C_2H_2	200	100
Ethylene, C_2H_4	200	100
Ethane, C_2H_6	250	100
Ethyl alcohol, C_2H_5OH	250	1
Butylene, C_4H_8	250	100
Butane, C_4H_{10}	250	100
Pentane, C_5H_{12}	250	20
Toluene, C_7H_8	250	10
Ethylbenzene, C_8H_{10}	250	0.2
Methanol, CH_3OH	250	3
Vinyl chloride, C_2H_3Cl	250	100
Nitrous oxide, N_2O	250	1.3

$$X_i = \frac{\ln\left[\dfrac{I_\lambda(0)}{\lambda_\lambda(L)}\right]}{f(\lambda,T,p)L} \tag{7.3}$$

The function $f(\lambda,T,p)$ is a constant if the wavelength, temperature, and pressure are fixed in the sampling cell, and this constant is determined by measuring the light attenuation across the reference cell that contains a known concentration, X_i, of the sample gas.

Table 7.1 lists some of the major gas species that can be detected using NDIR analyzers. Carbon monoxide and carbon dioxide are the most common species measured using NDIR analyzers, with resolutions down to 50 ppm levels. Carbon dioxide is a product of combustion and is also attributed to the greenhouse effect and associated global warming. Carbon monoxide is an extremely harmful gas and is also produced during combustion (usually in underventilated systems). Therefore, measurements of these gases are often useful, and NDIR furnishes a reliable, cost-effective method.

7.4 MEASUREMENTS OF NITRIC OXIDES

As shown in Table 7.1, although oxides of nitrogen can be measured using NDIR analyzers, the detection limit and interference are not always favorable. An alternative method, chemiluminescent analysis, has become the standard instrument for monitoring oxides of nitrogen, such as nitric monoxide (NO), in gas samples. Nitric dioxide can also be measured after being reduced to nitric monoxide under controlled conditions. Nitric oxides are extremely harmful gases, yet they are present in trace amounts in the urban atmosphere due to automotive and other anthropogenic sources. They cause so-called *brown clouds* due to their light absorption characteristics in the short visible wavelength range. Nitric oxides are formed in substantial amounts, particularly in combustion systems operating at relatively high temperatures, such as diesel engines. It is of environmental importance to curtail nitric oxide emissions from combustion systems, and reliable monitoring of nitric oxide thus becomes an important tool in such an effort.

Chemiluminescent analyzers (CLA) have a detection limit for nitric oxides down to one part per billion (ppb), a detection range of a factor of 10,000 (i.e., from 1 to 10,000 ppm, or from 1 ppb to 10 ppm), and a linear response over those ranges. Chemiluminescence is a process in which thermally excited molecules in a flame emit photons during downward transitions to stable states, and thus CLA is an application of emission spectroscopy. In a CLA, nitric monoxide is mixed with highly reactive ozone (O_3) in a controlled burner environment and burned. The chemical reactions during this flame process are

$$\begin{aligned} NO + O_3 &\rightarrow NO_2^* + O_2 \\ NO + O_3 &\rightarrow NO_2 + O_2 \\ NO_2^* &\rightarrow NO_2 + h\nu \\ NO_2^* + M &\rightarrow NO_2 + M \end{aligned} \tag{7.4}$$

In these chemical reactions, NO_2^*, $h\nu$, and M are the excited nitric dioxide molecule, the photon energy, and an arbitrary molecule, respectively. The first two reactions form either thermally excited or neutral nitric dioxide molecules, whereas the third reaction produces a photon during the transition to the ground state. The fourth reaction is a so-called quenching reaction where an arbitrary molecule collisionally deactivates the excited nitric dioxide molecule and therefore quenches the emission. In the reaction chamber of a CLA, where the conditions (temperature, pressure, and ozone feed rate) are kept constant, the number of photons depends linearly on the number of nitric monoxide molecules, and hence chemiluminescence gives a signal whose intensity is proportional to the nitric monoxide concentration in the gas sample.

Figure 7.6 shows a schematic of a CLA. A CLA consists of a reaction chamber (RC), a photomultiplier (PMT) with optical filters (OF), an amplifier to detect chemiluminescence, an ozone generator (OG), and a NO_2 converter ($NO_2 \rightarrow NO$) for nitric dioxide measurements. The NO–O_3 reactions produce a rather broadband chemiluminescence ranging from 600 nm to the infrared, with the peak occurring near 1.2 μm. The optical filter cuts off any light interference below 600 nm, and the photomultiplier detects the signal in the 600–900 nm range.

As shown in Figure 7.6, a CLA includes an NO_2 converter because the basic CLA relies on the NO–O_3 reaction. Thus, to measure NO_2, it must first be converted to NO. Three types of NO_2–NO converters are available: stainless steel, molybdenum,

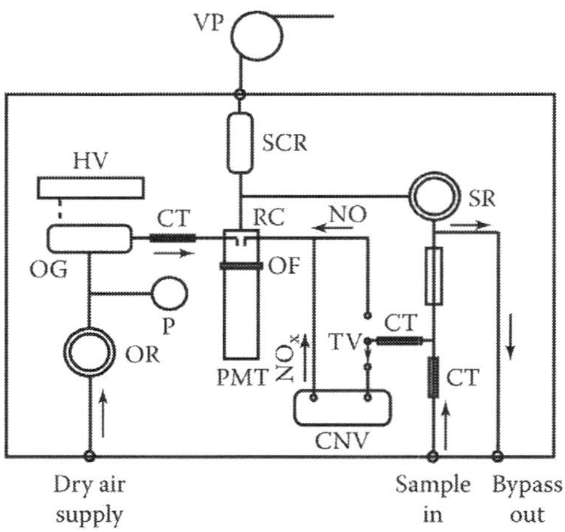

AR, air regulator; CNV, NO_2-to-NO converter;
CT, capillary tubes (0.005 in ID); HV, high-voltage source;
OF, optical filter; OG, ozone generator; P, pressure gage;
PMT, photomultiplier tube; RC, reaction chamber;
SCR, scrubber; SR, sample regulator;
TV, three-way valve; VP, external vacuum pump

FIGURE 7.6 A schematic of a chemiluminescent analyzer.

and carbon. In the stainless steel converter, NO_2 flows through a length of stainless steel tube, and the catalytic effect at high temperatures (~650°C) acts to promote the conversion reaction:

$$NO_2 \rightarrow NO + \frac{1}{2} O_2 \tag{7.5}$$

The converted NO is immediately analyzed in the CLA because NO can oxidize quickly and convert back to NO_2. In the molybdenum and carbon converters, the converter material actually participates in the reactions and is therefore consumed during the conversion processes, as shown in the following equation, although for molybdenum, the material loss can be recovered through a reversing reaction. The advantage of these converters is that high conversion rates are achieved at relatively low temperatures (350–400°C).

$$\begin{gathered} 3NO_2 + Mo \rightarrow MoO_3 + 3NO \\ NO_2 + C \rightarrow CO + NO \end{gathered} \tag{7.6}$$

The CLA units are nowadays packaged in compact, rack-mountable enclosures. As shown in Figure 7.6, a dry air supply is needed to operate the ozone generator, which feeds a constant flow rate of ozone to the reaction chamber through a capillary tube. The volumetric flow rate of the sample stream is also controlled by capillary tubes and the back pressure attained via an external vacuum pump. The excess amount of the sample is bypassed out. The sample stream goes directly into the reaction chamber in the NO analysis mode or through the NO_2-to-NO converter in the NO_x (NO_2 + NO) analysis mode. The mode is selected using a three-way valve. An optical filter and chopper are used to reject interference and noise. The combustion products from the reaction chamber are treated in the scrubber before being released from the analyzer.

7.5 HYDROCARBON ANALYSIS

Measurements of hydrocarbons are important in combustion systems for determining the fuel concentration during fuel preparation and unburned hydrocarbons after the combustion processes. The detection of hydrocarbons is predominantly accomplished using flame-ionization detectors (FID), which possess high sensitivity and linear response. A sample containing hydrocarbons is mixed with additional fuel gas (hydrogen or hydrogen diluted with helium or nitrogen). This mixture is burned with air in a controlled reaction chamber. At the temperatures of the flame produced in this manner, ions and electrons are formed. Under an electric field, the positively charged ions migrate to the cathode, whereas the free electrons migrate to the anode. The current established through this charge migration can be detected as a voltage change, which is proportional to the hydrocarbon content in the fuel stream. A schematic of an FID is shown in Figure 7.7. The chemical reactions of hydrocarbons that result in ion and electron production are

$$\begin{aligned} CH + O &\rightarrow CHO^+ + e^- \\ CHO^+ + H_2O &\rightarrow H_3O^+ + CO \\ H_3O^+ + e^- &\rightarrow H_2O + H \end{aligned} \tag{7.7}$$

It is apparent in the foregoing chemical reactions that CH and O are the initiators of the ion-producing reactions. CH is formed in the flame through thermal cracking of larger hydrocarbon molecules, i.e., high temperature and the associated kinetic energy of the molecules result in energetic collisions that break down the hydrocarbon

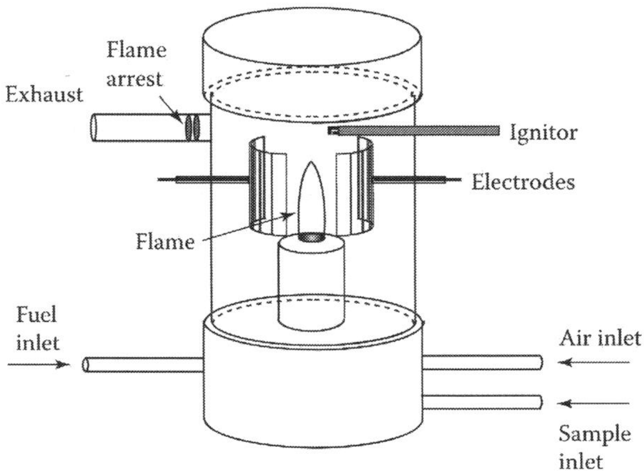

FIGURE 7.7 A schematic of a flame-ionization detector.

molecules, and one of the commonly occurring fragments is CH. Oxygen atoms are produced in reactions involving

$$\begin{aligned} H_2 + OH &\rightarrow H_2O + H \\ H + OH &\rightarrow H_2 + O \\ H + O_2 &\rightarrow OH + O. \end{aligned} \tag{7.8}$$

A large majority of hydrocarbons exhibit a nearly uniform response in the FID. The relative response with respect to the reference aliphatic hydrocarbons is 1.0 for aromatics, 0.95 for olefins (or alkenes), and 1.30 for acetylenic compounds. Aliphatics are hydrocarbons in which carbons are joined in straight or branched chains, such as methane (CH_4), ethylene (C_2H_4), and ethane (C_2H_6). Aromatics contain one or more planar sets of six carbon atoms joined in a benzene ring. Olefins (or alkenes) have a chemical formula of C_nH_{2n}, whereas acetylenic hydrocarbons have C_nH_{2n-2}. A group of hydrocarbons with a negative response in FID is alcohols, i.e., the ions are consumed by alcohols rather than being produced by them. Therefore, alcohol compounds represent a source of interference in FID measurements of hydrocarbon content. The variability in FID response listed earlier can be minimized to some extent by operating the FID burner at an optimum ratio of hydrogen-to-hydrocarbon flow rates (approximately 4.0). Then, the voltage generated in the FID is primarily a linear function of the hydrocarbon concentration in the sample gas.

7.6 MEASUREMENTS OF SULFUR OXIDES

Sulfur dioxide (SO_2) is formed when fuel containing sulfur is burned, as in coal-fired furnaces. As with other major pollutants, there are strict government restrictions on how much sulfur dioxide a power-generating station or an industrial plant can emit. In addition to being a toxic gas, sulfur dioxide can combine with atmospheric vapor to form sulfuric acid, a major contributor to acid rain. For these reasons, it is important to be able to monitor sulfur dioxide content in the exhaust flows of combustion systems and in the atmosphere.

Fluorescence detectors are used for gas-sampling analyses of sulfur dioxide content. As noted in Chapter 5, fluorescence is an optical process resulting from photon emission when an excited molecule undergoes a downward transition to a ground state. In gas-sampling detectors, the excitation and fluorescence of sulfur dioxide are achieved in a sample cell, as shown in Figure 7.8. For fluorescence analysis, an absorption band of sulfur dioxide near 210 nm is used for ultraviolet excitation, which results in broadband fluorescence in the 210–420 nm range. Out of the broadband fluorescence emitted, the bands in the short wavelength range from 210 to 230 nm are typically used for sulfur dioxide (SO_2) detection. A flashlamp with an ultraviolet light output is used with a filter selected to transmit light in the sulfur dioxide absorption bands. A photomultiplier tube is used for the detection of the fluorescence signal, again with a bandpass filter to accept only the sulfur dioxide signal in the 210–230 nm range. The optical processes involved in fluorescence generation are

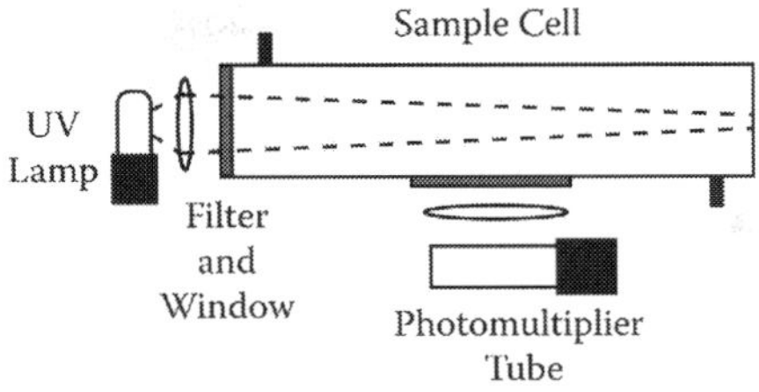

FIGURE 7.8 A fluorescence detector for sulfur dioxide measurements.

TABLE 7.2
Potential Interference from Various Species During SO_2 Fluorescence Measurements

Gas Species	Relative Concentration with Respect to SO_2 to Cause an Equivalent Response
NO	500:1
C_2H_4	4000:1
CS_2	500:1
H_2O	SO_2 signal decreases linearly with the percent increase of water vapor (at 2% H_2O, SO_2 signal decreases by 25%)
CH_4	Negligible if below 1%
CO_2	Negligible if below 3%

Source: From Okabe (1971).

$$\begin{aligned}\text{Excitation: } & SO_2 + h\nu_e \rightarrow SO_2^* \\ \text{Fluorescence: } & SO_2^* \rightarrow SO_2 + h\nu_f \\ \text{Quenching: } & SO_2^* + M \rightarrow SO_2 + M \\ \text{Dissociation: } & SO_2^* \rightarrow SO + O\end{aligned} \tag{7.9}$$

Excitation and fluorescence are the normal processes during which photons of frequencies ν_e and ν_f, respectively, are exchanged between excited (SO_2^*) and ground-state sulfur dioxide. Quenching means deactivation of the excited state through collisional energy exchange with an arbitrary molecule, M. Dissociation is the separation of sulfur dioxide into sulfur monoxide and an oxygen atom. Under controlled conditions (constant temperature and pressure) in the sample cell, all of the reaction rates in the processes listed are constant, and thus the sulfur dioxide concentration is proportional to the ratio of the fluorescence and input light intensities.

In the 210–230 nm wavelength range, few species can interfere with sulfur dioxide fluorescence measurements unless they exist in copious amounts. Table 7.2 gives an idea of the level required to produce appreciable interference (Okabe, 1971). The presence of typical gas species found in air, such as O_2, N_2, H_2S, NO_2, and CO, in small amounts leads to insignificant levels of interference for fluorescence measurements of SO_2.

7.7 GAS CHROMATOGRAPHY

Gas chromatography (GC) is a versatile method for analyzing the chemical compositions of the sampled gas or liquid. Most major gas and liquid species with boiling points below 300°C, including N_2, O_2, CO_2, H_2O, hydrocarbons, and many other gases, can in principle be detected down to ppb level using GC. GC is based on the differential diffusion of chemical species in a porous or dissolving medium, which results in the spatial separation of the constituent gas species. The sample gas or liquid is sent using an inert carrier gas through a column containing a stationary capillary, liquid, or solid medium. The diffusion speed of a particular chemical species through the medium will differ depending on the diffusivity of the species in the medium. A number of media are used in GC, such as capillary tubes, liquid columns, activated charcoal, silica gel, or other materials known by proprietary names (e.g., Porapak Q, Chromosorb 101, etc.).

A high adsorption rate (for solid media) or solubility (for liquid media), for example, will impede the progress of a gas sample through the medium, and therefore the diffusivity or diffusion coefficient will be small. Such a gas will lag behind other gases with higher diffusion coefficients. This spatial separation is translated into temporal separation, called the *retention time*, when the speed of the carrier gas motion is taken into account. The retention time is the spatial separation divided by the carrier gas speed. A detector is needed to quantify the amount of a particular gas species at a given retention time. Thus, in a gas chromatogram, the location on the time axis identifies the gas species, whereas the area under a peak gives the amount of that gas species.

GC has some important advantages over other analytical methods, such as relatively low cost and straightforward operation. By using different columns and detectors, GC can analyze a large variety of gases and liquids. Some amount of analysis time, typically 10–20 min, is needed for the separation of component gases, using a small amount of sample, typically 1–3 mL of gas.

A typical gas chromatograph setup and a chromatogram are shown in Figure 7.9a and 7.9b, respectively. A GC consists of a sample inlet, temperature and flow control, a column, and a detector. The column is rolled into a circular loop to economize the space needed to house its length and is placed in a temperature-controlled oven.

Depending on whether the sample is gas or liquid, a slightly different inlet arrangement is used; however, the important aspects of sample injection or inlet in all cases are speed, reproducibility, and a wide range of applicability. The most common method for sample injection into the GC is the use of a precision, small-volume syringe in the range of 1–500 μL. A more advanced and precise method, however, is to use sampling valves. Three basic designs of sampling valves are (a) spool, (b) sliding plate, and (c) diaphragm-actuated valve, two of which are shown in Figure 7.10. As shown in Figure 7.10a, the initial flow paths are separate between the carrier gas and the sample stream in the spool valve. The spool is rotated to connect the "loop" to the carrier gas line, and the sample contained in the loop is then sent into the GC column by the carrier gas. The sliding valve accomplishes the same task by having alternate routes in the plate that traverse across the carrier and sample ports, as shown in Figure 7.10b. The diaphragm-actuated valve is similar, except that it uses a diaphragm to control the flow paths.

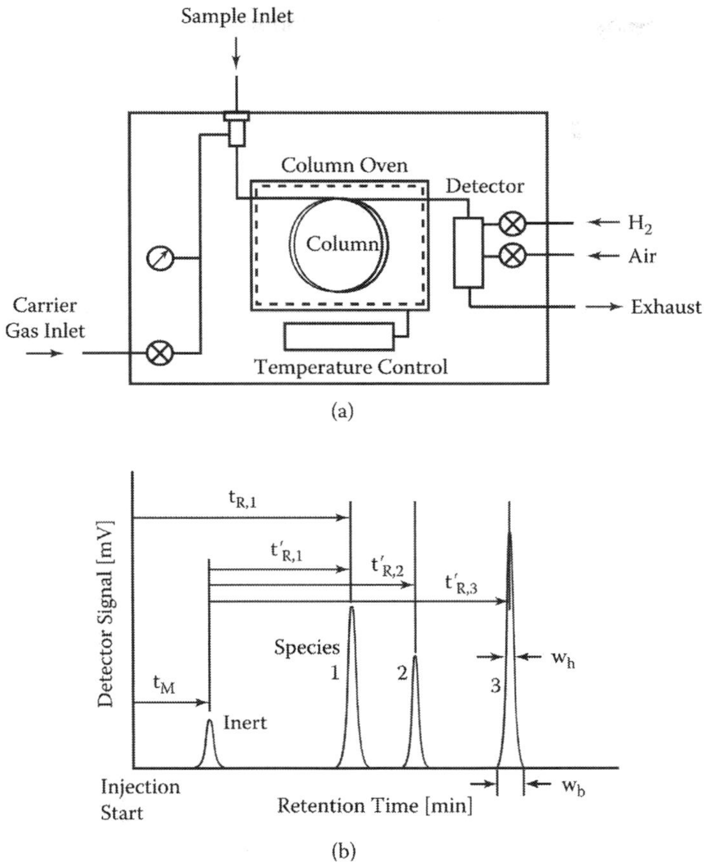

FIGURE 7.9 (a) A schematic of a gas chromatograph, and (b) a typical gas chromatogram.

In GC, the column or the porous medium only serves the function of separating the constituent components, and detectors are needed to quantify the amount of each separated species. Detectors commonly used in GCs include the thermal conductivity detector (TCD), FID, and electron capture detector (ECD). TCDs measure the change in thermal conductivity of the sample and can detect all gases with thermal conductivity different from the carrier gas. However, they tend to be used for less sensitive detection, from 0.01% to 100% content of the target gas. FIDs are similar to those used in hydrocarbon analyzers (Section 7.5) and thus require a hydrogen feed to sustain a flame in the detection chamber. They are highly sensitive to organic compounds but give insufficient or no response for the following compounds: He, Ar, Kr, Ne, Xe, O_2, N_2, HCHO, CS_2, COS, H_2S, SO_2, NO, N_2O, NO_2, HCOOH, NH_3, CO, CO_2, H_2O, $SiCl_4$, $SiHCl_3$, and SiF_4. ECDs are the most sensitive detectors, with sensitivities down to ppb levels, but require a radioactive source to ionize the gas sample to detect the change in electrical conductivity.

Table 7.3 lists some of the common arrangements for detectors used in commercial GCs and the corresponding carrier gas requirements.

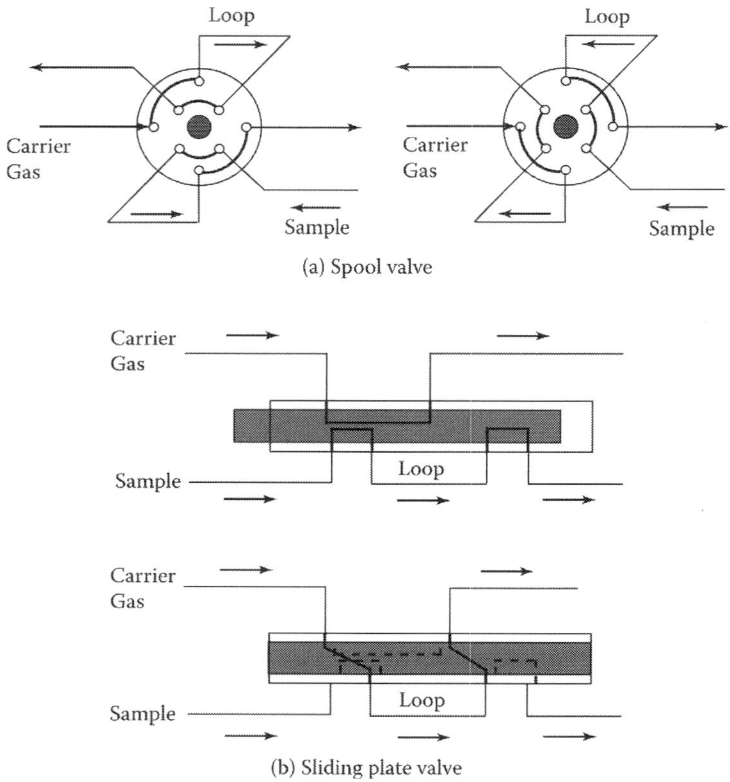

FIGURE 7.10 Sampling valves for gas chromatography.

TABLE 7.3
Typical Carrier Gas Arrangements for Gas Chromatography

Detector	Carrier Gas	Supply Pressure at Inlet (psig)	Flow Rate (cc/min)	Purity (%)
Thermal conductivity detector	He or N_2	80 (He), 77 (N_2)	30	99.995
Flame-ionization detector	He or N_2	80 (He), 77 (N_2)	30	99.998
	H_2	40	30	99.998
	Air	60	300	Best available
Electron capture detector	N_2 or Ar/CH_4 (90/10)	60	30	<0.02 ppm H_2O <1 ppm O_2

The choice of medium or column is important for each analyte (target species in the sample) and the desired resolution. There are two main categories of GC columns: capillary and packed. Capillary columns are essentially narrow-bore tubes ranging from 0.1 to 1 mm in inner diameter, made of metal or glass. The inner walls

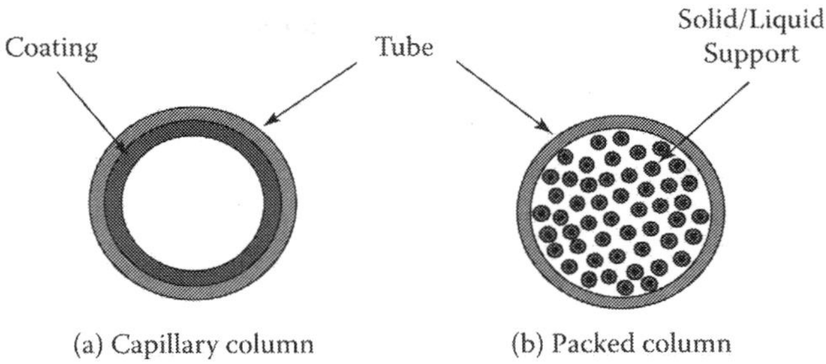

FIGURE 7.11 Two types of gas chromatography columns: (a) capillary column, and (b) packed column.

of the capillary columns are coated with an extremely thin liquid or adsorbent coating. Packed columns can contain either porous solids or stationary liquids supported by solid material. Some capillary columns are packed with porous solids, so these columns include aspects of both capillary and packed columns. Schematics of capillary and packed columns are shown in Figure 7.11.

There is a wide variety of commercially available GC columns, and based on measurement requirements, one needs to choose (1) column material, (2) stationary phase (the diffusion medium inside the column), (3) internal diameter, (4) film thickness, and (5) length. The important aspects to consider in choosing the GC column are the capacity (the volume of the sample that can be analyzed) and the resolution of the peaks representing different chemical species in the sample.

For column material, both fused silica and stainless steel offer structural integrity with flexibility (for looping) and chemical inertness. The length of the column is typically in the order of meters, so the columns are coiled into a small-diameter loop to fit into the GC unit. For packed columns, the stationary phase is critical for providing optimum separation of the analytes. Again, there is a wide range of stationary phase materials designed for specific analytes, which may include volatile hydrocarbons, solvents, pharmaceutical products, environmental samples, and even molecules responsible for food flavors. Due to the large variety of analytes and corresponding stationary phase materials, it is best to consult the manufacturers or their websites for available technical data and selection guides. The internal diameter of GC columns dictates the flow rate through the column. Thus, the sample volume and concentration need to be considered when selecting the internal diameter (from 0.1 to 1 mm). For example, if the sample concentration exceeds the column's flow rate capacity, i.e., if the internal diameter is too small, then resolution and reproducibility will suffer, along with possible peak distortions. Similar to the stationary phase for packed columns, the film type and thickness for capillary columns determine the separation characteristics. Finally, the column length increases the resolution at the expense of increased analysis time.

To briefly theorize the preceding qualitative descriptions of GC operation, the degree of separation of a given species, or *solute*, from the mobile phase to the stationary phase (liquid or solid medium) can be defined by the distribution constant K_c:

$$K_c = \frac{\text{mass of solute per unit volume of stationary phase}}{\text{mass of solute per unit volume of mobile phase}}$$
$$= \frac{\text{mass of solute in the stationary phase}}{\text{mass of solute in the mobile phase}} \times \frac{\text{volume of mobile phase}}{\text{volume of stationary phase}} \tag{7.10}$$
$$= k\beta$$

The mobile phase is the carrier gas plus all the solutes traveling through the GC column, so that a portion of the mobile phase, various solutes, can either adsorb to the solid stationary phase or dissolve in the liquid stationary phase. The relative mass concentration in each phase is described by the distribution constant. In Equation 7.10, k is referred to as the *retention factor*, whereas β is the ratio of the mobile to stationary volume and is called the *phase ratio*.

For a solute with a large K_c, the transit time (i.e., the retention time) will be large. In fact, the retention time is directly proportional to K_c, and as noted earlier, the retention time identifies a particular solute. For a finite length of the column, there will be a finite transit time for an inert gas such as the carrier gas, and therefore it is convenient to measure the solute retention time, t_R, with respect to the transit time of an inert species, t_M (see Figure 7.9b).

$$t'_{R,i} = t_{R,i} - t_M \tag{7.11}$$

The left-hand side of Equation 7.11, $t'_{R,i}$, is called the *adjusted retention time*.

The separation efficiency of a GC column is related to the ratio of the retention time to the width of the peak. If we use the standard deviation, σ, as a measure of the peak width, then the separation efficiency can be defined as

$$N = \left(\frac{t_R}{\sigma}\right)^2 \tag{7.12}$$

From the classical GC nomenclature, N is referred to as the *number of theoretical plates*, based on the notion that a given solute with a distribution constant, K_c, will occupy a hypothetical plate on the GC column. Further, if we assume the peak is Gaussian, then the standard deviation is related to the peak width at half peak height (w_h), by $w_h = 2.354\sigma$, or to the baseline peak width (w_b), by $w_b = 4\sigma$ (see Figure 7.9b).

Thus, Equation 7.12 can be rewritten as

$$N = \left(\frac{2.35 t_R}{w_h}\right)^2 = 5.54\left(\frac{t_R}{w_h}\right)^2 = 16\left(\frac{t_R}{w_b}\right)^2 \tag{7.13}$$

For closely spaced peaks, a degree of separation is needed for peak resolution. A measure of the separation of peaks n and $n+1$ is the separation factor, α:

$$\alpha = \frac{t'_{R,n+1}}{t'_{R,n}} = \frac{K_{c,n+1}}{K_{c,n}}, \tag{7.14}$$

and the resolution, R_S, to separate the peaks is quantified as

$$R_S = \frac{\left(t_{R,n+1} - t_{R,n}\right)}{\frac{\left(w_{b,n} + w_{b,n+1}\right)}{2}} = \frac{2\left(t_{R,n+1} - t_{R,n}\right)}{w_{b,n} + w_{b,n+1}} \tag{7.15}$$

Combining Equations 7.10, 12, 14, and 15, we obtain

$$R_S = \frac{1}{4}\sqrt{N}\,\frac{\alpha - 1}{\alpha}\,\frac{k_{n+1}}{k_{n+1} + 1}. \tag{7.16}$$

Typically, an R_S of 1.5 or greater is sought, at which point there will be a 99.7% or greater separation of baseline separations from two equal-area, symmetrical peaks. From Equation 7.16, the resolution is proportional to the square root of the number of theoretical plates, N, which in turn is proportional to the length of the GC column for a fixed carrier gas speed. Thus, to increase the resolution by a factor of two, a fourfold increase in the GC column length is required for a fixed carrier gas speed. Secondly, the resolution is determined by both the separation factor, α, and the retention factor, k_{n+1}. For example, packed columns are associated with a relatively small number of theoretical plates, N, and also a small α. Thus, to increase selectivity and resolution, R_S, one may opt for a capillary column with a larger α. However, due to a large phase ratio (β) in capillary columns, the retention factors (k_n) are typically small, thus offsetting the gains made by an increase in α. This is one of the main reasons for the very large number of GC column types and materials available to offer optimum combinations of separation and retention factors for a given set of analytes.

An example of a gas chromatogram is shown in Figure 7.12. Hydrocarbon components in a jet fuel sample were analyzed using GC. The horizontal axis represents the

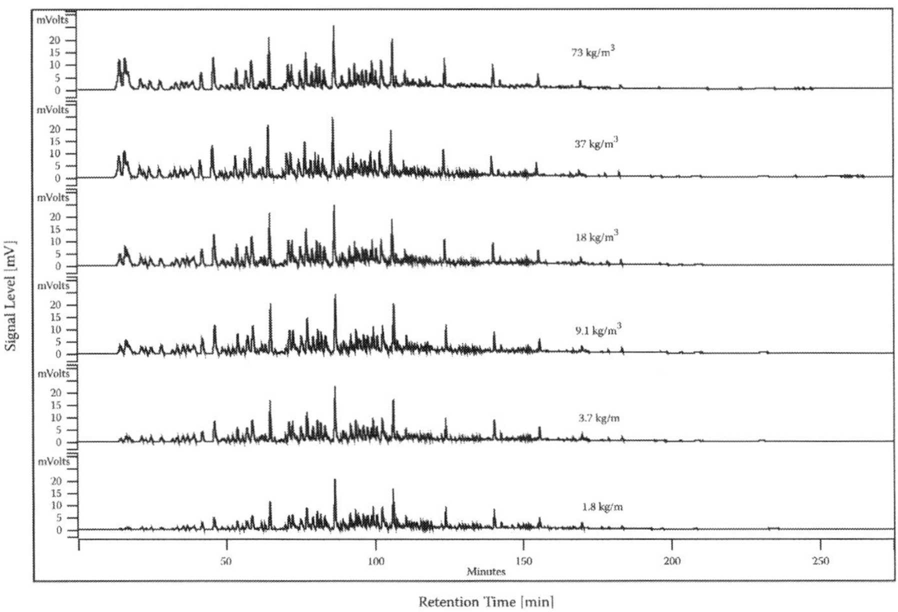

FIGURE 7.12 A gas chromatogram of hydrocarbon components in jet fuel.

retention time, so the peak location on this axis can be matched with known locations for different hydrocarbon species. To determine the exact locations of the peaks, a certified standard gas mixture consisting of equal parts of methane, ethane, propane, butane, and pentane, along with nitrogen, is run through the GC. For heavier hydrocarbons, a sample mix containing C_5 through C_{44} (ASTM D2887, Qualitative Calibration Solution, Inc.) is used. The vertical axis, in mV, is the detector response proportional to the amount of a particular hydrocarbon species. A capillary column with a 0.25-mm inner diameter and 100-m-length capillary column was used with an interior coating of nonpolar bonded methyl silicone at 0.5-mm film thickness. Carrier gas through the capillary column had a nominally 20 cm/min (linear velocity), and the maximum column temperature was 250°C. An FID was used for hydrocarbon species detection, whereas a thermal conductivity detector was used for measurements of air (for fuel-air ratio). There are six GC plots shown in Figure 7.12, each with a different fuel loading (the mass of liquid fuel divided by the total volume of the fuel chamber). It can be observed that at higher fuel loading, a larger vapor mass of all the hydrocarbon species is observed accentuated by higher amounts of the volatile hydrocarbons, i.e., those with smaller molecular weights observed at shorter retention times. As noted earlier, the association of the peaks at a given retention time with specific hydrocarbon species is made by running the calibration gases through the GC. Table 7.4 shows the identification of the peaks made in this manner. Due to the large number of peaks, however, several peaks remain unidentified in Table 7.4.

TABLE 7.4
A Listing of the Identified Peaks in the Gas Chromatogram Shown in Figure 7.12[a]

Peak Number	Retention Time (min)	Component Name[b]	Weight (%)	Mole (%)
1	11.00	Air	90.2913	97.3946
2	13.18	Propane, 2,2-dimethyl-	0.3518	0.1521
3	15.07	2-Propanol	0.1952	0.1009
4	16.07	2-Pentene, (E)-	0.1840	0.0817
5	17.90	C5?	0.0020	0.0016
6	19.25	2,3-Pentadiene	0.0011	0.0000
7	20.33	2-Pentene, 4-methyl-, (E)-	0.1239	0.0464
8	21.75	C5?	0.0542	0.0240
9	23.56	Methane, trichloro-	0.0988	0.0192
10	27.01	2-Butanol, 2-methyl-	0.0815	0.0288
11	30.78	1-Hexene, 3-methyl-	0.0383	0.0128
12	32.47	3-Hexene, 2-methyl-, (E)-	0.0830	0.0256
13	34.45	C6?	0.0956	0.0352
14	36.04	C6?	0.0796	0.0288
15	37.53	Cyclopentane, 1,3-dimethyl-, trans-	0.0512	0.0160
16	41.28	3-Heptene, (Z)-	0.1672	0.0528
17	45.48	Hexane, 2,2-dimethyl-	0.3048	0.0833

(Continued)

TABLE 7.4 (*Continued*)
A Listing of the Identified Peaks in the Gas Chromatogram Shown in Figure 7.12[a]

Peak Number	Retention Time (min)	Component Name[b]	Weight (%)	Mole (%)
18	48.12	Hexane, 2,4-dimethyl-	0.0431	0.0112
19	48.71	C7?	0.0250	0.0080
20	50.12	Cyclopentane,1,2,3-trimethyl-, (1a,2a,3b)-	0.0443	0.0128
21	51.76	C7?	0.0531	0.0160
22	53.31	C7?	0.1712	0.0528
23	55.10	2-Hexene, 2,5-dimethyl-	0.0456	0.0128
24	56.67	C8 Olefin?	0.2057	0.0576
25	58.45	Heptane, 3-methylene-	0.2927	0.0817
26	60.05	Cyclopentane, 1-ethyl-3-methyl-, trans-	0.0153	0.0048
27	60.96	1-Octene	0.0313	0.0080
28	61.66	C7?	0.0788	0.0240
29	62.83	C7?	0.0698	0.0224
30	64.57	3-Heptene, 2,2-dimethyl-, (E)-	0.3489	0.0865
31	66.06	C8?	0.0157	0.0048
32	67.44	Cyclohexene, 1,4-dimethyl-	0.0039	0.0016
33	68.33	Cyclopentane, propyl-	0.0196	0.0048
34	69.76	Cyclohexane, 1,1,4-trimethyl-	0.0513	0.0128
35	71.02	C9 Naphthene 4?	0.2446	0.0609
36	72.26	Hexane, 2,3,4-trimethyl-	0.1529	0.0368
37	72.69	C9 Olefin 2?	0.0659	0.0160
38	73.56	Benzene, 1,3-dimethyl-	0.0505	0.0144
39	74.97	Heptane, 3,4-dimethyl-	0.1357	0.0336
40	75.55	Heptane, 4-ethyl-	0.0732	0.0176
41	77.08	C8?	0.3205	0.0881
42	78.12	Pentane, 3,3-diethyl-	0.0427	0.0096
43	79.15	Ethane, 1,1,2,2-tetrachloro-	0.1581	0.0288
44	79.83	C8?	0.0204	0.0048
45	80.72	Cyclohexane, 1-ethyl-4-methyl-, trans-	0.1879	0.0464
46	81.79	Trimethylhexane(2)?	0.1678	0.0400
47	83.28	C8?	0.1464	0.0400
48	83.73	4-Nonene, (E)-	0.0571	0.0144
49	84.16	1,3-Cyclooctadiene	0.0115	0.0032
50	84.51	3-Nonene (Z)-	0.0222	0.0048
51	85.77	C9?	0.0094	0.0016
52	86.81	C9?	0.5441	0.1313
53	87.88	C9?	0.0050	0.0016
54	88.72	C9 Olefin 3?	0.0278	0.0064
55	89.50	C9?	0.0736	0.0176
56	89.79	5,7-Dimethyl-1–6-octadiene	0.0423	0.0096
57	90.32	Heptane, 2,4,5-trimethyl-	0.0340	0.0080

(*Continued*)

TABLE 7.4 (*Continued*)
A Listing of the Identified Peaks in the Gas Chromatogram Shown in Figure 7.12[a]

Peak Number	Retention Time (min)	Component Name[b]	Weight (%)	Mole (%)
58	90.66	1,5-Cyclooctadiene	0.0274	0.0080
59	92.01	Octane, 2,6-dimethyl-	0.1335	0.0288
60	92.76	C9?	0.0660	0.0160
61	93.80	Octane, 3,3-dimethyl-	0.1534	0.0336
62	94.75	C9?	0.1224	0.0304
63	95.24	C9?	0.0814	0.0192
64	96.35	Benzene, 1-ethyl-4-methyl-	0.1365	0.0352
65	96.69	C9?	0.0761	0.0192
66	97.69	Octane, 2,3-dimethyl-	0.1634	0.0352
67	98.28	C9?	0.0168	0.0048
68	98.79	Nonane, 5-methyl-	0.0449	0.0096
69	99.19	Heptane, 3,3,4-trimethyl-	0.0825	0.0176
70	99.68	Nonane, 2-methyl-	0.1477	0.0320
71	100.27	Octane, 3-ethyl-	0.0432	0.0096
72	101.01	Cyclohexane,1-methyl-4-(1-methylethyl)-	0.1301	0.0288
73	101.52	Benzene, (1,1-dimethylethyl)-	0.0355	0.0080
74	102.01	C10 Paraffin 3?	0.0083	0.0016
75	102.80	C10 Paraffin 6?	0.2137	0.0464
76	103.24	C9?	0.0267	0.0064
77	103.47	1-Decene	0.0736	0.0160
78	104.24	Benzene, (2-methylpropyl)-	0.0421	0.0096
79	105.03	Decane	0.0155	0.0032
80	105.43	C11 Paraffin 1?	0.0323	0.0064
81	106.64	Benzene, 1-methyl-3-(1-methylethyl)-	0.4807	0.1121
82	107.38	C10?	0.0014	0.0000
83	107.85	1H-Indene, 2,3-dihydro-{Indan}	0.1203	0.0320
84	108.70	Cyclohexane, (1-methylpropyl)-	0.0315	0.0064
85	109.85	3-Ethylnonane	0.0380	0.0080
86	110.29	C10?	0.0298	0.0064
87	110.89	C10?	0.1280	0.0272
88	111.45	C10?	0.0133	0.0032
89	111.90	Benzene, 1,3-diethyl-	0.0851	0.0192
90	112.63	Benzene, 1,4-diethyl-	0.0313	0.0080
91	112.92	Benzene, 1-methyl-4-proply-	0.0481	0.0112
92	113.53	Benzene, 1-ethyl-3,5-dimethyl-	0.0969	0.0224
93	114.36	Naphthalene, decahydro-, trans-	0.0478	0.0112
94	114.67	C10?	0.0703	0.0160
95	115.38	C11 Paraffin 2?	0.0087	0.0016
96	115.93	Cyclopentane, 1,2-dimethyl-4-propyl-	0.0557	0.0128
97	116.21	C11 Paraffin 3?	0.0442	0.0080

(*Continued*)

TABLE 7.4 (*Continued*)
A Listing of the Identified Peaks in the Gas Chromatogram Shown in Figure 7.12[a]

Peak Number	Retention Time (min)	Component Name[b]	Weight (%)	Mole (%)
98	116.87	Benzene, 2-ethyl-1,4-dimethyl-	0.0444	0.0096
99	117.42	C11 Paraffin 4?	0.0316	0.0064
100	118.06	Benzene, 4-ethyl-1,2-dimethyl-	0.0764	0.0176
101	118.38	Tricyclo[3.3.1.13,7]decane	0.0494	0.0112
102	119.22	Benzene, 2-ethyl-1,3-dimethyl-	0.0333	0.0080
103	119.38	C10?	0.0575	0.0128
104	120.06	C10?	0.0149	0.0032
105	120.41	C10?	0.0348	0.0080
106	120.85	Benzene,1-(1,1-dimethylethyl)-4-methyl-	0.0087	0.0016
107	121.12	1-Undecene	0.0079	0.0016
108	121.56	Benzene, 1-ethyl-2,3-dimethyl-	0.0366	0.0080
109	122.15	C10?	0.0214	0.0048
110	122.90	Butane, 2-cyclohexyl-2-methyl-	0.0665	0.0128

[a] Vapor sample taken at 115°C, 1 atm, and a fuel loading of 73 kg/m^3.

[b] Properties of unidentified peaks (CX?) were estimated with nearest *n*-alkane properties.

7.8 MASS SPECTROSCOPY

Mass spectroscopy (MS) is based upon the differential deflection of positively charged ions with different masses in an electromagnetic field to identify the mass, and therefore the identity, of the sample constituents. In optical spectroscopy, light can be dispersed depending on its wavelength content; in MS, the differential deflection of ions with different mass per electrical charge is used for spectral decomposition. As shown in Figure 7.13, first the sample needs to be ionized in MS and then accelerated into and through a so-called flight tube in which the ionized species are separated in an electromagnetic field by their mass. Finally, the separated ions are detected using ion detectors, as shown in Figure 7.13. Because ions and electrons used in MS need to have a sufficiently large mean free path to prevent ion collisions with extraneous molecules (these would lead to a loss or an erroneous signal for that molecule), MS systems are operated in vacuum (10^{-4} to 10^{-6} Torr).

There are several methods to ionize the sample: electron bombardment, high-intensity laser, or electrical discharge. A hot filament in a vacuum condition is used to generate electrons, which are then accelerated through an electric field. These electrons collide with sample molecules with sufficient kinetic energy (typically around 70 kV) to knock out the outermost electrons and create singly charged positive ions. The energy of photons from a high-intensity laser focused to a small spot can also ionize molecules. Electrical arcs are a stream of electrons between two electrodes, and they can be used to generate ions from the sample, although an electron beam

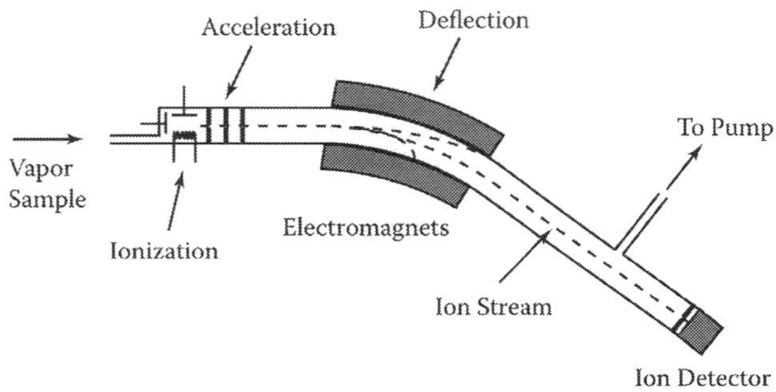

FIGURE 7.13 A schematic of a mass spectrometer.

augmentation is needed to achieve steady-state ionization. Electrical arcs are usually highly transient phenomena.

The ionized molecules with positive charges are then accelerated to a uniform kinetic energy through yet another electric field (typically in the order of 10,000 V) and then sent to the flight tube, where an electromagnetic field is applied to deflect the ion paths by amounts corresponding to their charge-to-mass ratio. For example, for an ion subject to a uniform electric field of strength E and length x the vertical displacement is

$$z = \frac{1}{2}\frac{Eq}{m}\left(\frac{x}{v}\right)^2 \tag{7.17}$$

E = electric field strength
q = charge of the ion
m = mass of the ion
v = velocity of the ion in the x-direction

Similarly, a magnetic field of strength B will cause a lateral deflection of

$$y = \frac{1}{2}\frac{Bqx^2}{mv} \tag{7.18}$$

B = magnetic field strength

By combining the preceding two equations, we can see that

$$y^2 = kz \tag{7.19}$$

where $k\left(=\frac{1}{2}\frac{q/m}{E/B}x^2\right)$ is a constant depending on the ion charge-to-mass ratio (q/m), applied field strengths, and geometry of the apparatus (x). Thus, for fixed applied field strengths and geometry, the ion trajectory will only depend on the charge-to-mass ratio. In MS, a magnetic field of varying strength can be used to separate ions with different q/m. Thus, the magnetic field strength as a function of time directs ion streams with a given q/m at a given time toward the detector, and the rest of the sample is evacuated using a steady-state vacuum pump. This sequential separation gives the ability to detect various molecular masses present in the sample.

Other methods of ion path separation are time-of-flight, quadrupole, and electrostatic. All of these methods are based on the differential path of ions subject to an electromagnetic field in either lateral or longitudinal directions and the ion charge-to-mass ratio.

The ion detector is based on the conversion of the ion kinetic energy to electrical energy, i.e., the generation of electrons at the detector surface. An ion detector using a microchannel plate is shown in Figure 7.14. Similar to the microchannel plates used in image intensifiers, these microchannel plates are an array of tubes electrically biased at approximately −2.2 kV. The collision of an ion at the top of the microchannel plate surface knocks one or more electrons from that surface. As these electrons bounce along the interior of the microchannel plates, multiple electrons are produced for each electron colliding with the wall. Figure 7.14 shows a schematic of an ion detector with two such microchannel plates and a funnel-shaped anode to capture the electrons released through the microchannel plates. This electron multiplication amplifies the signal generated by the incident ions, and the stream of electrons constitutes a voltage bias that is proportional to the number of ion collisions.

An example of a mass spectrum is shown in Figure 7.15 where the horizontal axis is typically in units of *m*/*q* (sometimes written as *m*/*z*, where *z* substitutes for the symbol for the charge, *q*). A pentane molecule (C_5H_{12}) is analyzed in this example. It can be observed that the mass spectrum of a molecule, particularly large ones, is not so simple to interpret, as it contains several prominent peaks along with minor satellite peaks. The peak at the largest *m*/*q* (72) is due to $C_5H_{12}^+$ ions. However, it can be observed that this is not the tallest peak in the spectrum. In fact, there are several other, more prominent peaks. This is because ions are unstable and tend to break up into molecular fragments. If the fragments are ions, they can also be accelerated, deflected, and detected in the MS. For neutral fragments, they simply become lost and are eventually evacuated by the vacuum pump from the MS. The peaks at $m/q = 57$ and 43, for example, are caused by the $C_4H_9^+$ and $C_3H_7^+$ fragments, respectively. The side products of these fragmentation reactions are methyl and ethyl radicals, which are neutral and not observed in the spectrum.

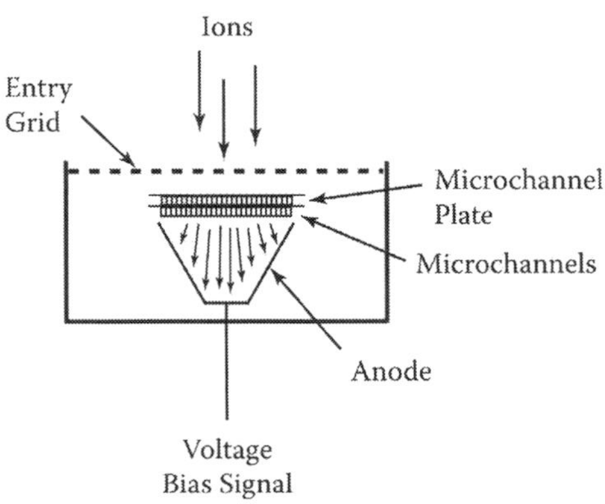

FIGURE 7.14 An ion detector based on the use of a microchannel plate.

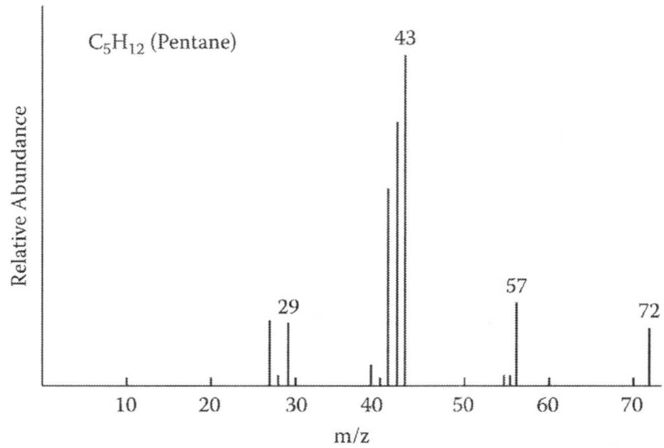

FIGURE 7.15 An example of a mass spectrum.

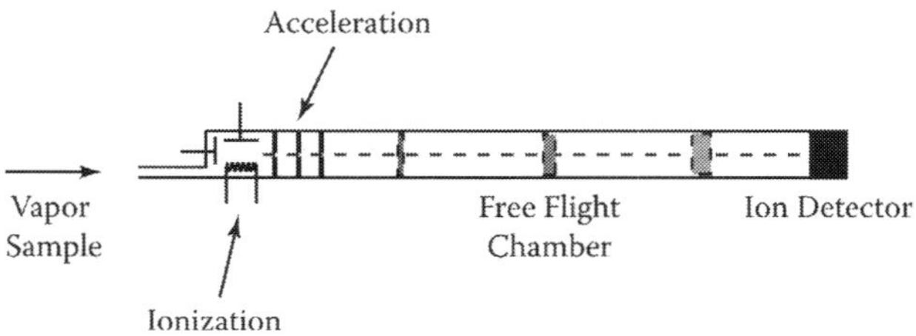

FIGURE 7.16 A schematic of a time-of-flight mass spectrometer.

The peaks immediately adjacent to $m/q=43$ are due to $C_3H_7^+$ isotopes. Similarly, a peak at $m/q=29$ is also observed for C_2H_5+ fragments. Thus, the interpretation of mass spectra is somewhat similar to FTIR in the sense that there are various contributing peaks, and some amount of post-analysis is needed to determine the exact origins of the peaks.

A time-of-flight MS operates on a slightly different principle. Instead of using a time-varying magnetic field to separate the components with different m/q, a time-of-flight MS uses a uniform electrical acceleration of the ions followed by a field-free flight tube. The idea is that, depending on the ratio m/q, the ions will be accelerated to different speeds so that those arriving at the ion detector will be separated by different times of flight. A schematic of the time-of-flight MS is shown in Figure 7.16. To increase the resolution, a large path-length flight tube is desirable, in which case an "ion mirror" is used to reflect the ions to make a double pass within a compact volume. The ion mirror is a positively charged plate that repels the ions. The ion detector in time-of-flight MS needs to have a high temporal resolution to discern the arrival times of different molecular ions.

PROBLEMS

7.1. In a nondispersive infrared analyzer for carbon dioxide, an attenuation of 15% of the incident infrared light is observed in a reference cell of length 0.25 m containing 150 ppm of carbon dioxide by volume. If an attenuation of 1.75% is detected in the sample cell (also 0.25 m in length), what is the carbon dioxide amount in ppm?

7.2. For a nondispersive infrared analyzer, plot the light attenuation as a function of the sample gas mole fraction for $f(\lambda,T,p)$ 2.5 m^{-1} and $L=0.35$ m.

7.3. For gas chromatography, compare the separation efficiency (or the number of theoretical plates, N, given by Equation 7.12) of (a) the maximum retention time, $t_R=1$ hr; baseline width, $w_b=15$ sec, and (b) $t_R=2.5$ hr; baseline width, $w_b=25$ sec.

7.4. In a gas chromatogram, two adjacent peaks have the retention times of 0.35 and 0.355 hr, while their base widths are 15 and 25 sec. What is the resolution for this set of peaks?

7.5. Show that the separation, R_S, in a gas chromatogram is given by Equation 7.16.

$$R_S = \frac{1}{4}\sqrt{N}\,\frac{a-1}{a}\,\frac{k_{n+1}}{k_{n+1}+1} \tag{7.16}$$

7.6. For a target analyte with a retention factor, k, of 0.025 and a separation factor of 1.127, what should be the number of theoretical plates, N, to achieve a separation of 1.5 or greater? Also, for this N, if the baseline width is 17 sec, what is the required retention time?

7.7. Plot the y and z coordinates of ion motion in a mass spectrometer as a function of the longitudinal distance, x. The electric field of 500 V is applied, while the magnetic field strength is 75 A/m. The ratio of the charge to mass of the ion is 1.760605×10^7 coulombs/kg, with the ion speed of 127 m/s. Also, how would the above charge-to-mass ratio convert to a dimensionless ratio, m/z, used in mass spectrometry?

BIBLIOGRAPHY

Grob, R. L., Ed., *Modern Practice of Gas Chromatography*, 3rd ed., Wiley Interscience, New York, 1995.

Okabe, H., Fluorescence and predissociation of sulfur dioxide, *Journal of the American Chemical Society*, 93(25), p. 7095, 1971.

Turns, S., *Lecture Notes: Species Measurements in Combustion Systems*, 1991.

8 Scaled Measurements

From Microscale Sensors to Atmospheric Diagnostics

8.1 INTRODUCTION

Thus far, we have discussed different approaches to measuring thermal and flow parameters. Nowadays, with the advent of nanotechnology and the importance of environmental effects, the requirements for obtaining thermal and flow characteristics extend far beyond the classical scales of measurement. Micro-electro-mechanical systems (MEMS) include the so-called microfluidic devices, for which thermal and flow characterization is important in design and applications. Large-scale processes, such as atmospheric circulations and transport, require diagnostic methods and data acquisition at vastly different scales. Micro- and nanoscale devices, in turn, have opened up a new discipline for sensor development and characterization. In this chapter, we will examine how some measurement methods can be scaled up or down for thermal and fluid measurements, and how emerging small-scale technology enables new or improved measurements.

8.2 MEMS DEVICES

MEMS refers to a wide class of devices that integrate mechanical and electronic functions on micron-scaled, solid-state platforms. They have the advantages of being amenable to mass production using semiconductor fabrication methods, reproducibility of electromechanical characteristics, direct interfacing of mechanical and electronic components at the chip level, and microscale dimensions requiring small sample volumes and minimal flow intrusion. However, as most semiconductor fabrication methods are surface processes, the construction of complex three-dimensional features can be a challenge, and precise control of fluids at microscales is sometimes difficult to achieve. Nonetheless, many measurement devices at "classical" scales can be miniaturized (along with some new devices feasible only at microscales) and directly interfaced with semiconductor circuitry. New functionality or improved performance of existing technology can be achieved using microfabrication methods. For these reasons, there is an increasing level of commercial and research applications of MEMS devices for thermal and flow process monitoring and control. Pressure sensors and accelerometers (in automobiles) are good examples of successful microsensor applications.

DOI: 10.1201/9781003611837-8

8.2.1 Microfabrication Methods

To begin, let us discuss some of the principal methods of microfabrication on a silicon-based platform. There are a large number of fabrication methods for semiconductor integrated circuit devices; however, for our purposes, we only need to understand some of the major categories of microfabrication methods for sensors and related devices. These are (1) photolithography (or patterning), (2) etching, (3) film deposition, and (4) doping. In addition, we may consider (5) bonding and (6) laser machining as auxiliary processes. Silicon is the most common platform or substrate material for these fabrication methods because of its favorable electrical and mechanical properties. Single-crystal wafers of silicon are readily available in diameters from 100 to 300 mm. Other substrate materials, such as glass, ceramics, or polymers, may also be used depending on the application.

The photolithographic process, shown in Figure 8.1, is the method used to create patterns on the substrate. A thin film of photoresist material is deposited on the substrate. Then, a mask with the desired pattern, appearing as opaque lines or other shapes, is aligned with the substrate. The mask material is typically glass with opaque features created using a chromium layer, so that the glass transmits light, while the chromium layer is opaque. The substrate is then exposed to ultraviolet light with the mask in place, resulting in the unmasked portions being exposed to the ultraviolet light. The photoresist material, once exposed to ultraviolet light, undergoes a chemical transformation, making it susceptible to etching by development chemicals. In the next step, the exposed photoresist material is removed through the development chemical process. This sequence corresponds to positive photoresist material, whereas for negative photoresist, the exposed area becomes immune to development

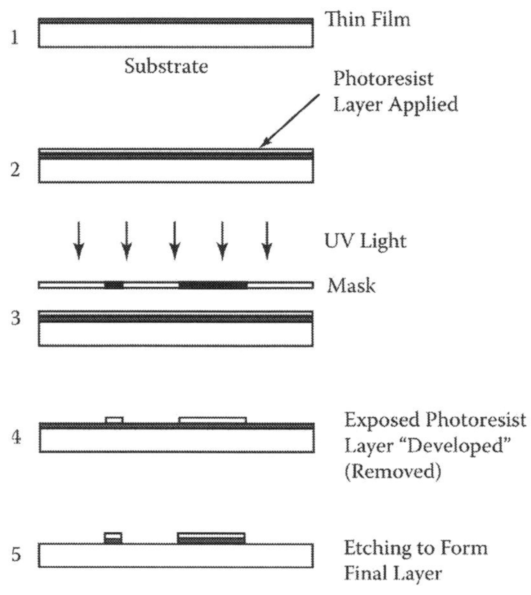

FIGURE 8.1 Photolithographic process to create a pattern on a substrate.

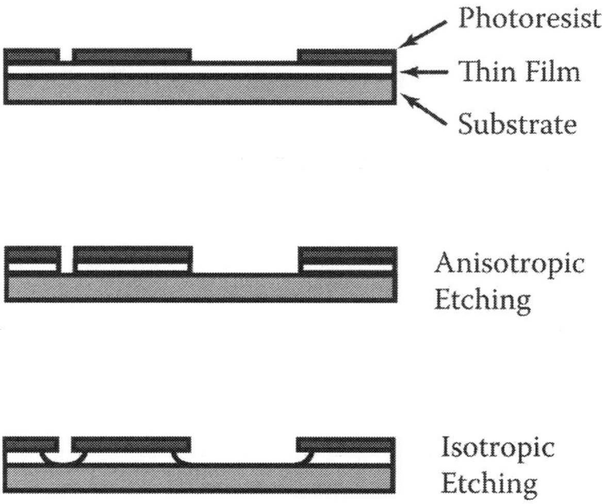

FIGURE 8.2 Anisotropic and isotropic etching.

chemicals, and it is the unexposed area that is removed during the development process. Through these steps, a patterned layer of photoresist material is created on the substrate, which acts as a mask to protect the substrate directly underneath it from etching or other processes. Once the etching process is completed, the photoresist material itself can be removed using a set of chemicals for the next photolithographic sequence.

Etching is a chemical process, such as the acidic dissolution of materials, used to remove material from the substrate. Figure 8.2 shows two different types of etching processes: isotropic and anisotropic. Typically, etching with liquid chemicals (wet etching) tends to be isotropic; however, gas-phase chemical reactions can also be used for etching, and they tend to be anisotropic. An example of a wet etchant is HNA, which is a combination of hydrofluoric acid (HF), nitric acid (HNO_3), and acetic acid (CH_3COOH). HNA dissolves silicon (Si) substrate material through the following chemical reaction:

$$18\ HF + 4\ HNO_3 + 3\ Si \rightarrow 2\ H_2SiF_6 + 4\ NO\ (gas) + 8\ H_2O \tag{8.1}$$

Silicon dioxide is commonly removed using diluted hydrofluoric acid (HF diluted with H_2O) or BOE (buffered oxide etch), which is a mixture of HF with NH_4F. The typical etch rate is in the order of 0.1 μm/min.

An example of dry etching is reactive ion etching (RIE), during which ions are generated in plasma and electromagnetically accelerated toward the substrate. Depending on the direction of the accelerated ions, the highest ion kinetic energy will be in that direction, leading to preferential, anisotropic removal of the substrate material. Anisotropic etching is important because it avoids mask undercutting and allows high-aspect-ratio features to be formed. There are also a large number of dry etching chemical reactions based on fluorine and chlorine. Table 8.1 shows various etchants and their characteristics.

TABLE 8.1
Various Etchants and Their Characteristics

	HNA[a]	KOH[b]	TMAH[c]	XeF_2	Deep-RIE[d]
Etch type	Wet	Wet	Wet	Dry	Dry
Directionality	Isotropic	Anisotropic	Anisotropic	Isotropic	Anisotropic
Etch rate for Si [μm/min]	3 at 25°C	2.5 at 95°C	0.6 at 80°C	1	2–3
Etch rate for SiO_2 [nm/min]	30–70 at 25°C	15 at 25°C	0.1 at 25°C	0.1	10–20

[a] HNA = mixture of HF, HNO_3, and CH_3COOH.
[b] KOH = potassium hydroxide.
[c] TMAH = tetramethyl ammonium hydroxide.
[d] dRIE = reactive ion etching.

Etching represents a removal process, and the addition of materials is accomplished by various film deposition processes. The two most common deposition methods are chemical vapor deposition (CVD) and physical vapor deposition (PVD). As the names suggest, CVD represents a chemical process during which gas-phase and surface reactions result in solid material being formed on the substrate. In contrast, PVD includes physical processes such as sputtering (using electron beams on a target material to produce particles or vapors of the material, for example) and evaporation. These processes deposit various reactive, sputtered, or evaporated materials onto the substrate, with resulting film thicknesses ranging from a few nanometers to a few hundred micrometers. An example of a surface chemical reaction is the oxidation of silicon. A silicon dioxide layer can be grown on the substrate using a high-temperature (900°C–1200°C) oxidation reaction in either an oxygen or steam environment:

$$\text{Oxygen ambient: } Si + O_2 \rightarrow SiO_2 \tag{8.2}$$

$$\text{Steam ambient: } Si + 2\ H_2O \rightarrow SiO_2 + 2H_2$$

Active vapor-phase chemical reactions are also used for SiO_2 deposition:

$$SiH_4 + O_2 \rightarrow SiO_2 + 2\ H_2 \tag{8.3}$$

Although metal materials are usually deposited through PVD, CVD processes are also available for depositing metals and metal compounds such as Cu, W, WSi_2, $TiSi_2$, and TiN:

$$2\ WF_6 + 3\ Si\ (\text{substrate}) \rightarrow 2\ W + 3\ SiF_4 \tag{8.4}$$

Metals are used for a number of applications on MEMS devices, such as electrical interconnects, electrodes, or reflective surfaces (for optical devices). The most common way to deposit metals or metal compounds is through PVD, such as sputtering or electron-beam evaporation. These processes involve metal elements being sputtered or evaporated from the target surfaces and then induced to migrate to the substrate.

Doping is primarily a process used to fabricate semiconductor devices, such as diodes and transistors, by modifying the electrical conductivity of silicon, gallium arsenide, or other electronic materials. For silicon, doping with phosphorus or arsenic produces an n-type semiconductor, whereas boron doping results in a p-type. These dopant atoms can be diffused into the substrate using gas, liquid, or solid sources. Ion implantation is a primary method to introduce a controlled concentration of dopant atoms into the substrate. Ionized dopant atoms are accelerated toward the substrate electromagnetically, with the mean penetration depth depending on the kinetic energy of the accelerated ions. High-temperature diffusion is also used following ion implantation to enhance the dopant concentration profile in the substrate.

Although two adjacent layers can be fabricated by sequential deposition, an alternative method is to directly bond two layers of different materials or substrates with another substrate. Direct bonding techniques, such as silicon fusion bonding and anodic bonding, are available. In silicon fusion bonding, two silicon wafers are heated to temperatures exceeding 1000°C until the two surfaces merge at the interface. Anodic bonding uses an electric field applied across the two surfaces, so that resistive heating causes bonding of the substrates. Laser machining is another method to form shapes on substrates by focusing laser energy onto a small spot for drilling, cutting, or bonding.

Using combinations and sequences of these fabrication techniques, both stationary and mobile structures, such as cantilevers, bridges, membranes, switches, actuators, and even gears and reactors, can be built on the substrate. Figure 8.3 shows the sequence used to fabricate a cantilevered mass in an accelerometer device. Both sides of the silicon substrate, covered with a SiO_2 layer, are used. First, the bottom-side features are drawn with a photolithographic exposure, followed by anisotropic etching using KOH. The top-side photolithography and etching create the final form, which has a mass connected by a beam. The deflection of the beam due to the acceleration applied to the mass is detected using a piezoresistor layered onto the silicon beam. Figure 8.4 shows another example of micro-device fabrication. A microfluidic channel is fabricated using a so-called sacrificial layer, which is removed after being used to build the fluid channel layer. The poly-xylene layers at the top and bottom of the fluid channel are used because this material is known to be compatible with biological compounds. The polyimide then serves as the top "wall" of the fluid channel. The sacrificial layer is removed by creating paths through the top polyimide and poly-x-xylene layers via photolithography and plasma treatment through a patterned mask, respectively, and then using acetone to remove the sacrificial layer.

8.2.2 Microfluidic Sensors and Devices

As noted earlier, many of the classical measurement devices or methods discussed in previous chapters can be miniaturized using microfabrication techniques. The principle used in hot-wire anemometers and in thermal mass flowmeters can be implemented in MEMS devices. A combination of resistor elements can be used to detect the change in flow temperature, which is a function of the convection heat transfer rate and therefore a function of the flow speed. Thus, the resistance change can be related to the flow speed. A MEMS flow rate sensor that incorporates this effect sends

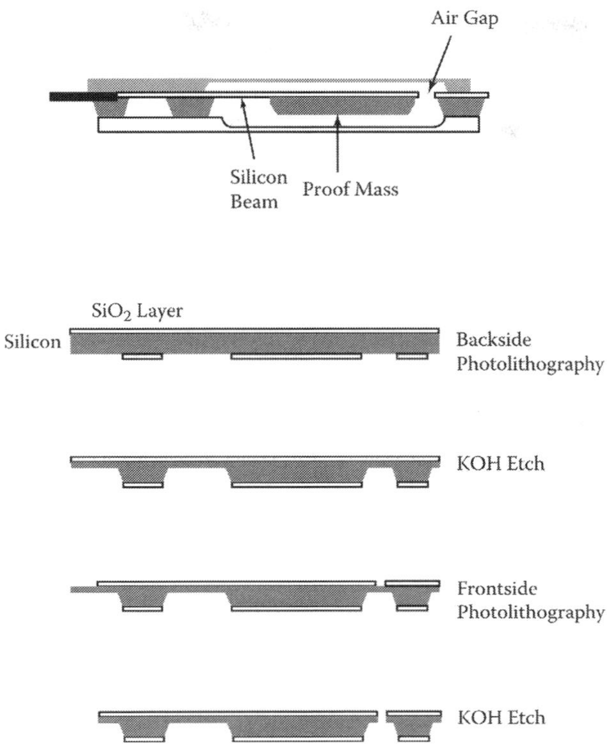

FIGURE 8.3 A sequence of microfabrication techniques for a cantilevered mass (used in an accelerometer). (Adapted from Kovacs, 1998.)

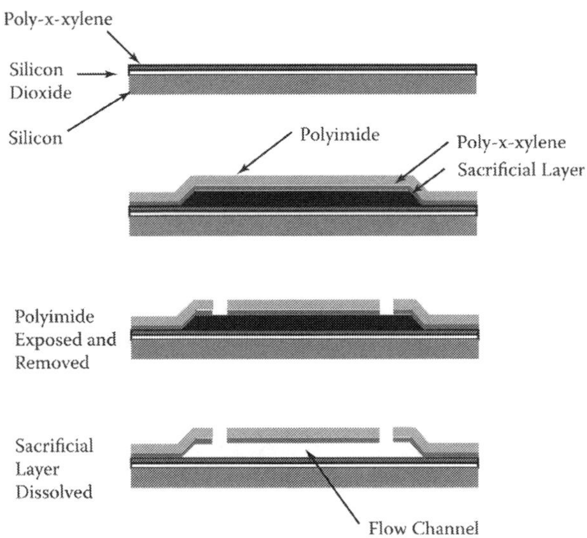

FIGURE 8.4 A micromachined fluidic channel.

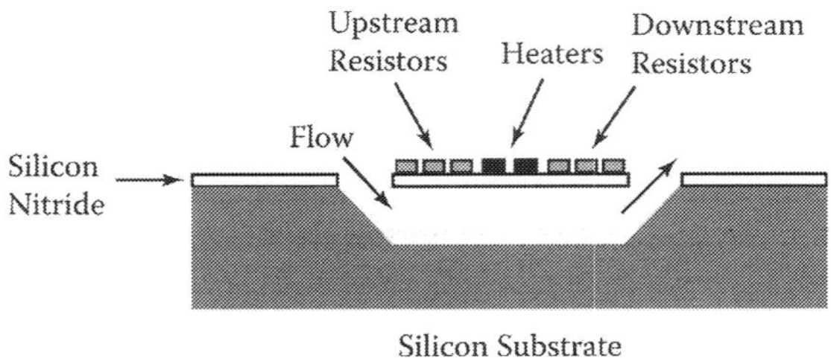

FIGURE 8.5 A MEMS flow sensor using the hot-wire anemometry principle.

liquid acetone through the paths to finally remove the sacrificial layer, as shown in Figure 8.5. The resistor elements are film-deposited platinum or permalloy (Ni-Fe alloy) on a silicon nitride surface. The flow is diverted into a microfluidic channel on the device, and the heating elements add thermal energy to the flow. The resultant change in temperature (a function of the flow speed) leads to a change in the resistivity of the resistor elements. The change in resistance is read using a Wheatstone bridge, similar to macroscale hot-wire anemometers, except that the circuit can be fabricated directly on the substrate.

A flow rate sensor based on the pressure drop across a microfluidic channel has also been suggested (see Oosterbroek et al., 1997). The idea is that the pressure drop across a flow channel depends on the flow velocity. For example, for laminar flow inside a round tube of radius R, the velocity profile is a function of the radius and the pressure gradient in the longitudinal direction, as given by

$$U(r) = -\frac{R^2}{4\mu}\left(\frac{dp}{dx}\right)\left(1-\frac{r^2}{R^2}\right) \tag{8.5}$$

r = radial distance from the centerline of the tube
μ = viscosity
dp/dx = pressure gradient in the longitudinal direction
R = radius of the flow tube

The negative sign means that the flow is in the positive direction for a negative pressure gradient, i.e., when pressure decreases in the direction of the flow. We can approximate the pressure gradient over a tube length L as $dp/dx = (p_2 - P_1)/L = \Delta p/L$. We can also integrate the velocity profile in Equation 8.5 across the tube cross-sectional area to find the volumetric flow rate, Q.

$$Q = \frac{\pi R^4}{8\mu L}\Delta p \tag{8.6}$$

From Equation 8.6, we can relate the pressure drop across a flow channel to the volumetric flow rate and, therefore, to the mean flow velocity. Microfabricated flow channels tend to have a rectangular cross-section, in which case Equation 8.6 is modified to

$$Q = \frac{wh^3}{12\mu L}\Delta p \tag{8.7}$$

w = channel width
h = channel height

For a rectangular channel with a low aspect ratio, an improved relationship is

$$Q = \frac{wh^3}{12\mu L}\left[1 - \frac{h}{w}\left(\frac{192}{\pi^5}\sum_{n=1}^{\infty}\frac{1}{n^5}\tanh\left(\frac{n\pi w}{h}\right)\right)\right]\Delta p \tag{8.8}$$

A microfluidic flow sensor based on the use of Equation 8.8 is shown in Figure 8.6.

In addition to flow sensors, many flow actuators can be fabricated, and these have uses in more complex microfluidic devices for sample handling and manipulation for downstream chemical or biochemical analyses (to be discussed in Section 8.2.4). A piezoelectric valve is shown in Figure 8.7. To improve the flexibility of the membrane and the seal between the substrates, a thin polymer is deposited to form the membrane that closes the gap in response to the activated piezoelectric stack. Such actuation of a membrane has also been used in microfluidic pump design. An interesting application of microfluidics is found in inkjet printers. An array of droplet generators fabricated on a glass substrate, as shown in Figure 8.8, forms the basis for patterned and sequential injection on scrolling paper for printing texts and graphics. Electrical signals are sent to a target array of thin-film resistors. Due to the small dimensions involved, the heating of the liquid ink on top of the resistor elements is rapid (on the order of 1 μs), and the heated ink vaporizes, pushing the liquid ink out

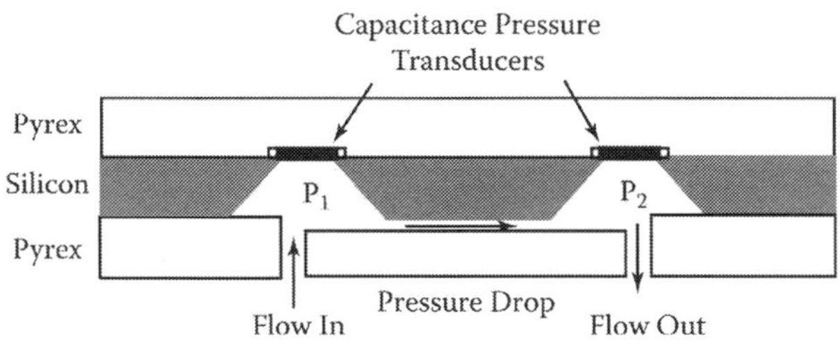

FIGURE 8.6 A flow rate sensor using the pressure drop across a microfluidic channel.

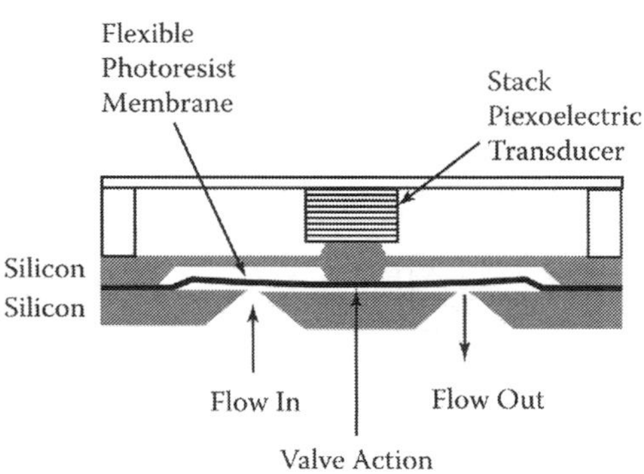

FIGURE 8.7 A piezoelectric valve using a polymer membrane.

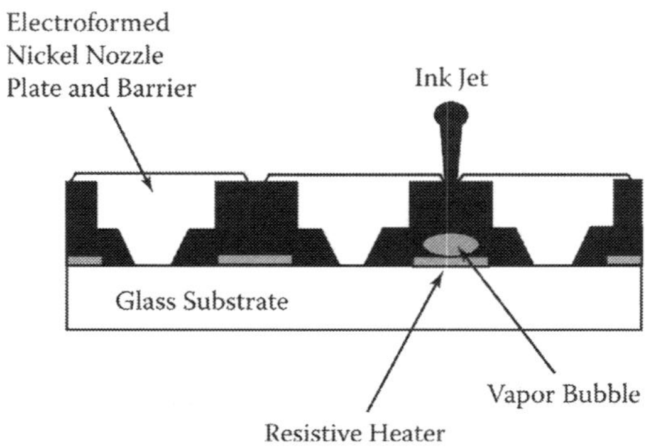

FIGURE 8.8 Microfabricated droplet generators used in inkjet printing devices.

of the nozzle, as illustrated in Figure 8.8. Turning off the electrical signal results in equally rapid cooling, which contracts the liquid ink, thereby drawing ink from the reservoir into the now-available volume above the resistor.

8.2.3 Pressure and Temperature Sensors

Pressure sensors represent a mature MEMS technology that offers improved performance and reduced size. Similar to the macroscale versions of pressure transducers, many MEMS pressure sensors rely on the measurement of strain on a microfabricated membrane. In Chapter 2, Section 2.3, the strain gauge based on resistance change was discussed. The resistance of a material depends on its resistivity (ρ_e), length (L), and cross-sectional area (A) of the element.

$$R = \frac{\rho_e L}{A} \tag{8.9}$$

We can differentiate the preceding equation and express the terms that contribute to the change in resistance.

$$\frac{dR}{R} = \frac{dL}{L}(1+2\nu)+\frac{d\rho_e}{\rho_e} \tag{8.10}$$

ν = Poisson's ratio (see Equation 2.40)

The first term on the right-hand side represents the geometrical effect, whereas the second term is referred to as the piezoresistive effect, i.e., the change in the fundamental resistivity due to strain on the material. Although metals in macroscopic strain gauges have a negligible piezoresistive effect, piezoresistive strain gauges are ubiquitously used in MEMS devices with large gauge factors. It may be recalled that the gauge factor (GF) represents the sensitivity of the strain gauge (Equation 2.47).

$$\text{Gauge Factor} = \text{GF} = \frac{dR/R}{dL/L} = 1+2\nu+\pi_e E \tag{8.11}$$

$$\pi_e = \frac{\frac{d\rho_e}{\rho_e}}{E\frac{dL}{L}}$$

$$E = \text{Young's modulus of elasticity}$$

Thin-film metals have a GF of about 2, whereas for diffused semiconductors, the gauge factors range from 80 to 200.

For a MEMS pressure gauge, a common design uses piezoresistive elements placed near the periphery of a diaphragm, as shown in Figure 8.9. In addition to the terms shown in Equation 8.11, temperature can also affect resistance (see Equation 2.38 in Chapter 2). Therefore, temperature-compensation circuitry is embedded in microfabricated pressure transducers. Microphones are essentially pressure transducers with a dynamic response, i.e., a high-frequency response. Piezoresistive microphones typically generate 25 μV/Pa output with a frequency response in the range of 100 Hz to 5 kHz. Using a capacitance-type device, both the sensitivity and frequency response can be improved to 0.2–25 mV/Pa and 10 Hz to 15 kHz, respectively. An example of a capacitance-based dynamic pressure transducer, i.e., a microphone, is shown in Figure 8.10.

Temperature sensors are conveniently fabricated on MEMS devices using the thermo-resistive effect.

$$R(T) = R_o\left[1 + \alpha(T - T_o)\right] \tag{8.12}$$

Where α is the temperature coefficient of resistance, and R_o is the reference resistance at the temperature T_o. Thin films of platinum, chromium, or other metals can

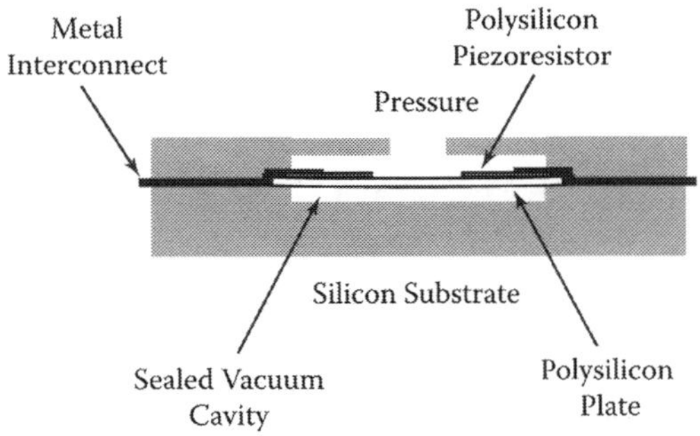

FIGURE 8.9 A microfabricated pressure transducer using piezoresistive elements.

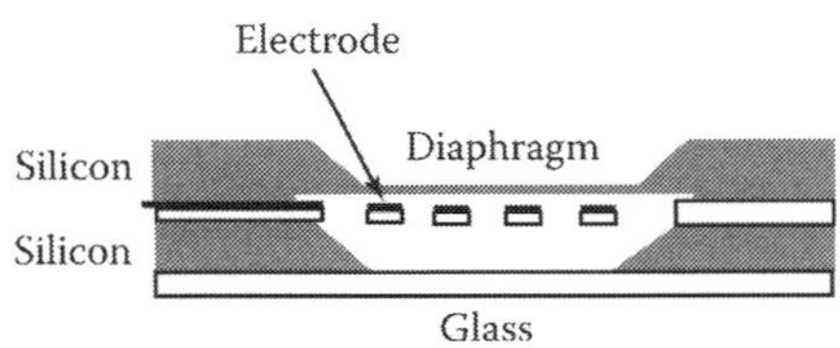

FIGURE 8.10 A capacitance-based microphone.

be used for highly accurate thermometers. In addition, thermistors made of metal oxides, selenides, or sulfides of Li, Cu, Co, Ti, Mn, Fe, or Ni can be used in microintegrated thermal sensors. Another obvious choice for a temperature sensor is based on the temperature dependence of the voltage across diode junctions. Typically, a −2.27 mV/K change in voltage occurs due to a number of effects such as carrier density and diffusion coefficients. This voltage across diode junctions can easily be measured and calibrated to yield temperature, with an accuracy of 0.1° or better.

Temperature devices can be integrated with catalytic surfaces as gas detectors. For example, platinum surfaces induce exothermic catalytic oxidation of combustible gases such as methane, hydrogen, propane, and carbon monoxide. The resulting thermal energy or temperature increase can be detected using micromachined thermocouple or thermistor devices as calorimetric sensors of combustible gases.

8.2.4 Bio- and Chemical Sensors

Chemical and biochemical sensing constitutes an immense opportunistic area due to its ubiquitous applications in industrial, environmental, and medical processes. The types of conversion methods include thermal, optical, electromagnetic, and electrochemical reactions. We will look only at some representative MEMS devices used for bio- and chemical sensors. Regardless of the type of transduction method, all microfabricated sensors need to have the following characteristics:

1. Interface with the environment to acquire and handle the sample, which sometimes requires microfluidic devices to be connected to the sensor elements.
2. Selectivity. The sensor needs to have a specific response only to target samples and must be able to overcome interference from other sample constituents.
3. Stability and repeatability. This includes preservation of the sensor characteristics in harsh environments, as well as zero-reset capability for repeated measurements of the same parameters.

In spite of the above requirements, microfabricated chemical and biochemical sensors have become useful in many areas of science and engineering due to their potential advantages, including small sample volume, mass production, and compact sensor dimensions.

Many organic polymers (molecules with a large number of repeating hydrocarbon structures) exhibit a change in their resistivity when exposed to a gas. For example, monolayers (thin layers with a thickness of one molecule) of Cu-substituted phthalocyanine are responsive to gases such as NH_3, NO_2, and CCl_4 at concentrations between 0.5 and 2 ppm. The monolayer is approximately 2.5 nm thick, and in actual devices, up to 45 multiple monolayers stacked on top of each other are used, with gold electrodes inserted in the layer stack to pick up the resistance change, as shown in Figure 8.11. This type of device is called a *chemiresistors*. One of the problems with these devices is the lack of selectivity, i.e., the sensor responds to an array of different gases, including NH_3, NO_2, and CCl_4.

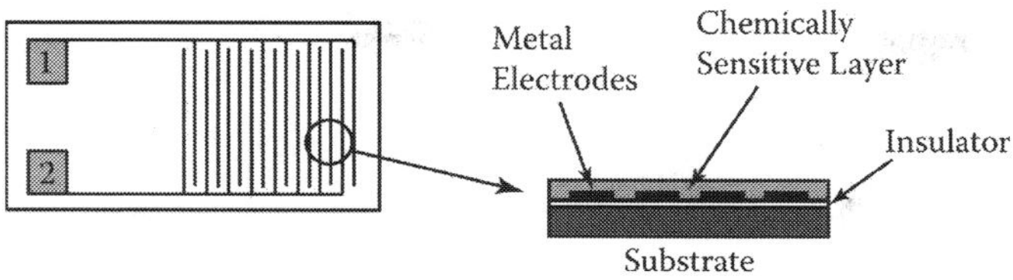

FIGURE 8.11 A chemiresistor design using organic polymer monolayers.

TABLE 8.2
Metal Oxide Gas Sensor Materials

Metal Oxide	Catalyst	Detectible Gas
$BaTiO_3/CuO$	La_2O_3, $CaCO_3$	CO_2
SnO_2	Pt + Sb	CO
SnO_2	Pt	Alcohols
SnO_2	La_2O_3	H_2, O_2, H_2S
SnO_2	CuO	H_2S
ZnO	V, Mo	Halogenated hydrocarbons
WO_3	Pt	NH_3
Fe_2O_3	Ti+Au	CO
Ga_2O_3	Au	CO
MoO_3	None	NO_2, CO
In_2O_3	none	O_3

Source: From Kovacs (1998).

A similar process is used in metal oxide gas sensors. Semiconducting metal oxides such as SnO_2, TiO_2, In_2O_3, ZnO, WO_3, and Fe_2O_3 undergo a change in resistivity in the presence of specific gases. This change in resistivity arises from the adsorption of gas molecules onto the metal oxide surface, where the gas molecules combine with electrons from the metal oxide to form ions. For example, oxygen binds with electrons in SnO_2 (as well as other metal oxide substrates) to form oxygen ions.

$$O_2 + 2e^- \rightarrow 2O^- \tag{8.13}$$

This loss of electrons in metals represents a loss of carriers of electricity and therefore an increase in resistivity. In many instances, catalytic additives are introduced into the metal oxide layers to increase the reaction rates, as shown in Equation 8.13. Table 8.2 shows the metal oxide and catalyst combinations used in the detection of various gases.

There is a class of devices that use the change in the work function due to the presence of chemicals near the material surface. Work function is the minimum energy required to remove an electron from a solid. The presence of certain chemicals affects the electric field of the solid, thereby altering its work function.

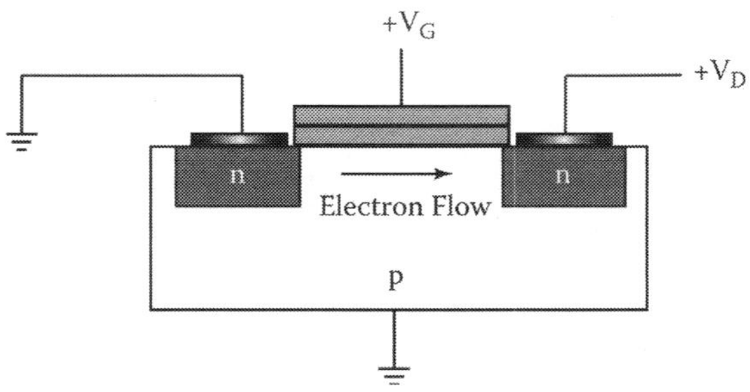

FIGURE 8.12 The operating principle of ion-sensitive field-effect transistors (ISFET).

The voltage output of microfabricated transistors, for example, is sensitive to a number of chemical parameters, including the pH level. This effect is illustrated in Figure 8.12. Field-effect transistors (FETs) are devices with n- and p-type semiconductors (silicon doped with either phosphorus for n-type or gallium for p-type) channels separated by a "gate" that is typically a silicon oxide layer. During normal operation of FETs, there is very little current from the source to the drain when the gate voltage (V_G) is zero, as the electrons in the n-type channel near the drain are attracted to the positive drain-source voltage, and the electron-depleted p-type channel does not furnish additional electrons for the current to flow. As the gate voltage is increased, however, electrons are drawn to the positive gate voltage and populate the p-type channel, thus allowing migration of electrons from the source to the drain, or current from the drain to the source. The operation of FETs for chemical and biosensors is based on the fact that the electron population due to gate voltage is strongly affected by the presence of ionic species at the gate, thereby affecting the FET current. The devices that utilize this phenomenon are called ion-sensitive field-effect transistors (ISFETs) or chemically sensitive field-effect transistors (CHEMFETs).

An ISFET for pH sensing is commercially manufactured and is capable of measuring pH levels with an accuracy of ± 0.01 pH units. Specificity, or the ability to detect specific chemicals, can be achieved using membranes that allow only the target ions to pass through. A further development of this selective transport or binding at the gate site involves enzymatic membranes that bind only with specific biochemical species. Devices using enzymatic membranes or layers are called enzyme-sensitive field-effect transistors (ENFETs). These devices have current and potential applications in biomedical (e.g., O_2 in blood), military (nerve gas and other toxic chemicals), mining (explosive gases, O_2), and industrial processes (CO_2, O_2).

ISFET devices can also be used as biosensors, detectors of complex biochemical molecules, including enzymes. To develop selectivity, various schemes are used so that only specific biomolecules adhere to the sensor substrate. Four major schemes are shown in Figure 8.13. A semipermeable membrane allows the passage of only specific molecules through the membrane wall. Once entrapped within the membrane, the ionic influence of the target biomolecules again alters the current output characteristics of the base FET, as described in the previous paragraph. A second method uses a membrane designed to adsorb the target molecules through combinations

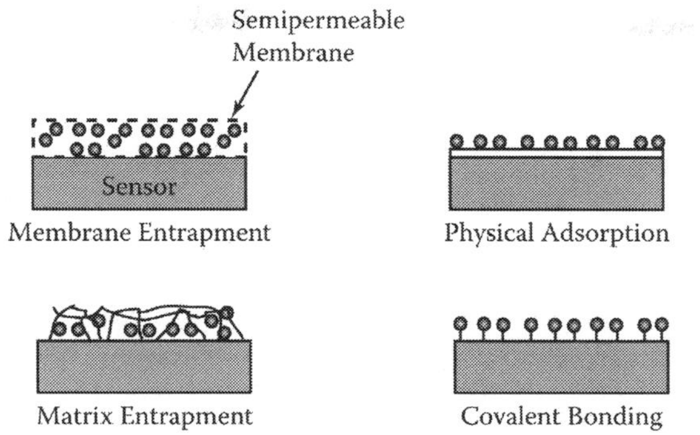

FIGURE 8.13 Methods to develop specificity in biosensors. (Adapted from Kovacs, 1998.)

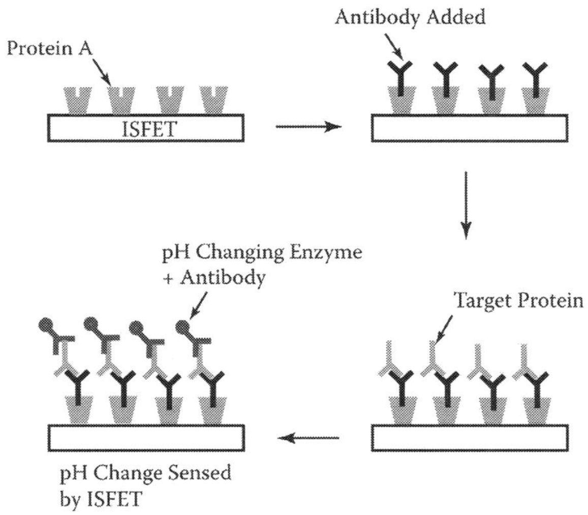

FIGURE 8.14 A CHEMFET using sequential covalent bonding to detect a specific target enzyme. (Adapted from Kovacs, 1998.)

of van der Waals forces (intermolecular forces arising from electrostatic attraction between molecules with asymmetric electric fields, i.e., dipole moments), hydrophobic forces, and ionic bond forces. Physical entrapment is also possible using specially designed matrix materials. Finally, the surface of the transducer can be treated to induce covalent bonding (chemical bonding resulting from the sharing of electrons in the orbitals of two different atomic nuclei) of the target species with the surface.

An example of the covalent bonding method for ISFET detection of enzymes is shown in Figure 8.14. The ISFET gate area is layered with a binding agent (typically a complex molecule, e.g., trimethoxysilane with glutaraldehyde) so that a "precursor" protein (Staphylococcus aureus) can bind to the surface (Colapicchioni et al., 1991). This precursor protein is needed to capture the target protein, in this case, human protein IgG (immunoglobulin G). The combination of these proteins, however, produces a negligible change in the local pH level. Thus, a secondary, pH-changing protein needs to be added to the sample to activate the ISFET, as shown in Figure 8.14. Detection limits of 0.1 μg/mL have been reported using this device.

8.3 NANOTECHNOLOGY SENSORS

Nanotechnology has become a ubiquitous term in many areas of science and engineering. Some applications represent truly unique and new technological advances. Nanotechnology is a frontier in science and engineering, enabling breakthroughs not possible at larger scales, and it is inevitable that some of these breakthroughs will lead to innovative and pervasive commercial and research applications. Because this is a rapidly evolving field, we will look at some of the underlying phenomena for representative nanoscale devices used in sensor applications.

Carbon nanotubes (CNTs) are tube structures consisting of carbon atoms bonded in hexagonal arrays. The covalent bonding between carbon atoms is facilitated by the sharing of electrons between two carbon atoms. If two electrons, one from each carbon atom, are shared in the molecular orbital as in diamond, it is called a single carbon bond or sp^3 bond (see Appendix D for molecular bonds). If four electrons, two from each carbon atom, are shared, the bond is a double bond or sp^2 bond. A triple carbon bond involves six shared electrons. The strength and ability of carbon bonds to form long chains of carbon and other atoms are evident in the ubiquity of carbon-based molecules, including organic and biochemical compounds. Carbon nanotubes involve sp^2 bonding of carbon atoms and therefore exhibit mechanical strength higher than that of diamond. CNT structures with diameters on the order of 1 nm and lengths ranging from several nanometers to as large as one centimeter can be fabricated using a variety of techniques, including CVD and arc discharge (to ablate carbon atoms from graphite rods). CNTs with a single layer of carbon atoms forming the cylindrical shell are called single-walled CNTs or SWNTs.

Single-walled CNTs (SWNTs) exhibit unique electrical and physical characteristics that can be transformed into sensor utilities. Due to the reactivity of the carbon atoms in the SWNT array, various chemical species can be adsorbed onto the SWNT outer surface or entrapped inside the SWNT channel. Similar to ion-sensitive FETs discussed in the previous section, the presence of chemical species attached to the SWNT influences the electrical conductivity. Exposure of SWNTs to nitrogen dioxide (200 ppm) and ammonia (1%) under ambient conditions results in a change in the nanotube's electrical conductivity by several orders of magnitude within a few seconds of response time. The underlying mechanism of the conductivity change is attributed to bonding with nitrogen dioxide, which tends to withdraw electrons from the nanotube structure and thereby create electron "holes." These electron holes act as hole carriers that allow electrons to hop from one site to another, thereby increasing electrical conductivity. Conversely, bonding with molecules such as ammonia tends to donate electrons to the nanotube lattice, filling the electron holes. This naturally leads to a decrease in electrical conductivity.

Due to the high reactivity of carbon atoms, SWNTs offer many possibilities for binding molecules or atoms to their surface to develop new functionalities.

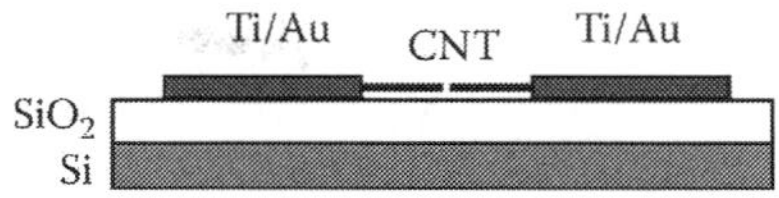

FIGURE 8.15 A sensor using a carbon nanotube.

Sputter-coated Pd nanoparticles on SWNTs, for example, have a propensity to absorb hydrogen molecules, thereby providing hydrogen-specific sensing capability for SWNTs. Multiwalled carbon nanotubes (MWNTs) have been used as biosensors for enzymes such as glucose by encapsulating these large molecules within the nanotube and sensing the resulting change in the electrical conductivity of the MWNT on a platinum substrate. Figure 8.15 shows a typical method for fabricating a carbon nanotubes on a silicon substrate. The carbon nanotube is actually cut in the device shown in Figure 8.15 so that DNA molecules can be electrostatically trapped.

Although it should be clarified that methods such as *surface-enhanced Raman scattering* (SERS) using nanosized particles existed long before nanotechnology became fashionable, this diagnostic method may now be classified under the nanoscale sensing category. As discussed in Chapter 5, Raman scattering is a useful technique, except that it suffers from low signal levels compared to interference from Mie scattering and spurious reflections. In fact, in flows with particles, Mie scattering can easily overwhelm even the best spectrometers and filtering devices. One method that overcomes this deficiency is SERS. SERS refers to the enhancement of Raman signal strength by up to 10^8-fold in the vicinity of nanosized particles or surface features. Although the theoretical basis of this enhancement is still being developed (and is somewhat beyond the scope of this book in any event), the signal enhancement occurs in part due to the electromagnetic resonance of the molecular dipole moment with the electric field of the nanoparticles. The Raman scattering signal is amplified to a degree easily observable over interference levels in SERS. A variety of metals can be used as the enhancement medium, such as gold, aluminum, lithium, copper, silver, zinc, sodium, germanium, and cadmium. These metals can be applied in metal tip surfaces, nanoparticle colloids, or metallic surfaces with controlled surface roughness. For example, silver colloids can easily be prepared by rapid mixing of $AgNO_3$ with $NaBH_4$ at low temperatures (~0°C). Alternatively, silver-coated polymer spheres can also be used in colloidal form. Aside from the provision of nanoscale metallic surfaces, the diagnostic arrangement is similar to standard Raman scattering measurements.

Quantum dots are nanocrystals, or simply nanoparticles, of semiconductor materials (silicon, germanium, tin, etc., elements whose electrical conductivity can be easily altered using dopants or external stimuli such as voltage or photon flux). Quantum dots typically range in size from 2 to 10 nm, consisting of 10–50 atoms. Given the small number of atoms, they tend to behave like molecules rather than bulk semiconductor materials. Quantum dots have interesting and useful optical properties in that their light absorption and emission wavelengths can be "tuned" by their size and composition. For example, the bandgap (the energy difference between the excited and ground states) of quantum dots is inversely proportional to the square of their size. This is due to the quantum mechanical structure of atoms and molecules in general, where large atoms or molecules have a large number of closely spaced high-energy states. Therefore, the energy difference at the top of the energy spectrum tends to be small, resulting in a small bandgap. This makes it possible to tune the light absorption and emission characteristics of quantum dots simply by changing their dimensions. Consequently, they offer the possibility of producing tailor-designed optical materials (e.g., lasing media or optical dyes) to suit a particular emission wavelength at high quantum yields (efficiency in emitting photons) typical of quantum dots. In fact, blue lasers using quantum dots are already commercially available.

Tunability and high quantum yield make quantum dots ideal for target-specific tracers. For example, CdSe-ZnS core-shell quantum dots coated with the organic molecule tri-n-octylphosphine oxide ligands are soluble in water, absorb at 548 nm, and emit at 563 nm. A target molecule, methyl viologen, binds with the organic coating, resulting in electron transfer from the quantum dot to the bound molecule. This, in turn, leads to a decrease in quantum dot luminescence intensity. Thus, the principle is to use surface ligands to form quantum dot conjugates with target bio- or other molecules, where binding to the target molecule modifies the electronic energy structure and alters the light absorption and emission properties. In contrast to nanodevices such as carbon nanotubes, quantum dots act as tracers that reveal target molecule concentrations upon external illumination at their absorbing wavelength range. Another interesting application of quantum dots is to use several of them with different sizes and materials, and therefore different colors and intensities, to tag target protein molecules with quantum dot "barcodes." With three different colors and ten intensity levels, this results in ~ 1000 combinatorial possibilities for such molecular barcodes.

8.4 MICROSCOPIC IMAGING TECHNIQUES

For microfluidics characterization or biochemical observations involving very small sample volumes, it is often necessary to use microscopic imaging methods. Optical microscopy is a classical optical method using a short-focal-length lens positioned close to the object. In this section, we will highlight three more advanced imaging methods applicable to thermal and flow processes at small scales: confocal microscopy, microscopic particle image velocimetry, and scanning electron microscopy (SEM). The last-named method is primarily used to scan microscopic surface features; however, in microfluidic devices and during flow-surface interactions, it is often necessary to determine surface characteristics using methods such as SEM.

Confocal microscopy is a method for creating sharp, three-dimensional reconstructions of a sample without the typical interference from objects between the target and the objective lens (which usually appear as blurs in standard microscopy). This method is widely used in biology, bioengineering, and medical sciences for relatively transparent samples that can be stained with fluorescent dyes. It is analogous to three-dimensional scanning of pointwise fluorescence signals from the object and, by necessity, requires digital processing of the set of pointwise signals to reconstruct the three-dimensional features. Figure 8.16 shows a schematic of a confocal microscopy setup. A set of rotating mirrors, similar to those used in three-dimensional

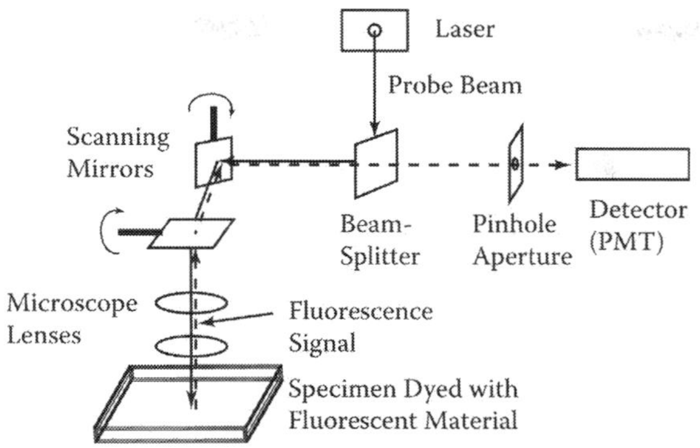

FIGURE 8.16 A schematic of a confocal microscopy setup.

scalar imaging (confocal microscopy *is*, in fact, a form of scalar imaging), scans the specimen dyed with fluorescent material (typically called fluorophores). Signal detection by a photomultiplier tube is synchronized with the mirror positions so that the signal at a given time is assigned a spatial coordinate. This data array is then used to construct a three-dimensional mapping of the fluorescent signals and, therefore, the specimen features.

The choice of fluorescent material, or fluorophores as they are called in biomedical imaging, is obviously important. They need to tag the targeted part of the specimen. For example, in cell imaging, many different fluorophores are commercially available that bind only to specific components or proteins in the cell, allowing specific regions to be imaged through confocal microscopy. The fluorophores typically used in confocal microscope systems need to emit a sufficient amount of light in response to the available laser light source. In addition, they need to be resistant to photobleaching, which refers to the loss of fluorescence due to oxidation of the fluorophores when exposed to air or an oxygen-containing environment over time.

For characterization of microscale flows, new challenges naturally arise in obtaining the requisite spatial resolution, appropriate seeding particles, and optical access. Standard particle image velocimetry (PIV), for example, can achieve spatial resolution (the spatial extent over which the velocity data is averaged and acquired) down to approximately 200 μm. A method to further enhance spatial resolution is called super-resolution PIV, which involves using the standard correlation-based PIV algorithm down to the 200 μm scale and then tracking individual particles within each interrogation window using the standard PIV results as a search guide. Keane et al. (1995) report that this method can achieve spatial resolution down to 50 μm in the interrogation plane and 200 μm in the transverse direction, limited by the thickness of the laser sheet.

Ultimately, as the size of the particle to be imaged becomes extremely small, the optical limit due to diffraction becomes relevant (see Chapter 4 and Equation 4.41). Diffraction refers to the property of light to spread when passing through an aperture. This can be visualized through the wave nature of light: as a planar light wave passes through a pinhole, the light from that point emanates as a spherical wavefront. Using this logic, the smaller the aperture, the larger the divergence. This effect of diffraction on imaging is described by the so-called *point spread function*, or the image diameter of an infinitesimally small particle:

$$d_s = 2.44(M+1)f\#\lambda \tag{8.14}$$

M = magnification
$f\#$ = f-number of the lens = (focal length)/(lens diameter)
λ = wavelength of the interrogation beam

For an illumination wavelength of $\lambda = 560\,\text{nm}$, a magnification of 60, and an $f\#$ of 0.36 (typical of microscopic lenses), the diameter of an infinitesimally small particle translated to the image plane is 30 μm. This is *not* the limit of spatial resolution but simply indicates the required pixel size and density at the image plane to visualize small particles in a small volume. The true indicator of spatial resolution is the resolving power, which is the ability of the imaging system to discern a feature, and it has a form similar to Equation 4.34:

$$d_r = 1.22f\#\lambda \tag{8.15}$$

Thus, for the foregoing system, 480 nm (2xdr) features can be discerned. This is a typical limit using visible-wavelength imaging and sets a hard limit on the spatial resolution for so-called micro-PIV.

The ability to track the motion of an apparent 480-nm-diameter particle depends on the pixel size and particle image diameter, as given by Equation 8.14. With a ratio of particle image diameter to pixel width of 3, approximately one-tenth of the particle motion can be detected, which sets the limit on the resolution with which velocity can be measured in micro-PIV. Santiago et al. (1998) presented a micro-PIV system with a reported spatial resolution of 6.9 × 6.9 × 1.5 μm over a 120 × 120 μm spatial domain. Polystyrene particles of 300 nm were used as flow tracers. Meinhart et al. (1998) reported an even better spatial resolution of 5.0 × 1.3 × 2.8 μm using fluorescent particles illuminated by an Nd:YAG laser with an oil-immersion microscopic lens. A schematic of a micro-PIV system is shown in Figure 8.17.

Scanning electron microscopy (SEM) is a method for obtaining very high magnifications (100,000× or higher) using electron beams, as opposed to photons in conventional microscopy. To understand how electrons can achieve magnification and resolution beyond optical microscopy, it is useful to consider the so-called de Broglie wavelength. The de Broglie postulate states that particles can also behave as waves. Consider photons, for example. Photons are light particles that also exhibit wave-like behavior, such as diffraction and interference. The energy of matter, on the other hand, can be expressed using Einstein's special theory of relativity.

$$E^2 = (mv)^2c^2 + m^2c^4 \tag{8.16}$$

c = speed of light
m = mass of the particle
v = velocity of the particle

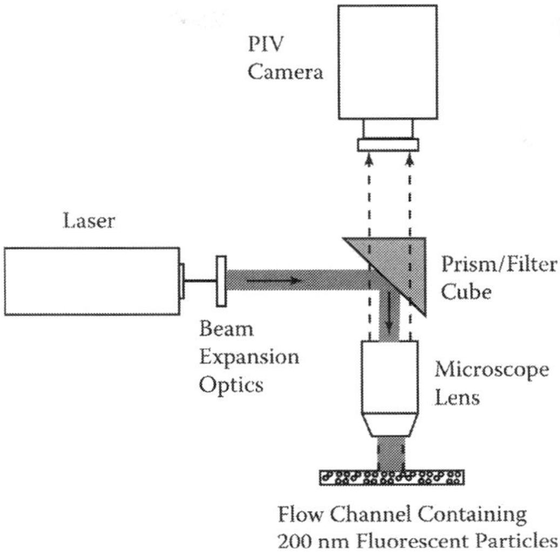

FIGURE 8.17 A schematic of a micro-PIV system.

In the absence of momentum, this expression for energy reduces to the famous form, $E = mc^2$. However, for photons, it is the "rest energy" term that disappears because photons have zero mass (but finite momentum).

$$E = (mv)c \tag{8.17}$$

We also know from photoelectric effects, such as fluorescence, that the photon energy is proportional to its frequency.

$$E = hv \tag{8.18}$$

h = Planck's constant = 6.63×10^{-34} J s

Equating Equations 8.17 and 8.18, we obtain the so-called de Broglie wavelength, which applies to all particles exhibiting wave properties, known as "matter waves."

$$E = hv = (mv)c \rightarrow \lambda = \frac{h}{(mv)} \tag{8.19}$$

Thus, if we apply the de Broglie wavelength to electron beams with charge q and mass m, subject to an electric field V, the electrons will be accelerated to a velocity v. The resulting wavelength of the electrons is

$$\lambda_e = \frac{h}{mv} = \frac{h}{\sqrt{2qmV}} = \frac{1.21}{\sqrt{V}}[\text{nm}] \tag{8.20}$$

So, for an applied electrical potential of $V = 10{,}000\,V$, the electron wavelength is 0.012 nm, much smaller than the visible wavelength of 400–700 nm. As shown in Equation 8.15, resolution is proportional to wavelength, and therefore, electron beams can achieve much higher resolution.

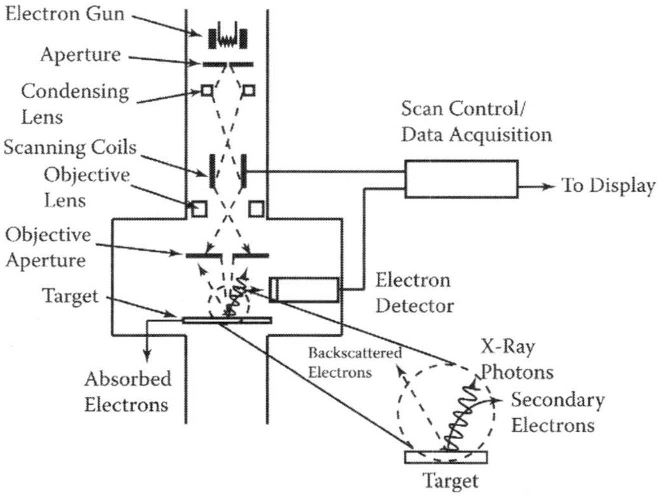

FIGURE 8.18 A schematic of a scanning electron microscope.

Although electron beams are used in transmission, scanning, and emission modes in electron microscopy, we will limit our discussion to SEM. A schematic of SEM is shown in Figure 8.18. Electrons are thermionically generated in a tungsten or LaB_6 (lanthanum hexaboride) filament; that is, the thermal energy of the electrons at the surface of the heated filament overcomes the work function and ejects the electrons from the material surface. The electrons are focused using electromagnetic fields (also called lenses, in analogy to optical microscopy) to a very fine focal spot size of ~1 nm. The focal point is scanned across the target object, and the intensity of the scattered electrons corresponding to each point is measured and digitally processed to display a two-dimensional map. The electrons interact with the matter at the focal point and produce backscattered electrons, X-rays (from the photoelectric effect), and secondary electrons. All these particles can be used for imaging; however, secondary electrons are most commonly used because their trajectories depend on the surface tilt angle, revealing surface features. For example, secondary scattering is produced due to the interaction of incident electrons with the surface matter down to a depth of ~10 nm. If the sample surface is tilted, the sample volume swept by the electron beam to this depth increases by $1/\cos\theta$ (see Figure 8.19). Thus, different surface angles—and therefore surface features—produce varying degrees of secondary scattering, and electron detectors capture the scattered electrons and convert them to an electrical signal for the given focal point coordinates. Figure 8.20 shows some example images produced by SEM.

8.5 LIDARS

Lidars (light detection and ranging) are the optical analog of radars (radio wave detection and ranging). Many of the optical diagnostic methods discussed in Chapter 5 can be used for remote sensing of large-scale processes, collectively known as lidar

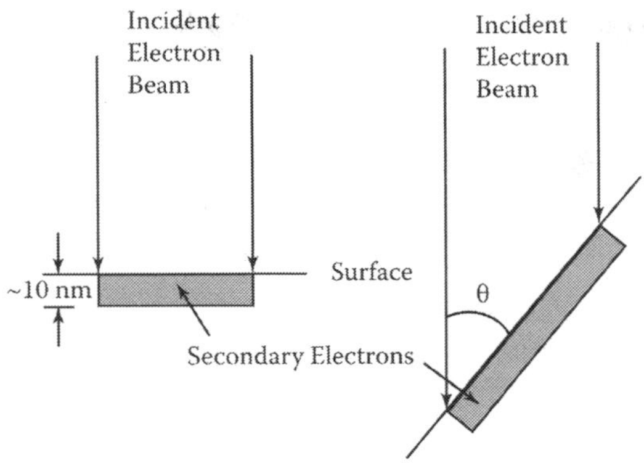

FIGURE 8.19 The dependence of secondary scattering on the tilt angle of the sample surface.

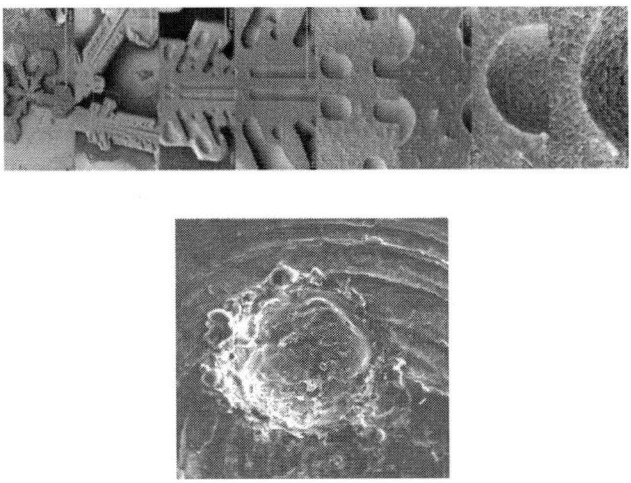

FIGURE 8.20 Examples of scanning electron microscopy (SEM) images. The top photos are SEM images of a snowflake at a sequentially higher magnification (image courtesy of the US Department of Agriculture). The bottom image is the aluminum surface damaged by an electrical spark.

methods. Much of the current knowledge base for atmospheric processes, including fluid dynamics and chemistry, has been acquired through lidar experimentation. Many satellite-borne systems also use either passive or active lidar concepts. Some primary examples are atmospheric turbulence, the water cycle, planetary boundary layer processes (the region of the atmosphere influenced by surface friction), ozone fluxes, monitoring of gaseous and particulate pollutants, and various meteorological processes, including precipitation and frontal passages. In the previous sections, applications of the diagnostic methods to very small scales were discussed; in this section, we move in the opposite direction in the spectrum of length scale.

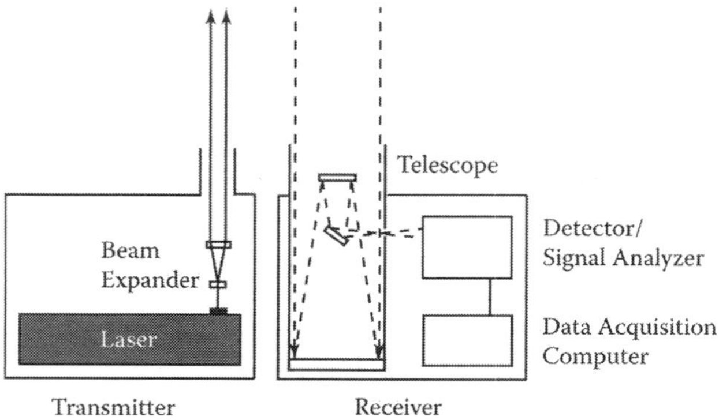

FIGURE 8.21 A schematic of a generic lidar system setup.

As already noted, lidars can be implemented using a variety of optical processes such as Mie scattering, Rayleigh and Raman scattering, fluorescence, and gas or particle absorption. A generic setup for lidar systems is shown in Figure 8.21. Similar to optical diagnostic systems at laboratory scales, lidars consist of transmitting and receiving optics, along with lasers, detectors, and data processing units. Due to the large distances probed with lidars, usually the backscattering component is detected—that is, the signal component that returns to the lidar unit is observed. The detector, in principle, can be located far from the light source, but this will inevitably make the system more difficult to manage. The light source or laser used in lidars typically covers the ultraviolet, visible, and infrared wavelength ranges. Corresponding laser types include ruby, copper vapor, CO_2, excimer, and Nd\:YAG lasers, with output wavelengths ranging from 250 nm to ~11 μm. A convenient choice, as in laboratory-scale diagnostics, is the Nd:YAG laser with a fundamental wavelength output of 1064 nm, which can be frequency doubled (532 nm), mixed (355 nm), or quadrupled (266 nm) to cover infrared to ultraviolet. For gas absorption measurements requiring tunable lasers, dye lasers, solid-state lasers, or optical parametric oscillators are also used.

In spite of the collimating optical components in the transmitting optics, the laser beam undergoes a small divergence, typically ~100 m rad. Thus, the receiving optics are designed to capture signals from a cone with a divergence angle of a few hundred μrad. Because the apertures of the primary optical components in lidars range from 0.1 to about 1 m, it is most practical and cost-effective to use reflective mirrors, as shown in Figure 8.21. It is convenient to describe lidar scanning of space in terms of a spherical coordinate system: spatial coverage is achieved by pointing the transmitting or receiving optics over a range of polar (ϕ) and azimuthal (θ) angles. The third component, radius or distance (r), is resolved by timing the backscattered return signal relative to the laser pulse.

$$r = c(t - t_o) \tag{8.21}$$

c = speed of light in air (~ 300,000 km/s)
t = the time at which the signal is detected
t_o = the time at which the laser pulse is sent out

It is evident from Equation 8.20 that, in order to resolve r, the data acquisition speed and temporal resolution must be very high. For example, if Δt denotes the time interval during which the signal is collected, then the spatial resolution in r is

$$\Delta r = c\frac{\Delta t}{2} \tag{8.22}$$

The factor of 1/2 is needed because the light must travel up and down the sample bin of Δr. Thus, to resolve a spatial bin of $\Delta r = 10\,\text{m}$, a minimum 67 ns data sampling interval or a 17 MHz data acquisition rate is required.

The lidar signal strength can be related to the target parameter to be measured using the following lidar equation:

$$I(r) = KG(r)\beta(r,\lambda)T(r,\lambda) \tag{8.23}$$

K = constant that depends on the performance of the optical components
$G(r)$ = geometrical factor (see the following text)
$\beta(r, \lambda)$ = backscatter coefficient
$T(\text{r}, \lambda)$ = transmissivity of the medium to and from the distance r

The optical geometry leading to the lidar equation, Equation 8.23, is depicted in Figure 8.22. The factor K depends on the laser pulse duration and intensity, the receiving mirror or lens aperture, and the optical component efficiency, and therefore is a fixed parameter for a given optical system. The geometric factor, $G(r)$, is illustrated in Figure 8.22. The solid angle extended from the aperture of area A to the distance r is $\Delta\Omega = A/r^2$. Thus, the solid angle decreases quadratically as a function

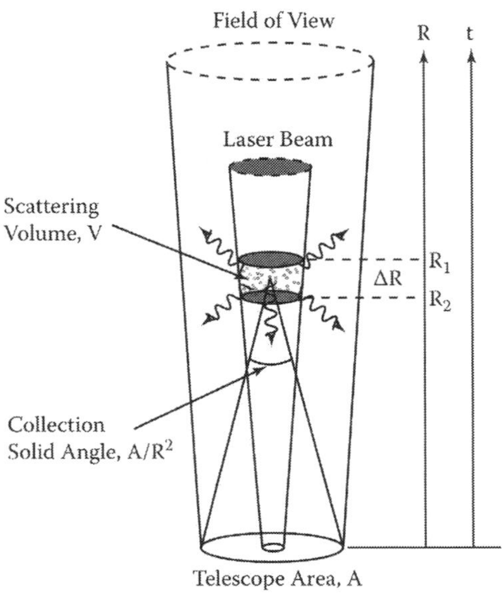

FIGURE 8.22 Lidar geometry.

of r, and therefore the signal strength will also decrease likewise, i.e., $G(r) \sim 1/r^2$. However, for a given optical system, it is again a fixed function of *r*.

$\beta(r, \lambda)$ is then the primary atmospheric parameter to be resolved by lidar, whereas $T(r, \lambda)$ is the transmissivity of the medium. The backscattering coefficient may have several contributing species, both gaseous and particulate, in the atmosphere, and therefore is written as a summation:

$$\beta(r,\lambda) = \sum_j N_j(r)\left(\frac{d\sigma(\lambda)}{d\Omega}\right)_j \tag{8.24}$$

N_j = number concentration of species j
$(d\sigma(\lambda)/d\Omega)_j$ = backscattering coefficient of species *j* at wavelength λ

Similarly, the transmissivity function $T(r, \lambda)$ depends on the constituent species and their overall extinction (absorption + scattering) coefficient, α_e, and is described by the Beer-Lambert law for light extinction:

$$T(r,\lambda) = \exp\left[-2\int_0^r \alpha_e(r,\lambda)\,dr\right] \tag{8.25}$$

Again, a factor of two inside the square brackets is needed because the light must travel to and from the sample volume at *r*. In instances where there is appreciable absorption and scattering of various species at the probe wavelength, it requires knowledge of the number concentrations and optical properties (absorption and backscattering coefficients) for all of these species, and data extraction can quickly become hopelessly complex. For that reason, it is judicious to choose the laser wavelength that poses minimum coupling of the absorption and backscattering of the species with the target species.

Some of the major optical processes used in modern lidars are:

Elastic backscattering (Mie and Rayleigh scattering)
Differential optical absorption spectroscopy
Raman backscattering
Fluorescence
Doppler shift

Elastic backscattering uses Mie and Rayleigh scattering from particles. Mie scattering refers to elastic scattering (emitted light having the same wavelength as the incident light) from particles with sizes on the order of, or larger than, the incident light wavelength, whereas Rayleigh scattering involves particles much smaller (say 1/10th) than the wavelength. For visible and nearby wavelengths, this means that Rayleigh scattering occurs from air molecules (N_2 and O_2), and Mie scattering from atmospheric aerosols such as ice crystals, soot agglomerates, mineral dust, and sea-salt particles. Although both Rayleigh and Mie scattering are described by Mie theory, Rayleigh scattering in the limit of small particle size exhibits a unique $1/\lambda^4$ dependence that highly biases the short-wavelength components. The Mie scattering intensity, on the other hand, depends on the particle size and index of refraction, and for spherical and other relatively simple geometrical shapes, Mie theory gives exact solutions for scattering and extinction properties. However, atmospheric particles tend to be anything but spherical: ice crystals have a variety of crystalline shapes, soot agglomerates are at best fractal aggregates of nearly spherical primary particles, and mineral dusts tend to be highly irregular in shape and sometimes compounded by attached particles. Therefore, although aerosol particles can be easily detected due to the strong Mie scattering and the conversion of Mie-scattering intensity to quantitative data such as particle size, composition, and morphology, this process is not straightforward, and much ongoing work aims to quantify the Mie-scattering signatures of nonspherical particles.

Due to the elastic nature of both Mie and Rayleigh scattering, aerosols and air molecules contribute to the signal at the detection wavelength. One method used to separate the two components is high spectral resolution lidar (HSRL), which allows monitoring of aerosols separately from atmospheric molecular scattering, and vice versa. The principle is based on the fact that molecules exhibit much higher levels of random motion due to thermal energy than aerosol particles. At typical atmospheric temperatures, the velocities of gas molecules' random motion range approximately $\pm 300\,\text{m/s}$, which translates to a Doppler frequency shift of ~ 1 GHz (see Equation 4.2 in Chapter 4). In contrast, aerosol particles are not subject to such random motion and instead move in response to wind (~ 10 m/s) and atmospheric turbulence (~ 1 m/s), producing Doppler shifts of only ~ 3 to ~ 30 MHz. The result is that if the probe laser has a very narrow linewidth, the resulting Rayleigh scattering will be broadened to a much greater extent than Mie scattering. By using a very high spectral resolution device on the detector side, such as a Fabry-Perot interferometer or atomic/molecular absorption filters, the Mie and Rayleigh-scattering components of the elastic scattering signal can be separated.

Fabry-Perot interferometers are devices with a so-called *etalon*, which consists of a pair of partially reflecting surfaces separated by a precise distance. For multiple internal reflections, as shown in Figure 8.23, each successive pass of light produces a path length difference of

$$d = 2nl \cos \theta \tag{8.26}$$

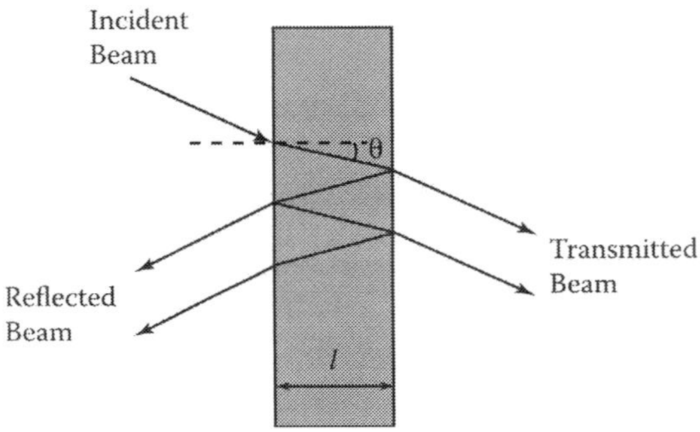

FIGURE 8.23 The principle of the Fabry-Perot interferometer.

n = refractive index of the material between the reflecting surfaces
l = thickness of the etalon or the distance between the two surfaces
θ = incident angle of light into the etalon

Maximum transmission through the etalon occurs when the path length difference is an integer multiple of the wavelength, or when

$$\lambda = 2mnl \cos\theta \qquad m = 1, 2, 3, \ldots \tag{8.27}$$

Also, in the absence of absorption by the etalon material, (transmission) + (reflection) = 1. Therefore, the components not transmitted are reflected by the etalon. By tuning the Fabry-Perot interferometer to the precise center wavelength of the probe laser in HSRL, the narrow Mie-scattering component will be transmitted, whereas the remaining Rayleigh-scattering component is reflected, allowing both to be separately detected.

Atomic or molecular filters are essentially gas cells containing specific atomic or molecular vapor or gas. For example, barium has a strong atomic absorption line at 553.7 nm and can be converted to the vapor phase upon heating to 700–800°C. Although this simplifies the filtering scheme, atomic filters require the laser to be tuned to the absorption line wavelength, which is usually achieved using narrow-linewidth dye lasers. Atomic or molecular filters reject the narrowband Mie-scattering component due to aerosols and are therefore suitable for monitoring Rayleigh scattering from atmospheric gas molecules. Iodine is also used as an atomic filter vapor, requiring lower heating temperatures (~25°C–100°C) and offering absorption lines that match the 532 nm output of frequency-doubled Nd:YAG lasers.

Mie backscattering may be used to determine the aerosol distribution in the atmosphere through measurements of the aerosol extinction coefficient, α_e. Using Equations 8.23 and 8.25, we can write

$$S(r) \equiv \frac{I(r)}{KG(r)} = \beta(r,\lambda)T(r,\lambda) = \beta \exp\left[-2\int_0^r \alpha_e(r)\,dr\right] \tag{8.28}$$

The r dependence of the scattering coefficient, β, is omitted for convenience. Taking the logarithm of the left-most and right-most sides of the above equation and then taking a partial derivative with respect to r, we get

$$\frac{\partial(\ln S(r))}{\partial r} = \frac{1}{\beta}\frac{\partial \beta}{\partial r} - 2\alpha_e(r) \tag{8.29}$$

The solution of the above differential equation is of the form

$$\alpha_e(r) = \frac{S(r)}{\dfrac{S(r_{\max})}{\alpha(r_{\max})} + 2\displaystyle\int_r^{r_{\max}} S(\xi)\,d\xi} \tag{8.30}$$

The preceding equation is solved iteratively by first assuming a value for the unknown $\alpha_e(r_{max})$ until the left- and right-hand sides are matched within acceptable margins, using the signal values, $I(r)$, divided by KG(r).

Once the extinction coefficient due to atmospheric aerosols is thus obtained, the data can then be converted to optical depth or visibility.

$$\text{Optical thickness} \equiv \int_0^{r_{max}} \alpha_e(r)\,dr \tag{8.31}$$

$$\text{Visibility} \equiv \frac{3.912}{\alpha_e}[\text{km}] \tag{8.32}$$

Measurement of the extinction coefficient from Equation 8.30 can also be used to *estimate* the aerosol size distribution. By assuming a certain form for the aerosol size distribution and the sphericity of the aerosol particles, the integrated extinction coefficient can be calculated and fitted to the measured α_e. For atmospheric aerosols, the modified gamma distribution can be used:

$$n(R) = aR^{\varepsilon}\exp\left(-bR^{\gamma}\right) \tag{8.33}$$

$$R = \text{aerosol particle radius}$$

The values a, b, ε, and γ are model parameters and are all real and positive. The preceding distribution has the most probable radius, R_c, and total number density, N, of

$$R_c = \left(\frac{\varepsilon}{b\gamma}\right)^{\frac{1}{\gamma}} \tag{8.34}$$

$$N = a\gamma^{-1}b^{-\frac{a+1}{\gamma}}\Gamma\left(\frac{a+1}{\gamma}\right) \tag{8.35}$$

$$\Gamma = \text{gamma function}$$

Typically, the parameters ε, γ, R_c, and N are varied until a match with the observed α_e is achieved; then a and b can be calculated using Equations 8.34 and 8.35. Table 8.3 shows typical parameter values for atmospheric aerosols.

Figures 8.24 and 8.25 show, respectively, a photograph of a LIDAR unit being deployed in the field and some aerosol backscatter data. The data swaths are obtained by sequential data acquisition of angularly swept laser pulses and show the movement of aerosol clouds due to wind conditions.

Differential optical absorption spectroscopy (DOAS) is a form of absorption spectroscopy in which two sequential beams with offset wavelengths are used as reference and probe beams. One beam (probe) is tuned to the absorption wavelength of the target species, whereas the reference beam is offset slightly from this absorption wavelength so that it undergoes similar scattering and extinction processes, except that, at that wavelength, it has zero or at least different absorptivity due to the target species. The acronym DIAL (differential absorption lidar)

TABLE 8.3
Typical Parameter Values for Atmospheric Aerosols

Aerosol Type	A	ε	B	γ	R_c [μm]	N [m^{-3}]
Heavy fog	0.027	3	0.3	1.0	10.0	20×10^6
Moderate fog	0.066	3	0.375	1.0	8.0	20×10^6
Culumus cloud	2.373	6	1.5	1.0	4.0	100×10^6
Corona cloud	0.01085	8	1~24	3.0	4.0	100×10^6
Moderate haze	16×10^5	1	8.943	0.5	0.05	100×10^6

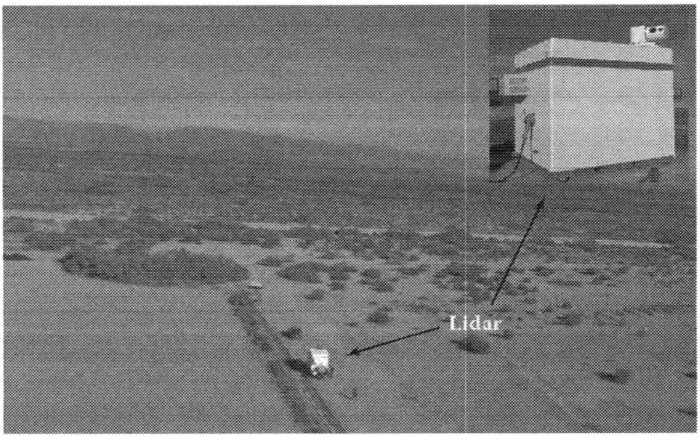

FIGURE 8.24 A photograph of a LIDAR being deployed for aerosol backscatter measurements in Arizona desert (Photo courtesy of Prof. Ron Calhoun, ASU).

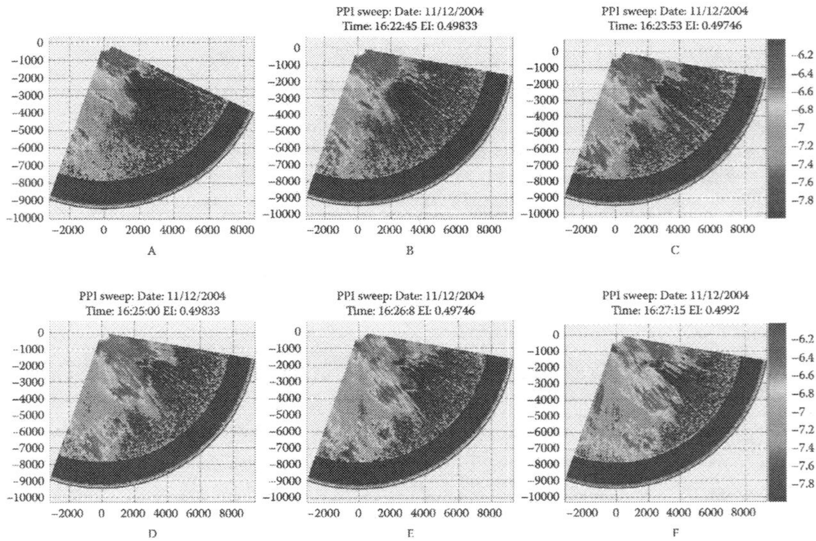

FIGURE 8.25 Aerosol backscatter levels as measured by Lidar (Time advances linearly from A through F. Notice advancement of front of strong winds over the sequence). (Photo courtesy of Prof. Ron Calhoun, ASU.)

is also used for the lidar application of this principle. DIAL has been used in ultraviolet and infrared wavelengths for atmospheric monitoring of ozone, SO_2, NO_2, NH_3, HCl, CO, hydrazine, and Hg.

We can write the difference in the absorption coefficients at the probe and reference wavelengths.

$$\Delta\alpha_e = \alpha_{e\lambda,\text{probe}} - \alpha_{e\lambda,\text{ ref}} = N\left(\tilde{\alpha}_{e\lambda,\text{probe}} - \tilde{\alpha}_{e\lambda,\text{ ref}}\right) \equiv N\Delta\tilde{\alpha}_e \tag{8.36}$$

where the tilde (~) over the absorption coefficient denotes the per-molecule absorption coefficient, and N is the number density of the target molecule. Again, using the preceding expression in the lidar equation, Equations 8.23 and 8.25, and after some algebraic manipulations, we get

$$N = \frac{1}{2\Delta\tilde{\alpha}_e}\left[\frac{d}{dr}\ln\left(\frac{I_{\text{probe}}(r)}{I_{\text{ref}}(r)}\right)\right] \tag{8.37}$$

Of course, in lidar measurements, the data are not continuous in *r*, so a discretized form of Equation 8.37 is needed.

$$N = \frac{1}{2\Delta\tilde{\alpha}_e\Delta r}\left[\frac{I_{\text{ref}}(r+\Delta r)}{I_{\text{ref}}(r)}\frac{I_{\text{probe}}(r)}{I_{\text{probe}}(r+\Delta r)}\right] \tag{8.38}$$

Figure 8.26 shows a schematic of a DIAL system, based on two NdYAG-pumped dye lasers, used to measure atmospheric pollutants O_3, NO_2, and SO_2. Probe/reference wavelengths of 296.16/297.35 nm were used for SO_2, whereas the 280.92/282.72 nm combination was used for O_3.

Optical parametric oscillators are replacing dye lasers in lidars as tunable light sources, with a tuning range from 220 nm to 4 μm.

Raman scattering, discussed in Chapter 5, is also used in lidars. Both vibrational and rotational Raman signals are employed. Water vapor, ozone, and trace gas concentrations have been measured using vibrational Raman signals with ranges up to 30–40 km, whereas rotational Raman backscattering has been used to profile atmospheric temperature. Similar to laboratory-scale Raman measurements (discussed in Chapter 5), short-wavelength lasers are optimal for Raman lidars, including the frequency-tripled output at 355 nm of Nd:YAG lasers or ultraviolet excimer lasers such as XeCl and XeF at 308 and 351 nm, respectively.

For example, water vapor concentration with respect to dry air density can be determined by comparing the vibrational Raman lidar signals of water vapor and nitrogen. Because the Raman backscatter coefficient is proportional to the number density, a ratio of the lidar equation (Equation 8.23) written for water vapor and nitrogen gives

$$\frac{m_{wv}}{m_{\text{air}}} \equiv \mu(z) = C\frac{I_{wv}(z)\exp\left[-\int_0^z \alpha_{e,N2}(\zeta)\,d\zeta\right]}{I_{N2}(z)\exp\left[-\int_0^z \alpha_{e,wv}(\zeta)\,d\zeta\right]} \tag{8.39}$$

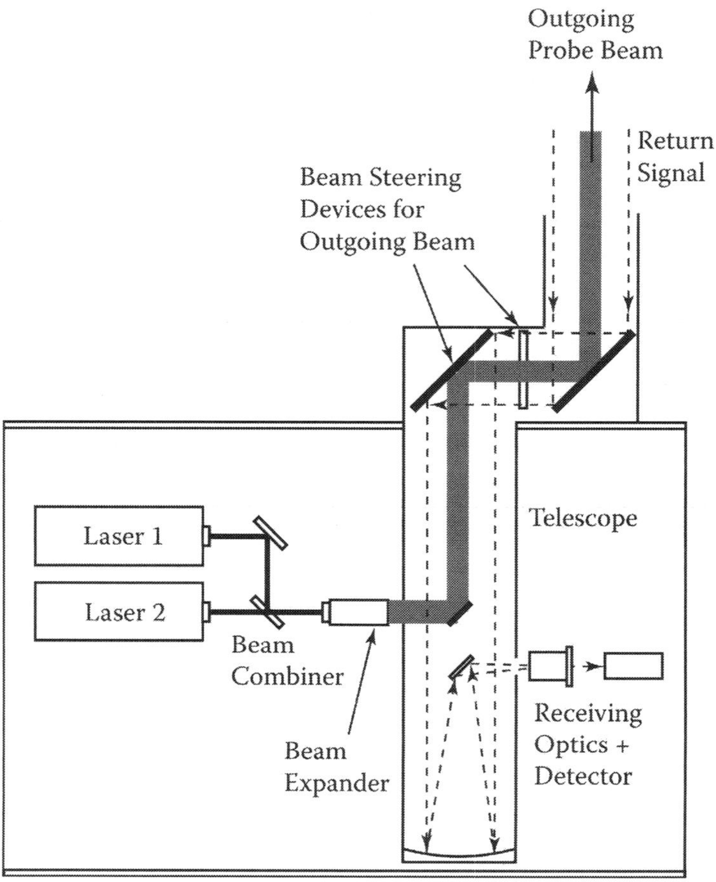

FIGURE 8.26 Schematic of a DIAL system.

$\mu(z)$ = ratio of water vapor to air densities at distance
z = "mixing ratio"
wv = water vapor

An innovative application of Raman backscattering is differential absorption for ozone measurements. The Raman frequency shift from abundant nitrogen molecules is chosen to coincide with the ozone absorption band. A slight up- or downshift of the laser wavelength then provides the reference beam in this Raman-DIAL setup. The lidar equation is then applied twice—for the absorption (probe) and reference beams—and the logarithms of the ratio of the two signals are differentiated.

$$N = \frac{1}{\Delta\tilde{\alpha}_e}\left[\frac{d}{dz}\ln\left(\frac{I_{\text{probe}}(z)}{I_{\text{ref}}(z)}\right) - \Delta\alpha_{\text{e,aerosol}}(z) - \Delta\alpha_{e,\text{mol}}(z)\right] \tag{8.40}$$

$$\Delta\tilde{\alpha}_e \equiv \tilde{\alpha}_{e\lambda,\text{probe}} - \tilde{\alpha}_{e\lambda,\text{ref}}$$

$\Delta\alpha_{e,\text{aerosol}}$ = difference in aerosol extinction coefficients at λ_{probe} and λ_{ref}
$\Delta\alpha_{e,\text{mol}}$ = difference in air extinction coefficients at λ_{probe} and λ_{ref}

In comparison to Equation 8.37, the two differential extinction coefficients for aerosols and air molecules are included, whereas the factor 1/2 is removed because the Raman-shifted signal makes a one-way trip from the scatterer to the detector.

The *Doppler wind lidar* operates on the same principle as laser Doppler velocimetry, discussed in Chapter 4, except that a dual-beam setup is not feasible due to the large probe distances involved. Because aerosol backscatter is much stronger than molecular scattering, it is more convenient to use aerosols as wind tracers wherever possible. The single-beam arrangement again requires an extremely high-resolution device to detect the small Doppler frequency shifts, such as a Fabry-Perot interferometer. Another alternative is to mix the returned Doppler-shifted signal with the unshifted laser signal (heterodyne detection). In field implementations, a local oscillator laser provides the unshifted frequency source, while a pulsed, narrowband, high-power laser is used as the transmitting source.

The fact that a ground-based lidar scans a conical volume above it requires a set of transformations to recover the velocity components. In principle, a single Doppler lidar measures only the radial component of the velocity, as shown in Figure 8.27 and described in terms of the angles θ and ϕ.

$$v_r = -u \sin\theta \cos\phi - v \cos\theta \cos\phi - w \sin\phi \tag{8.41}$$

v_r = radial component of the velocity measured by Doppler wind lidar
u = west-east component
v = south-north component
w = vertical component
θ = azimuthal angle, clockwise from the north
ϕ = elevation angle

However, the lidar is scanned in the azimuthal plane, and assuming that the atmospheric wind conditions are homogeneous in the measured plane, all the parameters

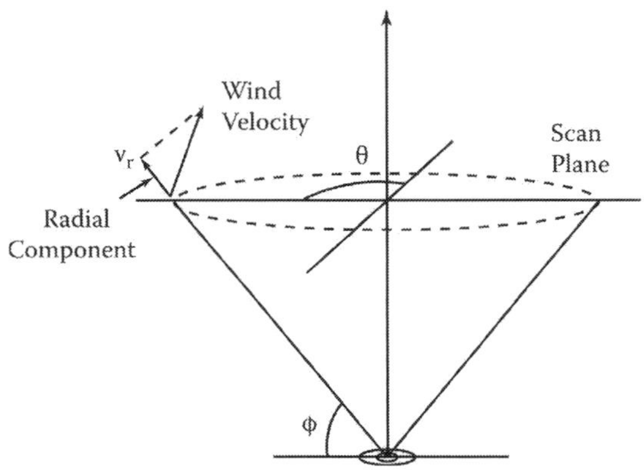

FIGURE 8.27 Velocity components in atmospheric Doppler wind lidar measurements.

in Equation 8.41 can be set as constant except for the scanned angle, θ, and the measured v_r can be fitted to a sinusoidal function of θ.

$$v_r = a + b\cos(\theta - \theta_{max}) \tag{8.42}$$

a = offset
b = amplitude
θ_{max} = phase shift

Equation 8.42 represents the effect of pointing the lidar at an angle θ and observing the radial component, at that angle, of a homogeneous velocity vector. The amplitude b corresponds to the magnitude, and the phase shift at which the maximum v_r occurs gives the azimuthal direction of the wind velocity vector. Thus, using a, b, and θ_{max}, we can recover the three velocity components.

$$\begin{aligned} u &= -\frac{b \sin \theta_{max}}{\cos \phi} \\ v &= -\frac{b \cos \theta_{max}}{\cos \phi} \\ w &= -\frac{a}{\sin \phi} \end{aligned} \tag{8.43}$$

8.6 SATELLITE-BASED SENSORS

For studying the Earth's atmospheric flows and global thermodynamics (e.g., global warming), satellite-based remote-sensing data are used. As the global climate system represents a critical thermal and flow problem, the importance of satellite-sensing methods is increasing. Many of the sensors on satellites operate using the diagnostic methods discussed in this book, particularly in Chapter 5 on optical diagnostics. Satellite-based sensors can be classified as active or passive. Active sensors use an on-board or coordinated source of electromagnetic radiation and then detect the return signal, similar to laser diagnostics. For example, LIDAR, discussed in the previous section, can be placed on board for air quality profiling. Passive sensors receive reflected light or emission as the signal source. LANDSAT data are an example of passive sensing. Target objects of satellite-based remote sensing include the atmosphere, ocean, vegetation, soil/earth surface, and anthropogenic surfaces such as pavements and buildings. Below are some of the widely used satellite sensors.

Some of the signals used in satellite sensors include visible/infrared (for vegetation and urban area monitoring), thermal infrared (for temperature monitoring of land and sea surfaces), lidar (for wind speed and ice sheet thickness), synthetic aperture radar (SAR, for Earth surface topography), and microwave (for sea surface temperature, water vapor content, precipitation, and snow cover thickness). The sensors are mounted on satellites that orbit the Earth in geosynchronous (constant observation of the same area) or polar orbits (global coverage through multiple passes). Below are some examples of satellite-sensing platforms (Figure 8.28).

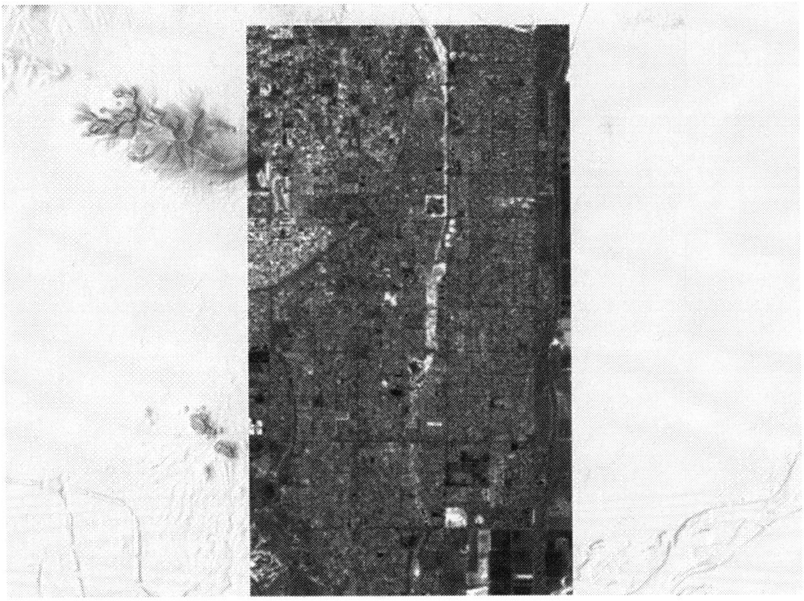

FIGURE 8.28 An example of a LANDSAT image, used for urban surface monitoring.

A. LANDSAT (Land Satellite) 4, 5, 7, 8: LANDSAT is a passive imaging platform with a vast amount of archived data that is readily available. LANDSAT has 30 m spatial resolution with a 16-day time period of orbit. On LANDSAT 8 (launched in 2013), the following data bands are acquired.
Band 1: Coastal Aerosol (0.43–0.45 μm)
Band 2: Blue (0.450–0.51 μm)
Band 3: Green (0.53–0.59 μm)
Band 4: Red (0.64–0.67 μm)
Band 5: Near-Infrared (0.85–0.88 μm)
Band 6: SWIR 1(1.57–1.65 μm)
Band 7: SWIR 2 (2.11–2.29 μm)
Band 8: Panchromatic (PAN) (0.50–0.68 μm), 15-m resolution
Band 9: Cirrus (1.36–1.38 μm)
Band 10: TIRS 1 (10.6–11.19 μm), 100-m resolution
Band 11: TIRS 2 (11.5–12.51 μm), 100-m resolution

B. MODIS (Moderate Resolution Imaging Spectro-radiometer): MODIS observes the Earth using 36 spectral bands (0.41–14.2 μm), with a 2,330-km wide viewing swath. Bands 1 and 2 have a resolution of 250 meters, bands 3–7 have 500 meters, and bands 8–36 have 1000 meters (therefore, the name "moderate resolution"). With this instrument, global climate parameters such as cloud cover, cloud aerosol size distributions, columnar (vertical) distributions of atmospheric water vapor content and temperature, and photosynthetic activity (vegetation indices) can be measured.

C. ERBS (Earth Radiation Budget Satellite): The Earth's climate system is arguably the most important problem in science and engineering. The Earth exchanges thermal energy primarily through radiation, and NASA has a sensor platform dedicated to studying incoming and outgoing radiation, called ERBS. Incoming radiation from the Sun is measured using an active-cavity radiometer with a precision of 0.01%. Outgoing radiation is measured separately using total wavelength (0.2–50 μm), long total wavelength (0.5–50 μm), and short total wavelength (0.2–5 μm) thermistor bolometers. The Earth's radiation budget is analyzed in combination with other sensors such as the Upper Atmosphere Research Satellite (UARS), Solar and Heliospheric Observatory (SOHO), and Variability of Solar Irradiance and Gravity Oscillations (VIRGO).

PROBLEMS

8.1. Sketch the sequence of microfabrication methods (i.e., photolithography, etching, deposition, etc.) to produce the following microsensor for flow rate measurements.

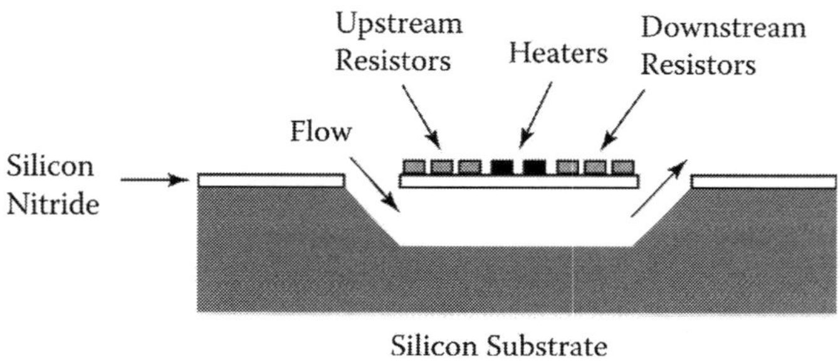

8.2. Show that for $dp/dx = (p_2 - p_1)/L = \Delta p/L$, Equation 8.5 integrates to give the volumetric flow rate of $Q = \int_A U(r)\,dA = \frac{\pi R^4}{8\mu L}\Delta p$ (Equation 8.6).

$$U(r) = -\frac{R^2}{4\mu}\left(\frac{dp}{dx}\right)\left(1 - \frac{r^2}{R^2}\right) \tag{8.5}$$

8.3. Plot the volumetric flow rate of air at $T = 300$ K and $p = 1$ atm through a rectangular channel as a function of the pressure drop, Δp, using Equations 8.7 and 8.8. Vary w/L from 1, 0.5, 0.2, 0.1–0.05.

8.4. A semiconductor material has a Young's modulus of $8.59\cdot10^{11}$ dyn/cm^2 and Poisson's ratio of 0.31. The factor, $\pi_e = \dfrac{\frac{d\rho_e}{\rho_e}}{E\frac{dL}{L}}$, in Equation 8.10 is estimated as $7.59\cdot10^{-10}$ cm^2/dyn for this material. If a resistive strain gauge element made of this material has an unstrained resistance of 25 Ω. What is the resistance change for a strain of 0.0001, 0.001, 0.01, and 0.1?

8.5. For a temperature sensor using thermo-resistive effect (Equation 8.11), what is the required temperature coefficient of resistance, α, to achieve a 10% change in resistance for a temperature variation of 25 K. What are some of the materials that can give this kind of thermo-resistive response?

8.6. What is the diffraction-limited image diameter of a particle using a microscopic imaging system with a magnification of 50, *f*# of 0.55, and using a wavelength of 540 nm?

8.7. Equation 4.40 also gives a relationship between the particle size and the image diameters. For a particle of 480 nm, use Equation 4.40 to find the image diameter.

8.8. Using Equations 8.14 and 8.19, plot the resolving power of an electron microscope as a function of the applied voltage from $V = 5{,}000$ to 20,000 V. The f# can be determined using the distance from the objective "lens" to the target (2.5 cm) and the objective aperture of 5 mm, as shown in Figure 8.18.

8.9. A focal spot size in a scanning electron microscope has a diameter of 50 nm. The interaction of the electrons with the target material produces 10^2 secondary electrons per 1 nm^3 of the material volume for a uniform electron penetration depth of 10 nm. Plot the number of secondary electrons released from the target as a function of the orientation angle, θ, of the target surface (see Figure 8.19).

8.10. Plot the particle size distributions of heavy fog and cumulus clouds using the parameter values in Table 8.4. What are the mean particle diameter and the total number density?

8.11. For a Doppler wind lidar, show that the velocity components are given by Equation 8.42.

$$u = -\frac{b \sin \theta_{max}}{\cos \phi}$$

$$v = -\frac{b \cos \theta_{max}}{\cos \phi} \tag{8.42}$$

$$w = -\frac{a}{\sin \phi}$$

BIBLIOGRAPHY

Colapicchioni, C., Barbaro, A., Porcelli, F., and Giannini, I., Immunoenzymatic assay using CHEMFET devices, *Sensors and Actuators*, B4, p. 245, 1991.

Keane, R.D., Adrian, R.J., and Zhang, Y., Super-resolution particle imaging velocimetry, *Measurement Science and Technology*, 6, p. 754, 1995.

Kovacs, G.T.A., *Micromachined Transducers Sourcebook*, McGraw-Hill, New York, 1998.

Meinhart, C.D., Wereley, S.T., and Santiago, J.G., Micro-resolution velocimetry techniques, in *Developments in Laser Techniques and Applications to Fluid Mechanics* (Adrian, R.J., Ed.), Springer-Verlag, Berlin, 1998.

Oosterbroek, R.K., Lammerink, T.S.J., Berenschot, J.W., van den Berg, A., and Elwenspoek, M.C., Designing, realization and characterization of a novel capacitive pressure/flow sensor, *Proceedings of Transducers* 97, 1, p. 151, 1997.

Santiago, J.G., Wereley, S.T., Meinhart, C.D., Beebe, R., and Adrian, R.J., A particle imaging velocimetry system for microfluidics, *Experiments in Fluids*, 25(4), p. 316, 1998.

Weitkamp, C., Ed., *Lidar: Range-Resolved Optical Remote Sensing of the Atmosphere*, Springer Series in Optical Sciences, Springer-Verlag, Berlin, 2005.

Appendix A
Electronic Devices

Electronic devices and circuitry are frequently used in experimentation. Power supplies, signal conditioning, timing circuitry, frequency filters, data acquisition systems, and displays are some of the devices in which electronic circuits are used. Although most of these devices can be purchased "off the shelf," it is useful to know some of the fundamentals of electronic operations for the judicious selection of equipment and for the laboratory fabrication of simple devices where possible. It is not possible to cover all the topics in electronic circuits and devices, but we will provide some basic principles as starters.

We saw an example of electric circuitry in measuring the resistance change in resistance-based temperature detectors (or RTDs) in Chapter 2. RTDs undergo a change in resistance that needs to be detected to recover the temperature data. Resistance can be measured in a classical Wheatstone bridge, as shown in Figure 2.16a and reproduced here in Figure A.1. This example involves the use of the basic Ohm's law ($V = IR$). In the Wheatstone bridge, the variable resistance, R_3, is adjusted until the voltage across the bridge (V_{BD}) is zero. The fact that V_{BD} is zero means that the voltage drops across the elements must be the same for both sides of the bridge, $V_{AB} = V_{AD}$ and $V_{BC} = V_{DC}$. Using Ohm's law, this condition translates to

$$\begin{aligned} I_{AB}R_1 &= I_{AD}R_3 \\ I_{BC}R_2 &= I_{DC}R_{RTD} \end{aligned} \tag{A.1}$$

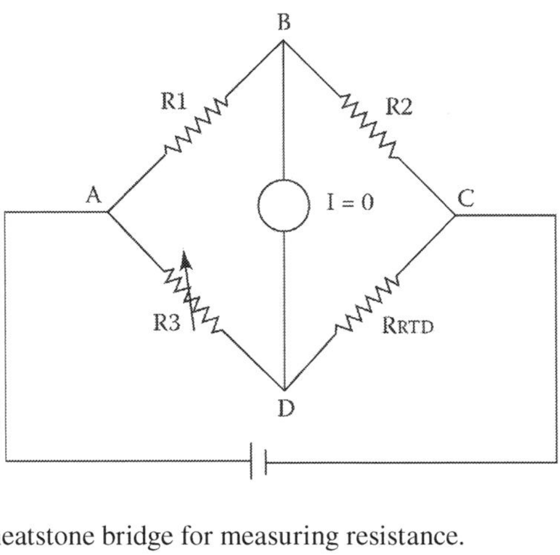

FIGURE A.1 A Wheatstone bridge for measuring resistance.

Also, the fact that $V_{BD}=0$ means that $I_{BD}=0$, leading to $I_{AB}=I_{BC}$ and $I_{AD}=I_{DC}$. Using this fact in Equation A.1, we obtain a relationship between the known resistances and the RTD resistance:

$$\frac{R_2}{R_1} = \frac{R_{RTD}}{R_3} \tag{A.2}$$

Next, we move from the preceding example of the Wheatstone bridge to op-amp-based electronic devices commonly used in laboratories.

A.1 OPERATIONAL AMPLIFIERS

Operational amplifiers, or op-amps, are packaged solid-state devices that can be used as "off-the-shelf" components in many simple and useful circuits. Although op-amps were initially designed to perform analog mathematical operations such as addition, subtraction, integration, and differentiation, their usage in circuitry is much more diverse. Op-amps are integrated circuits with approximately 20 transistors along with resistors and capacitors, but for our purposes, they can be considered as a "black box" that performs various functions depending on how circuit elements are connected to it. Low-cost op-amps are commercially available with a wide range of performance characteristics such as gain, supply voltage, voltage offset, slew rates, and other parameters. Some of the basic characteristics are common to all op-amps and op-amp circuits; they provide gains of 10^5 to 10^6 and are powered by a nominal 15-V DC supply. An op-amp typically has an eight-pin design, as shown in Figure A.2, and is drawn as a triangle in a circuit diagram with + and − inputs and an output terminal. Usually, the power supply to the op-amp is omitted in op-amp circuits. A wide range of circuits involving op-amps can be analyzed without too much difficulty by applying the following op-amp rules:

Op-amp Rule 1: The voltage difference between the + and − inputs is pushed to zero by the op-amp's internal circuit.

Op-amp Rule 2: The inputs draw zero current.

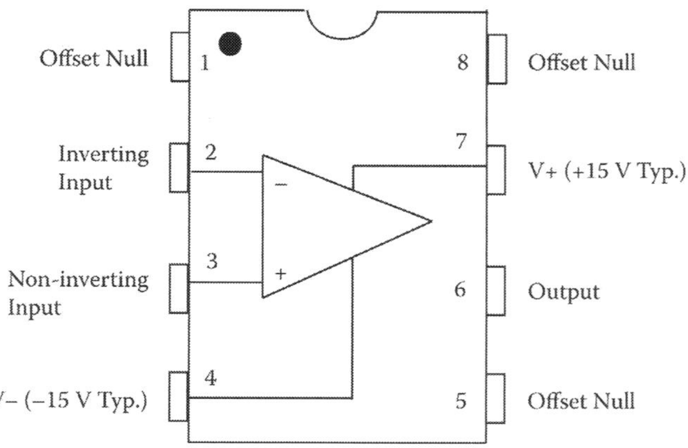

FIGURE A.2 An operational amplifier.

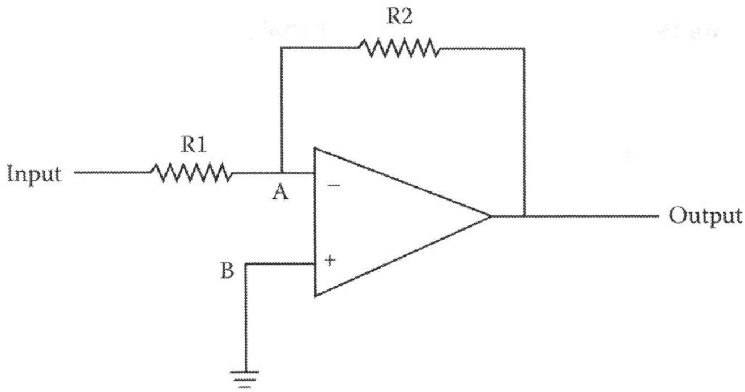

FIGURE A.3 An inverting amplifier circuit using an op-amp.

A.1.1 Inverting Amplifier

Figure A.3 shows an op-amp circuit that outputs an amplified voltage inverted with respect to the input voltage. In the circuit shown, the positive input is grounded, so the voltage at that point is zero ($V_B=0$). Applying Rule 1 (the voltage difference across the input terminals is zero) means that the voltage at the negative terminal is also zero ($V_A=0$). Now, if we apply Rule 2, which states that the inputs draw no current, meaning that the current going into the "–" input terminal is zero, we can set the current at point *A* equal in magnitude but opposite in sign because both the input and output voltages are oriented to send current into the node at *A*.

$$I_1 = -I_2 \tag{A.3}$$

I_1 = current across the resistor R_1
I_2 = current across the resistor R_2

Using Ohm's law, we have

$$\frac{V_{out}}{R_2} = -\frac{V_{in}}{R_1} \tag{A.4}$$

And the gain from this circuit is

$$\text{Gain} = \frac{V_{out}}{V_{in}} = -\frac{R_2}{R_1} \tag{A.5}$$

A.1.2 Noninverting Amplifier

An amplifier with the same output polarity as the input can be designed, as shown in Figure A.4. Again, applying the two rules, $V_A = V_{in,}$ and the minus input terminal draws no current. Therefore, the current going through R_1 is the same as the current going through R_2. So, we can write

$$\frac{V_A}{R_1} = \frac{V_{out}}{R_1 + R_2} \tag{A.6}$$

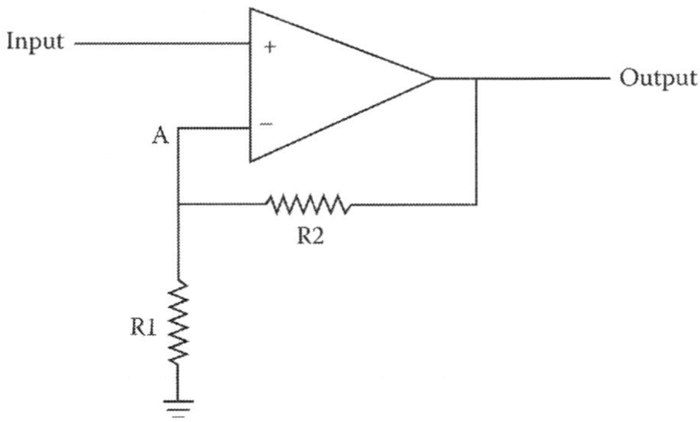

FIGURE A.4 A noninverting amplifier circuit using an op-amp.

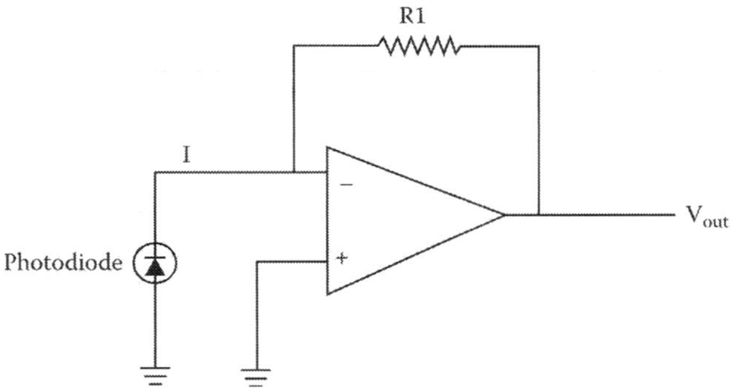

FIGURE A.5 A current-to-voltage converter.

Thus, the gain is

$$\text{Gain} = \frac{V_{\text{out}}}{V_{\text{in}}} = 1 + \frac{R_2}{R_1} \tag{A.7}$$

A.1.3 Current-to-Voltage Converter

Another useful device is a current-to-voltage converter, as shown in Figure A.5. Optical devices such as photodiodes generate current in response to light incident on the photodiode surface. One way to read the signal from the photodiode is to convert this current into voltage using the current-to-voltage circuit. The analysis is as follows: the input draws no current, so all the current from the photodiode flows through the resistance, R_1, and the output voltage is proportional to the input current, I.

$$I = \frac{V_{\text{out}}}{R_1} \tag{A.8}$$

A.1.4 Differential Amplifier

An output proportional to the difference between two input voltages can be generated using the differential amplifier circuit shown in Figure A.6. The output is given by

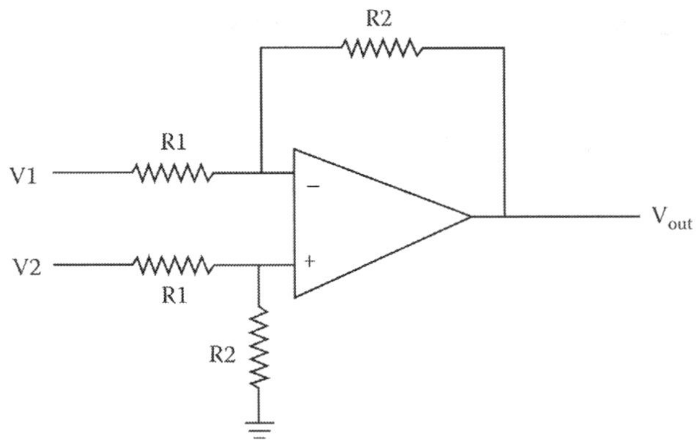

FIGURE A.6 A differential amplifier.

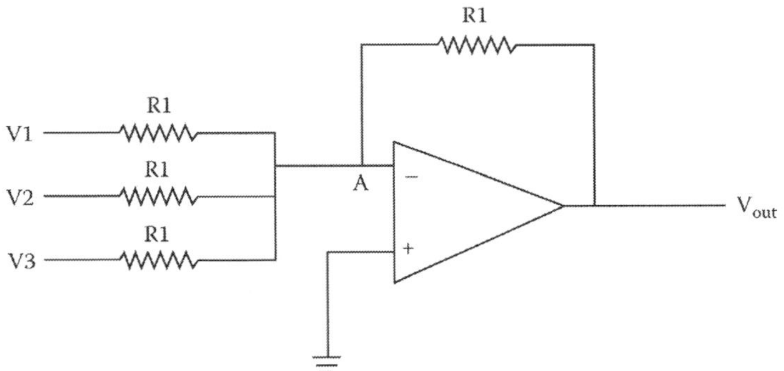

FIGURE A.7 A summing amplifier.

$$V_{out} = \frac{R_1}{R_2}(V_2 - V_1) \tag{A.9}$$

As will be shown later, the differential, or more exactly, difference operation, achieved by this device is useful for comparing the incoming signal, V_2, with respect to some reference signal, V_1. Because op-amps can produce gains on the order of 100,000, the voltage difference, $V_2 - V_1$, needs to be a fraction of a few millivolts before the output reaches either negative or positive saturation (peak output voltage). Thus, R_1 and R_2 can be chosen so that the output is either the positive or negative maximum output, depending on whether V_2 does or does not exceed V_1. This is a useful operation in laboratory instrumentation for triggering electronic components, and there is a class of dedicated devices called *comparators* that perform this operation with high-frequency response, to be discussed in the next section.

Figures A.7 and A.8 show a couple of other interesting and useful devices that use an op-amp: a summing amplifier (Figure A.7) and an integrator (Figure A.8). The summing amplifier gives an output of $V_{out} = -(V_1 + V_2 + V_3)$, whereas the integrator gives a time integral of the input voltage ($V_{out} = \frac{1}{RC}\int V_{in}\,dt +$ constant).

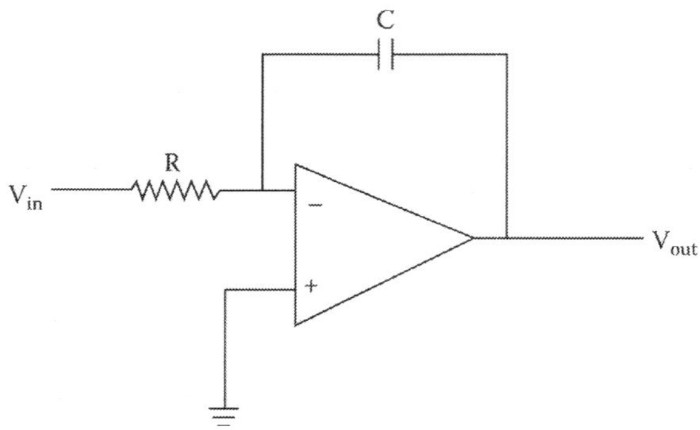

FIGURE A.8 An integrator.

A.2 TRIGGER ELECTRONICS

One of the simple yet useful electronics in thermal and flow measurements is an electronic trigger. Comparators are similar to the differential amplifiers discussed earlier, except that they go from ground (when V_{in} is less than some threshold voltage) to +5 V (when V_{in} is greater than the threshold voltage). The ground-to-+5 V signal is used in many electronic devices and is referred to as the Transitor-to-transitor logic (TTL) signal. TTL signals are used to trigger many laboratory electronic devices. *Schmitt triggers* also generate up or down signals, except that two threshold levels can be set for the upswing and down swing, respectively. Typical input and output responses of comparators and Schmitt triggers are shown in Figure A.9. It can be seen that comparators, due to their single set threshold, can generate rapid swings in the output on noisy input signals. On the other hand, Schmitt triggers provide a margin for fluctuations in the input signal before the output is reset to ground. A *timer* is yet another solid-state device that can be purchased off the shelf, and it generates an output signal with a certain time delay that can be adjusted using an external resistor and capacitor. These and numerous other devices are manufactured by many electronic device manufacturers (e.g., NEC, Motorola, Texas Instruments, Samsung, and so on) and, in most instances, can be purchased at low cost along with technical specifications to design simple circuits around them.

The functions of these devices, the Schmitt trigger and timer, are combined in a useful device called a *multivibrator*. Because it combines these functions and more, we will use it as an example of how we can build a simple triggering device for laboratory use. A multivibrator typically comes in a 16-pin package, as shown in Figure A.10, and we will use the Texas Instruments SN54122 retriggerable monostable multivibrator as an example (other devices share very similar characteristics and are well described in the attendant data sheets). Depending on the high, low, rising, or falling input signal, either a positive or negative "gate" output can be generated, with the gate width that can be adjusted with external resistors and capacitors. Table A.1 shows the function table for this device.

A good number of different input combinations can be used to tailor the output, and the output gate width can also be controlled using an external resistor and

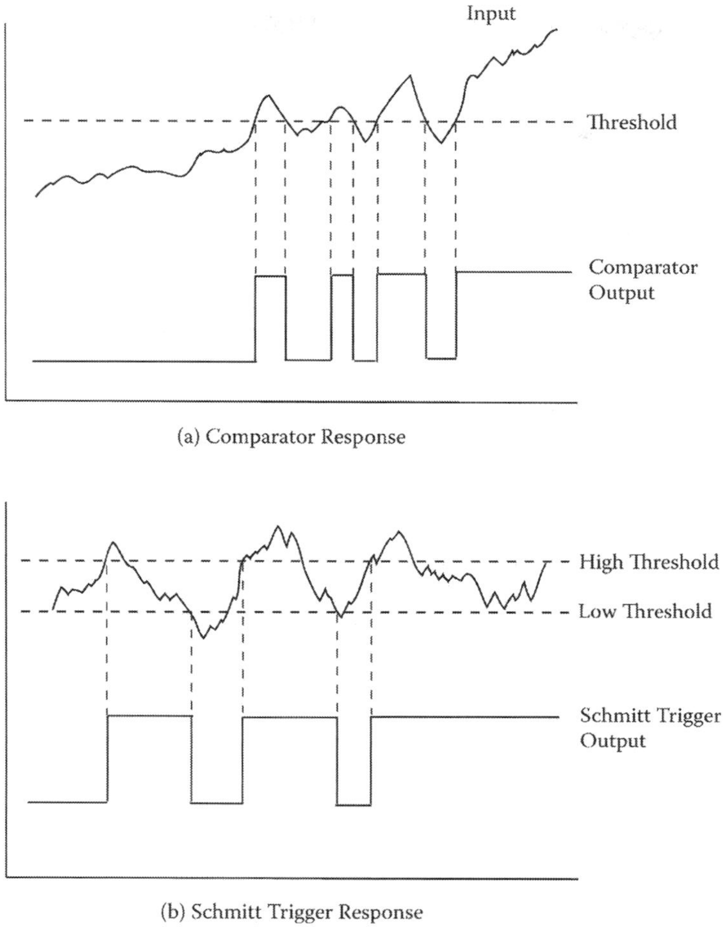

FIGURE A.9 Signal response of comparators and Schmitt triggers.

capacitor connected to the R_{ext}/C_{ext} terminal, as shown in Figure A.10. The resistor and capacitor connection is also shown in Figure A.10. The output gate width, t_W, for this device is determined as follows (for other devices, this functional dependence is provided in the technical specifications):

$$t_W = KR_{ext}C_{ext}\left(1+\frac{0.7}{R_{ext}}\right) \quad [\text{ns}] \tag{A.10}$$

K = constant = 0.32
R_{ext} = external resistance in kΩ
C_{ext} = external capacitance in pF

Thus, a multivibrator generates a gate output pulse of an adjustable length, depending on the combination of inputs. The following example shows how this device can be used to generate a delayed trigger pulse in response to an input.

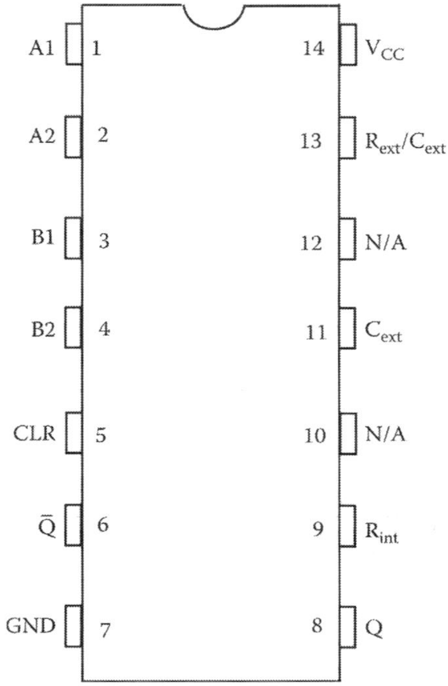

FIGURE A.10 A multivibrator pin configuration (Texas Instruments SN54122 retriggerable monostable multivibrator).

Example A.1

We want to trigger an imaging system based on an external event. For example, we want to image flame development 200 µs after ignition is observed by a photodiode. The photodiode produces a signal as shown in the following text. The imaging system requires a gate input of 10 µs.

First, we draw a timing diagram, as shown in Figure A.11. To generate a delay, we use the SN54122 multivibrator discussed earlier. To obtain a delay pulse width of 200 µs, we set the external resistance and capacitance using Equation A.10.

$$t_W = KR_{\text{ext}}C_{\text{ext}}\left(1+\frac{0.7}{R_{\text{ext}}}\right)\ [\text{ns}] \tag{A.10}$$

$K = 0.32$

Because we have t_W of 200,000 ns, we use a 500-pF capacitor and a resistor with 1249.3 kΩ resistance. Actually, a resistor of 1249.3 kΩ is usually not available, so instead we use a variable resistor (or a potentiometer) to adjust the resistance until we see the 200,000 ns = 200 µs pulse width on an oscilloscope. Using Table A.1, by connecting the photodiode signal to port B1 of the multivibrator (while holding CLR = H = + 5 V and A1 = L = 0 V) and using the Q output, we will get a top-hat pulse that is triggered by the rising edge of the photodiode signal.

The next step is to generate the actual gate signal for the imaging device after the earlier-described delay. So, we use a second multivibrator with C_{ext} = 500 pF and R_{ext} = 61.8 kΩ, which will give us a gate width of 10 µs. Because we are gating

TABLE A.1
Function Table for a Multivibrator

CLR	A1	A2	B1	B2	Output Q	Output $\bar{Q}$
L	N/A	N/A	N/A	N/A	L	H
N/A	H	H	N/A	N/A	L	H
N/A	N/A	N/A	L	N/A	L	H
N/A	N/A	N/A	N/A	L	L	H
H	L	N/A	↑	H	⊓	⊔
H	L	N/A	H	↑	⊓	⊔
H	N/A	L	↑	H	⊓	⊔
H	N/A	L	H	↑	⊓	⊔
H	H	↓	H	H	⊓	⊔
H	↓	↓	H	H	⊓	⊔
H	↓	H	H	H	⊓	⊔
↑	L	N/A	H	H	⊓	⊔
↑	N/A	L	H	H	⊓	⊔

Note: L = low; H = high; N/A = not used.

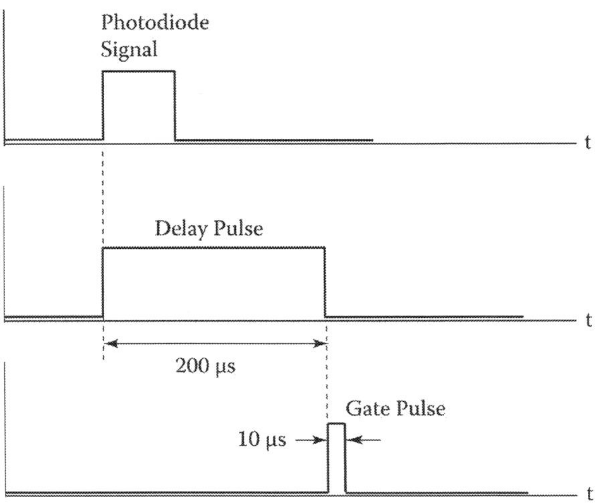

FIGURE A.11 The timing diagram for generating a gate pulse using a multivibrator.

off the falling edge of the delay pulse, we send the delay pulse signal to A2 while holding CLR = A1 = B1 = B2 = H = + 5 V. This will lead to the final gate pulse as shown in Figure A.11.

A.3 FREQUENCY FILTERS

A.3.1 FILTER BASICS

In dealing with frequency signals (signals that contain time-dependent characteristics), it is often necessary to "clean" the signals by applying analog or digital filters. The most basic filter circuit, consisting of a resistor and a capacitor, is shown in Figure A.12. The capacitor is a device that stores charge when a voltage is applied. It is useful, for the sake of understanding filter operations, to review the basics of the transient behavior of electrical circuits. The capacitance is defined as

$$C = \frac{Q}{V} \quad \text{or} \quad Q = CV \tag{A.11}$$

Q = charge [Coulomb]
C = capacitance [F, μF, or pF]
V = voltage [V]

Differentiating both sides of Equation A.11, we get

$$I = \frac{dQ}{dt} = C\frac{dV}{dt} \tag{A.12}$$

The current is defined as the time rate of change of charge. Using Equation A.12, we can see the result of a time-dependent input voltage, $V(t)$. First, we write $V(f)$ as a sinusoidal function with a circular frequency, $\omega = 2\pi f$, using the Euler notation, where f is the frequency of the input signal.

$$V(t) = \text{Re}\left(V_o e^{j\omega t}\right) = \text{Re}\left\{V_o\left(\cos j\omega t + \sin j\omega t\right)\right\} \tag{A.13}$$

V_o = amplitude
j = complex number

Substituting Equation A.13 into A.12, we obtain

$$I(t) = C\frac{d}{dt}\text{Re}\left(V_o e^{j\omega t}\right) = C\text{Re}\left(V_o j\omega e^{j\omega t}\right) = j\omega C\text{Re}\left(V_o e^{j\omega t}\right) = \frac{V(t)}{-\dfrac{j}{\omega C}} \tag{A.14}$$

We write the equation in the preceding form so that it becomes analogous to Ohm's law, $I = V/R$. In fact, we can write a general form of Ohm's law where the denominator is the "reactance" of a resistor, a capacitor, or an inductor.

$$I = \frac{V}{X} \tag{A.15}$$

X = reactance

As we can see in Equation A.15, the reactance for a capacitor is then

$$X_C = -\frac{j}{\omega C} \quad \text{for capacitors} \tag{A.16a}$$

For other circuit elements, the reactance can be written.

$$X_R = R \quad \text{for resistors} \tag{A.16b}$$

R = resistance [Ω]

$$X_L = j\omega L \quad \text{for inductors} \tag{A.16c}$$

L = inductance [H]

Now, we can apply the generalized Ohm's law to the filter element shown in Figure A.12. The relationship between the input voltage and the current is

$$I = \frac{V_{\text{in}}}{X_{\text{total}}} = \frac{V_{\text{in}}}{R - \frac{j}{\omega C}} = \frac{\left(R + \frac{j}{\omega C}\right)V_{\text{in}}}{R^2 + \frac{1}{\omega^2 C^2}} \tag{A.17}$$

Because the resistor and the capacitor are in series, we simply add the two reactances to find the total reactance, X_{total}. We can also write the relationship between the output voltage and the current (which is the same as the current in Equation A.17).

$$V_{\text{out}} = IX_R = IR = \frac{V_{\text{in}}}{R - \frac{j}{\omega C}} R = \frac{R}{R - \frac{j}{\omega C}} V_{\text{in}} \tag{A.18}$$

Essentially, the circuit in Figure A.12 is a voltage divider using one resistor and one capacitor. Now, we are only interested in the amplitude, not the phase, of the output voltage. Therefore, we convert the output voltage in Equation A.18 into an amplitude equation.

$$|V_{\text{out}}| = \left(V_{\text{out}} V_{\text{out}}^*\right)^{1/2} = \frac{R}{\left[R^2 + \frac{1}{\omega^2 C^2}\right]^{1/2}} V_{\text{in}} \tag{A.19}$$

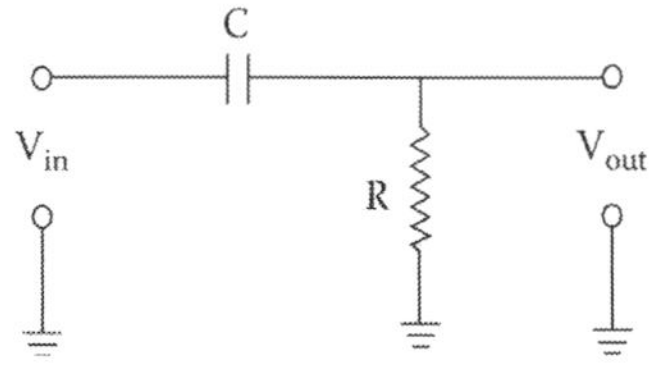

FIGURE A.12 A basic filter circuit using a resistor and a capacitor.

Using $\omega = 2\pi f$, we can again find the response V_{out}/V_{in}, which gives us the frequency response.

$$\frac{V_{out}}{V_{in}} = \frac{2\pi fRC}{\left[1+(2\pi fRC)^2\right]^{1/2}} \tag{A.20}$$

The frequency response is entirely determined by the choice of R and C, and the circuit action is to cut off the low-frequency component while passing the high-frequency component (the output increases with increasing f). This circuit action is called a high-pass filter.

In filter terminology, we often use the so-called −3 dB cutoff points. This is where the output signal starts to fall off due to the filtering action of the circuit. The decibel, dB, is defined in terms of the logarithm based on 10.

$$\mathrm{dB} = 20\ \log_{10}\frac{A_2}{A_1} \tag{A.21}$$

For example, +6 dB means that the signal ratio, A_2/A_1, is 2, since as $20\log_{10}(2) = 20(0.3010) \approx 6$. The signal ratio of 10 gives +20 dB, whereas 1/10 gives −20 dB, and so on. Thus, the −3 dB cutoff point means that V_{out}/V_{in} has reached a ratio of 0.7079. The −3 dB point in the circuit represented by Equation A.20 is then

$$f_{3\mathrm{dB}} = \frac{1}{2\pi RC} \tag{A.22}$$

R = resistance in kΩ
C = capacitance in μF

A.3.2 Active Filters

The preceding example is referred to as a passive filter. Filters using op-amps are called *active filters*, and they are widely used in signal processing due to their well-defined and precise performance. Figure A.13 shows fundamental active filter circuit elements for (a) a low-pass and (b) a high-pass filter. There are different types of filters that can be built by cascading the elements shown in Figure A.13. Different filters give different output characteristics; however, they are based on the same filter element, except that the R and C values are adjusted. For example, a *Butterworth* filter is built by setting the R and C values as $R_1 = R_2 = R$ and $C_1 = C_2 = C$. This gives the cutoff frequency defined by R and C.

$$RC = \frac{1}{2\pi f_C}$$
$$f_C = f_{3\mathrm{dB}} \tag{A.23}$$

The Butterworth filter gives the following response:

$$\frac{V_{out}}{V_{in}} = \frac{1}{\left[1+(f/f_C)^{2n}\right]^{1/2}} \tag{A.24}$$

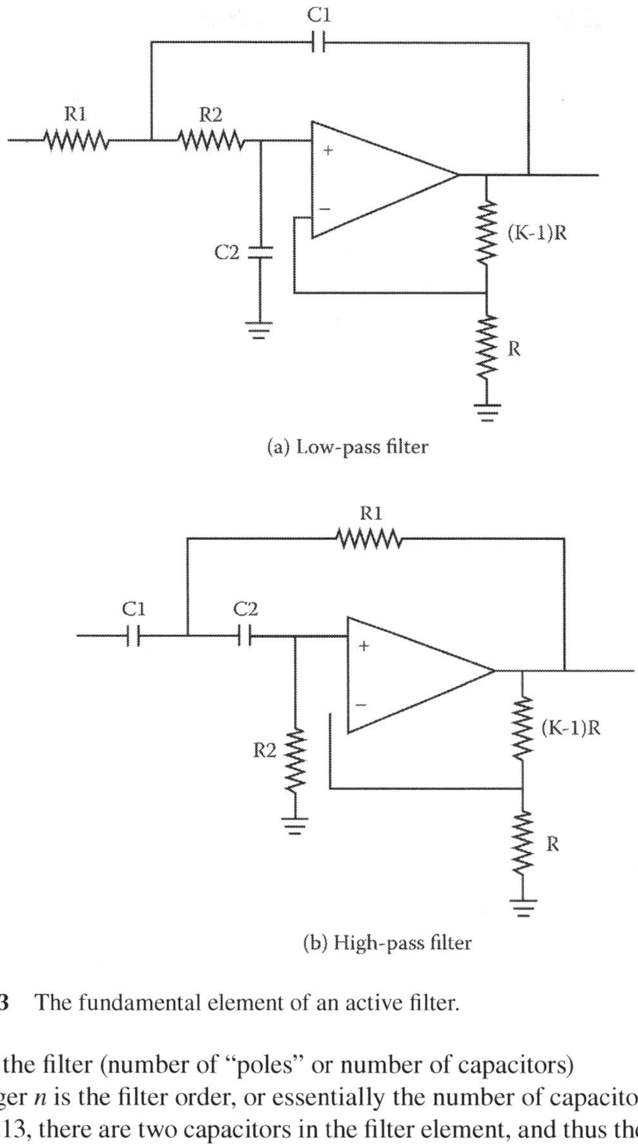

(a) Low-pass filter

(b) High-pass filter

FIGURE A.13 The fundamental element of an active filter.

n = order of the filter (number of "poles" or number of capacitors)

The integer n is the filter order, or essentially the number of capacitors. As shown in Figure A.13, there are two capacitors in the filter element, and thus the exponent in Equation A.24 is $2n$. Higher-order filters are achieved simply by cascading the filter element in Figure A.13 with identical R and C values. The higher the filter order, the steeper the cutoff will be, with the cutoff frequency given by Equation A.23. The resistance on the output side of the filter is given by R and $(K-1)R$, where the value of K is the filter gain and is listed in Table A.2 for Butterworth and other types of filters. It can also be seen in Figure A.13 that a low-pass filter can be converted to a high-pass filter by simply replacing R_1 and R_2 with C_1 and C_2, and vice versa. The cutoff frequency and the filter gain are then exactly the same in low- and high-pass filters, except, of course, that one passes low-frequency components and the other passes high-frequency components. A *bandpass* filter is constructed by cascading a low and a high-pass filter with an appropriately set pass band in between.

TABLE A.2
Filter Parameters

Total Number of Poles	Pole Number	Butterworth K	Bessel f_n	Bessel K	Chebyshev (0.5 dB) f_n	Chebyshev (0.5 dB) K	Chebyshev (2.0 dB) f_n	Chebyshev (2.0 dB) K
2	2	1.586	1.272	1.268	1.231	1.842	0.907	2.114
4	2	1.152	1.432	1.084	0.597	1.582	0.471	1.924
	4	2.235	1.606	1.759	1.031	2.660	0.964	2.782
6	2	1.068	1.607	1.040	0.396	1.537	0.316	1.891
	4	1.586	1.692	1.364	0.768	2.448	0.730	2.648
	6	2.483	1.908	2.023	1.011	2.846	0.983	2.904
8	2	1.038	1.781	1.024	0.297	1.522	0.238	1.879
	4	1.337	1.835	1.213	0.599	2.379	0.572	2.605
	6	1.889	1.956	1.593	0.861	2.711	0.842	2.821
	8	2.610	2.192	2.184	1.006	2.913	0.990	2.946

Other commonly used filter designs include *Bessel* and *Chebyshev* filters, the parameters for which are given in Table A.2. The *R*s and *C*s are constant in all of the cascaded elements in the Butterworth filter. However, in Bessel and Chebyshev filters, we alter the *RC* values according to

$$RC = \frac{1}{2\pi f_n f_C} \tag{A.25}$$

f_n = normalizing factor given in Table A.2

Within each pole element, again $R_1 = R_2 = R$, and $C_1 = C_2 = C$, and they are just incremented in the next pole elements. High-pass and bandpass filters can again be constructed in a manner identical to Butterworth filters, i.e., by reversing *R*s and *C*s and cascading high- and low-pass filters. The two different types of Chebyshev filters in Table A.2 (0.5 and 2 dB) refer to the magnitude of ripples that occur in the pass-band for these filters. Butterworth and Bessel filters do not have ripples.

A.4 DATA ACQUISITION

Most of the measurement methods discussed in this book are interfaced with a computer (for data storage and control) through data acquisition (DAQ) devices. DAQ converts analog to digital, and vice versa for control output. Transducers (for pressure, temperature, flow rate, etc.) generate electrical signals that are converted to digital for display, storage, and data analysis. DAQ serves this purpose and, as noted above, also enables reverse operations of converting digital into voltage outputs for a switch or relay on-off control. DAQ also has installed circuits for real-time signal conditioning and analysis, although the same can be accomplished through a

TABLE A.3
Display Options

On-Board Function	Purpose
Oscilloscope	View real-time voltage signals
Function Generator	Output sine, square, and triangle waveforms
Digital Multimeter	Measure voltage, current, and resistance
Bode Analyzer	Analyze system frequency response
Variable Power Supply	Provide DC voltage for circuits
Digital Reader/Writer	Interface with digital logic
Analog Input/Output	Multiple channels for inputs and outputs

computer control interface. In addition to analog-to-digital conversion, modern DAQ includes other useful functions (Table A.3). An example of a DAQ device is shown in Figure A.14, with a convenient USB connection to a computer (laptop or PC) and an input/output module.

A similar set of factors discussed in Section 1.3 is important for the selection and use of DAQ. These are:

Accuracy. Digitized (read) value by DAQ relative to the true value of the signal.
Resolution. The smallest increment in the measured quantity. Since the signal is "digitized," the resolution is given by 2^n. For example, for a 16-bit DAQ covering +/− 5 V, the resolution is $10\ \text{V}/2^{16} = 10\ \text{V}/65{,}536 = 0.0001526\ \text{V}$.
Sampling Rate. This is the rate at which the signal is acquired. For accurate temporal representation, the Nyquist criterion should be met: sampling rate must be at least twice the signal frequency. Figure A.15 is an example of an under-sampled signal.
Repeatability. Duplicability of the data through DAQ.
Linearity. Although linearity calibrations are factory-set in DAQ, drifts can occur depending on the operating conditions.
Noise. Noise can adulterate the signal at both the sensor and, to a lesser extent, through DAQ. Although linearity calibrations are factory-set in DAQ, drifts can occur depending on the operating conditions.

Once the data are acquired through DAQ, they can be analyzed at the software level (external programming). Dedicated software interfaces are also frequently used for data display and analysis. LabVIEW[R] ("Laboratory Virtual Instrument Engineering Workbench") is an example of interface software that provides a graphical programming environment. It is widely used for DAQ, instrument control, and experimental/industrial automation. Instead of writing lines of code, LabVIEW users create programs by connecting graphical icons and wires called "virtual instruments." The graphical interface consists of a Front Panel and a Block Diagram, as shown in Figure A.16. Table A.4 shows a typical set of software functions.

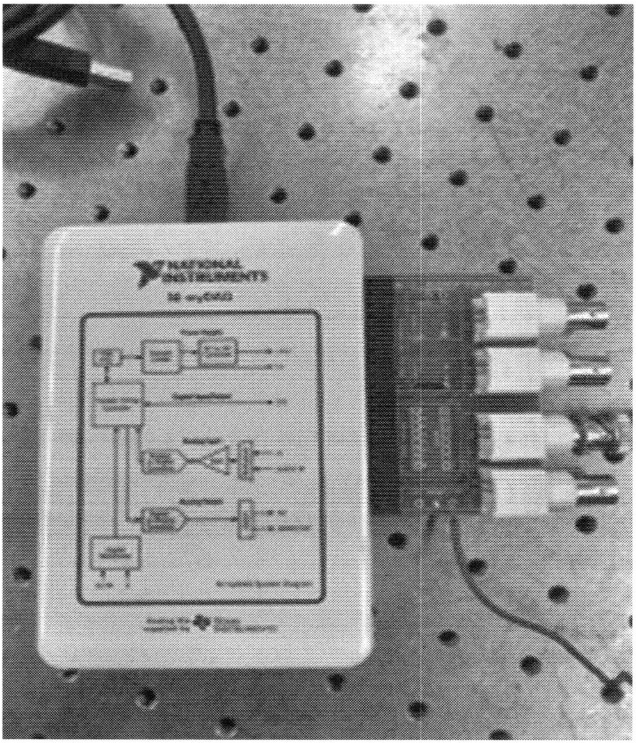

FIGURE A.14 An example of DAQ.

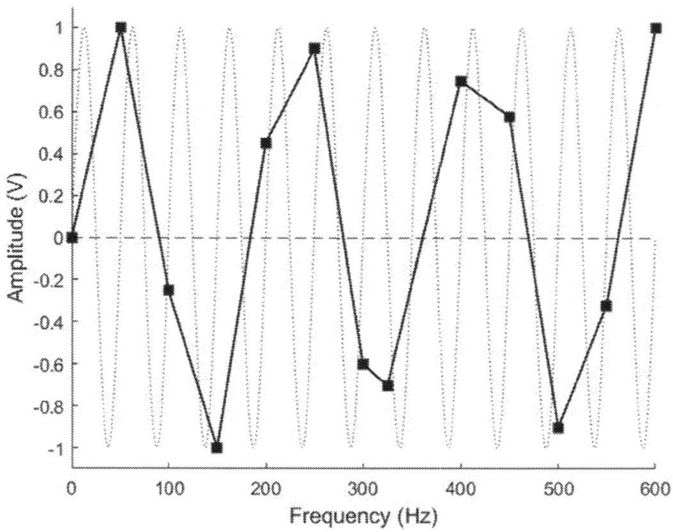

FIGURE A.15 An example of a low sampling rate, relative to the signal frequency. The dotted line is the actual signal, while the squares are sampled data points, leading to an inaccurate representation (solid line).

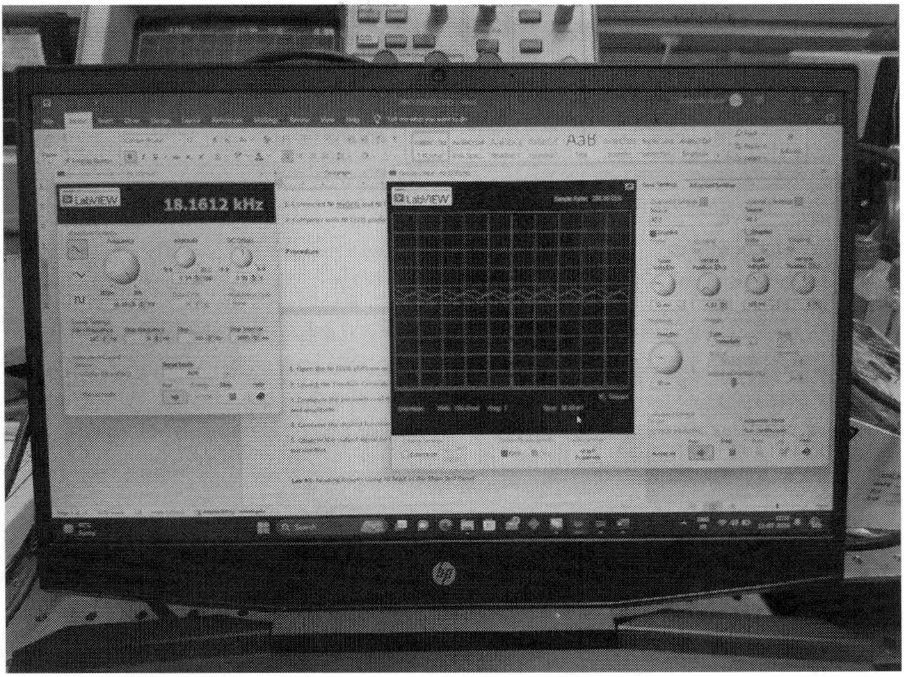

FIGURE A.16 "Front panel" on LabVIEW®. The graphic display is designed to resemble an oscilloscope.

TABLE A.4
Function Options

Function Category	Examples
Mathematics	Add, Subtract, Multiply, Divide
Array Operations	Build Array, Index Array, For Loop
File I/O	Write to File, Read from File
DAQ	Data blocks for hardware I/O
Timing	Wait (ms), Elapsed Time
Conversion	Number to String, String to Number

BIBLIOGRAPHY

Horowitz, P. and Hill, W., *The Art of Electronics*, 2nd ed., Cambridge University Press, New York, 1989.

Howatson, A. M., *Electrical Circuits and Systems: An Introduction to Engineers and Physical Scientists*, Oxford University Press, New York, 1996.

Appendix B
Optics

Optical components are used in many thermal and flow measurement systems. Most, if not all, of the flow visualization methods discussed in Chapter 2 require lenses and other optical devices to transmit the probe beam and to image the resulting signal. Flow velocity measurement methods, such as laser Doppler velocimetry and particle image velocimetry, involve sophisticated optical components to precisely align the probe laser beams and to collect and filter the signals. Additionally, some of the other diagnostic methods discussed in this book are based on optical processes, such as species concentration, particle size, and lidar measurements. For this reason, we will present some of the basic elements of optics while referring to the detailed treatment of optical phenomena in dedicated reference books listed at the end of this appendix.

The most fundamental quantity in optics is probably the speed of light. The speed of light in a vacuum is a universal constant and is 2.997925×10^8 m/s, or approximately 300,000 km/s. The speed of light is the propagation speed of a traveling electromagnetic wave of frequency (f) and wavelength (λ), so that

$$c = f\lambda \tag{B.1}$$

f = frequency of the light [s^{-1}]
λ = wavelength [μm or nm]

In different media, however, the speed of light *does* change appreciably. The ratio of the speed of light in a vacuum to that in a medium is called the refractive index and is the next most fundamental quantity in optics.

$$n = \frac{c_o}{c} = \text{ refractive index} \tag{B.2}$$

c_o = speed of light in a vacuum
c = speed of light in a medium

It turns out that the speed of light in air is quite close to its value in a vacuum, so for practical intents and purposes, the refractive index of air is considered to be one. In fact, for most gases, the refractive index does not deviate appreciably from one, as shown in Table B.l. The table gives the refractive index of common materials at a wavelength of 589 nm.

The refractive index itself is not a constant; it varies depending on the wavelength of the light interacting with the material, as well as the material's density and temperature. Table B.l gives representative values for different media at a wavelength of 589 nm. The wavelength-dependent refractive index actually causes some problems in color imaging, called chromatic aberration (color dispersion, particularly

TABLE B.1
Refractive Index of Common Materials at 589 nm

Material	Refractive Index
Air	1.00029
Carbon dioxide	1.00045
Methane	1.00044
Nitrogen	1.00030
Oxygen	1.00027
Water vapor	1.00025
Chlorine	1.385
Ethyl alcohol	1.36
Liquid water	1.33–1.32 (15°C–100°C)
Glass, crown	1.50–1.55
Glass, light flint	1.55–1.61
Glass, heavy flint	1.62–1.71
Fused quartz	1.52–1.68
Ice	1.31

TABLE B.2
Refractive Index of Optical Materials at Different Wavelengths

Optical Material	n_C (at 656.3 nm)	n_D (at 589.2 nm)	n_F (at 486.1 nm)	n_G (at 434.0 nm)
Glass, borosilicate crown	1.49776	1.50000	1.50529	1.50937
Glass, dense flint	1.61216	1.61700	1.62901	1.63923
Glass, extra dense flint	1.71303	1.72000	1.73780	1.75324
Glass, light flint	1.57100	1.57500	1.58500	1.59400
Fused quartz	1.45640	1.45845	1.46318	1.46690
Glass, spectacle crown	1.52042	1.52300	1.52933	1.53435
Glass, telescopic flint	1.52762	1.53050	1.53790	1.54379

noticeable at edges). Table B.2 gives the color (wavelength)-dependent refractive index for some optical materials.

The refractive index gives rise to a useful optical phenomenon called *refraction,* as the speed of light changes when light traverses from one medium to another. For a *specular* surface (a surface that reflects the light in a specular manner, as defined in the following text), the angle of incidence is equal to the angle of reflection, as shown in Figure B.1. All the angles are measured from a line perpendicular to the interface.

$$\phi_r = \phi_i \tag{B.3}$$

ϕ_r = reflected angle
ϕ_i = incident angle

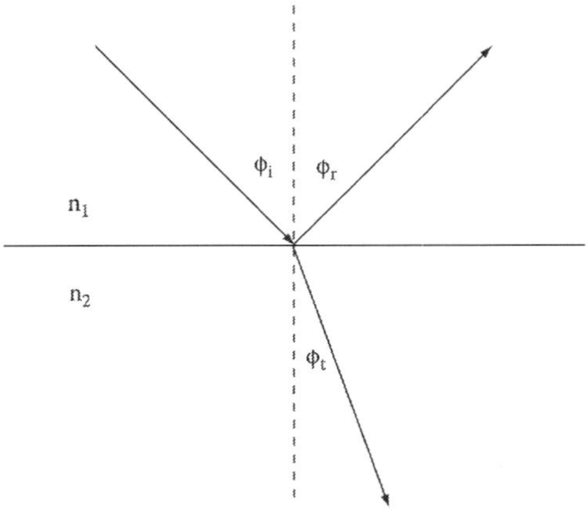

FIGURE B.1 Laws of reflection and refraction.

For the refracted ray of light, the angle is determined by Snell's law.

$$n_2 \sin \phi_t = n_1 \sin \phi_i \quad \text{or} \quad \frac{\sin \phi_t}{\sin \phi_i} = \frac{n_1}{n_2} \tag{B.4}$$

n_1 = refractive index of medium 1
n_2 = refractive index of medium 2

One action of refraction is to collect light through a lens from a distant object, and then form an image on a plane, such as a photographic plate or a CCD chip. For an object at an object distance s, the image will be formed at an image distance s', as determined by the focal length of the lens f, as illustrated in Figure B.2.

The relationship between the object distance, image distance, and focal length is known as the *lens equation*.

$$\frac{1}{s} + \frac{1}{s'} = \frac{1}{f} \quad \text{The Lens Equation} \tag{B.5}$$

s = object distance
s' = image distance
f = focal length of the lens

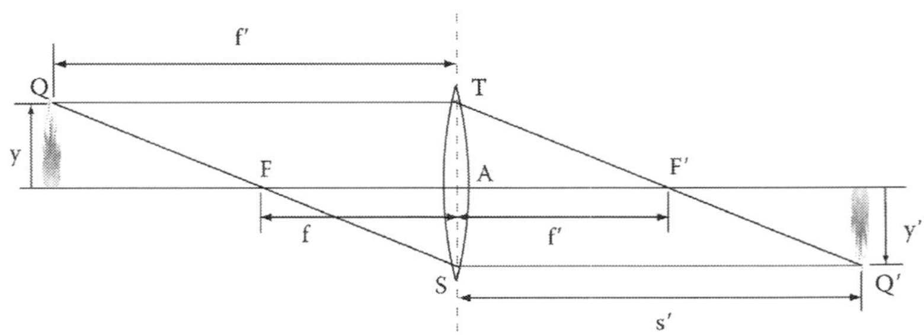

FIGURE B.2 Geometry for focusing of an object using a lens (the lens equation).

The lens equation is easily proven geometrically. We use similar triangles, Q'TS and F'TA, in Figure B.2, from which we obtain the ratio

$$\frac{y-y'}{s'}=\frac{y}{f} \tag{B.6}$$

In Equation B.6, the image distance is considered negative if it is below the centerline, i.e., $y'<0$. This is the reason for the minus sign for y' on the left-hand side. Similarly, we can use two similar triangles on the other side of the lens, QTS and FAS, to write

$$\frac{y-y'}{s}=-\frac{y'}{f'} \tag{B.7}$$

Adding B.6 and B.7,

$$\frac{y-y'}{s}+\frac{y-y'}{s'}=\frac{y'}{f'}-\frac{y}{f} \tag{B.8}$$

If the refractive index of the medium is the same on both sides of the lens, $f=f$.

Then, we obtain the lens equation (B.5).

$$\frac{y-y'}{s}+\frac{y-y'}{s'}=\frac{y'}{f}-\frac{y}{f}=\frac{y'-y}{f}\Rightarrow\frac{1}{s}+\frac{1}{s'}=\frac{1}{f} \tag{B.9}$$

This is an example of the geometrical proof of an optical equation, and from this point on, we will simply accept these results from optics references and apply them to different geometries.

Example B.1

If the object is at a distance of 6 cm from the lens (focal length 4.0 cm), then at what distance from the lens would the image form?

We can use the lens equation, B.5. The image distance is s'.

$$\frac{1}{s'}=\frac{1}{f}-\frac{1}{s}=\frac{s-f}{fs}\Rightarrow s'=\frac{fs}{s-f}$$

$$s'=\frac{(+6)(+4)}{6-4}=12\text{ cm}$$

The image distance is 12 cm, and it is positive. This means that the image is real, meaning that the image can be viewed on a screen located at this distance. Only real images, as opposed to *virtual* images, can be projected onto a screen in such a manner.

The difference in the image formation of real and virtual images is shown in Figure B.3. For the real image, the light paths bring the image to the right side of the lens. For the virtual image, the light paths make the image appear as if it is projected from the real object.

Here, it is useful to summarize the sign convention used in the optical relationships, where the direction of light rays is from left to right.

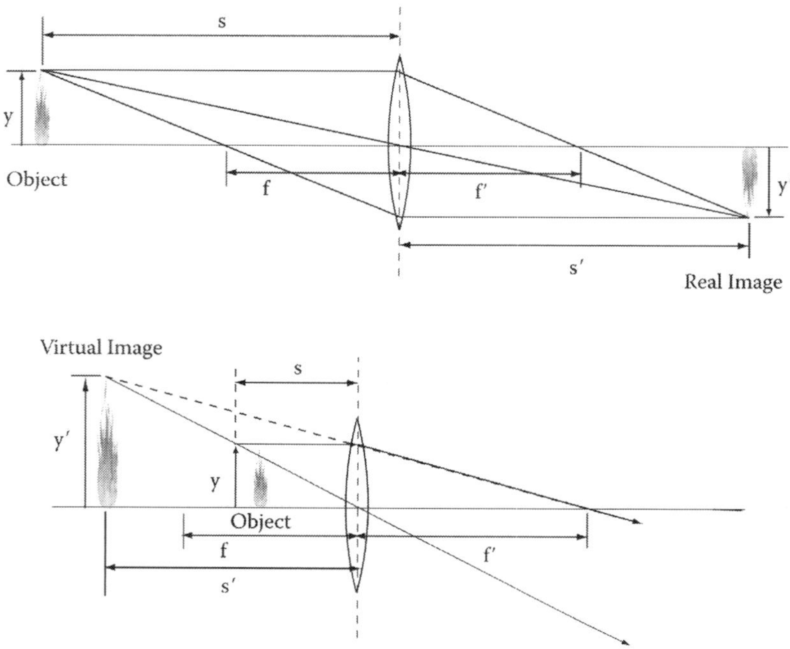

FIGURE B.3 Real and virtual images.

1. The object distance s is positive if the object is to the left of the lens.
2. The image distance s' is positive for *real* images and negative for *virtual* images (see Figure B.3).
3. Focal lengths f are positive for converging lenses and negative for diverging lenses.
4. Object and image dimensions, y and y', are positive upward. If y' is negative, then the image is inverted.
5. The magnification is negative for inverted images.
6. The radius of curvature of optical surfaces is positive for convex surfaces toward the incident light, → (. Similarly, the radius of curvature is negative for concave surfaces, →).

Another useful equation is Lensmaker's formula, which is used to calculate the focal length of lenses.

$$\frac{1}{f} = \frac{n - n_m}{n_m}\left(\frac{1}{r_1} - \frac{1}{r_2}\right) \tag{B.10}$$

n = refractive index of the lens material (see Table B.2)
n_m = refractive index of the medium (see Table B.1)
r_1 = radius of curvature of the front surface of the lens
r_2 = radius of curvature of the back surface of the lens

Again, the sign convention is such that the radius of curvature is positive for a convex surface toward the incident light, → (, and negative for concave, →).

This results in a positive focal length for a convex or converging lens and a negative f for a diverging lens in Equation B.10. n_m is the refractive index of the medium, and thus $n_m = 1$ is substituted for a lens placed in air.

Example B.2

For a plano-convex lens, where the first surface is flat (the plano part) and the second surface is convex, determine the radius of curvature of the convex surface for the lens to have a focal length of 25.0 cm, with the lens material having a refractive index of 1.52.

This is an application of the Lensmaker's equation, B.10. We solve for the radius of curvature of the second surface, r_2. We assume that the medium is air, so $n_m = 1$.

$$\frac{1}{r_2} = \frac{1}{r_1} - \frac{1}{n-1}\frac{1}{f} = \frac{1}{\infty} - \frac{1}{1.52-1}\frac{1}{25} \rightarrow r_2 = -13.0 \text{ cm}$$

The radius of curvature is negative, so the second surface is concave toward the light ray coming in from the left.

Some of the useful parameters to characterize the lens are (1) the focal length, (2) the f-number, and (3) the magnification. The focal length is calculated using Equation B.10, and the f-number is the ratio of the focal length and aperture (diameter).

$$f - \text{number} = \frac{f}{D} \tag{B.11}$$

D = aperture (lens diameter)

An alternate form of the focal length is the *dioptric power*, which is the inverse of the focal length measured in meters.

$$\text{Dioptric power} = \frac{1}{f}\left[\text{m}^{-1}\right] \tag{B.12}$$

The magnification is given by

$$m = \frac{y'}{y} = -\frac{s'}{s} \tag{B.13}$$

Example B.3

For an object distance, s, of 6.0 cm and a focal length, f, of +10.0 cm, find the image distance and the magnification.

Again, we use

$$s' = \frac{fs}{s-f} = \frac{(+6)(+10)}{6-(+10)} = -15 \text{ cm}$$

$$m = -\frac{s'}{s} = -\frac{(-15)}{(+6)} = +2.50$$

The image distance is negative, so the image is said to be virtual. Also, the magnification is positive, and the image is upright.

Example B.4

For a concave (diverging) lens of focal length −6.0 cm and an object distance of +12.0 cm, find the image distance and the magnification.

For concave or diverging lenses, the focal length is negative.

$$s' = \frac{fs}{s-f} = \frac{(-6)(+12)}{+12-(-6)} = -4.0 \text{ cm}$$

$$m = -\frac{s'}{s} = -\frac{(-4)}{(+12)} = +\frac{1}{3}$$

Example B.5

A two-lens combination is in use, as shown in Figure B.4. The object is located 4 cm to the left of the first lens, whose focal length is 3 cm. The second lens has a focal length of 4 cm, and the distance between the two lenses, *d*, is 2 cm. Find the final image distance s_2'.

$$s_1' = \frac{f_1 s_1}{s_1 - f_1} = \frac{(+4)(+3)}{(+4)-(+3)} = +12 \text{ cm}$$

We take the image at a distance of s_1' from the first lens to be the object for the second lens. So, the object distance for the second lens is calculated by taking into account that the second lens is 2 cm downstream of the first lens.

$$s_2 = d - s_1' = 2 - 12 = -10 \text{ cm}$$

$$s_2' = \frac{f_2 s_2}{s_2 - f_2} = \frac{(+4)(-10)}{(-10)-(+4)} = +2.86 \text{ cm}$$

The final image forms 2.86 cm to the right of the second lens and is considered to be a real image. If $d=0$, i.e., the two lenses are in contact, then it can be shown that the two-lens combination has a total focal length of

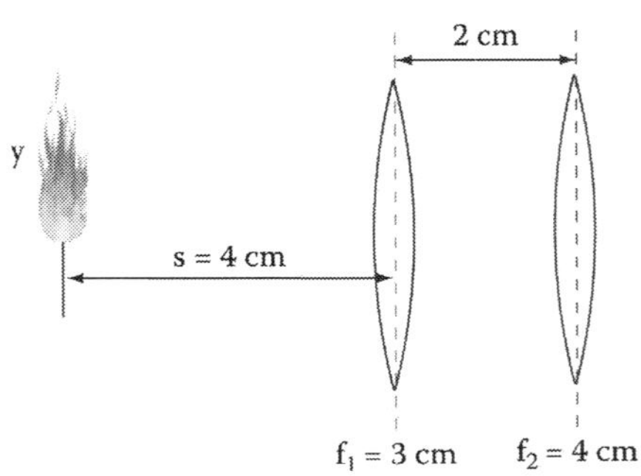

FIGURE B.4 A two-lens combination.

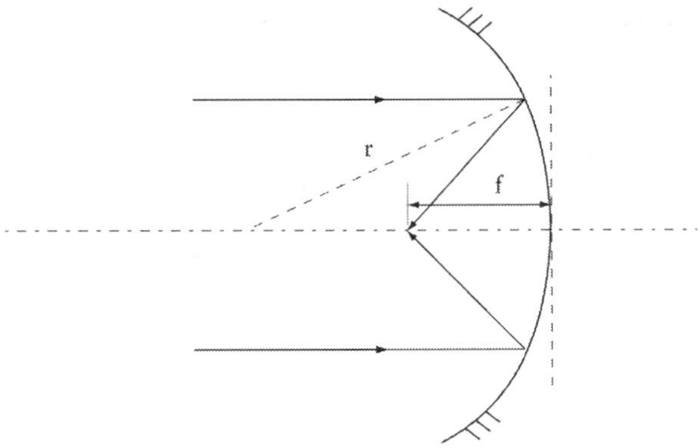

FIGURE B.5 A spherical mirror geometry.

$$\frac{1}{f} = \frac{1}{f_1} + \frac{1}{f_2} \tag{B.14}$$

Another commonly used optical component is a mirror. A spherical mirror is a reflective surface with an equi-centered surface (i.e., a spherical arc), as shown in Figure B.5. The focal length of a spherical mirror is

$$f = -\frac{1}{2}r \tag{B.15}$$

r = radius of curvature of the mirror surface

This is an analog of the Lensmaker's equation, Equation B.10. Once we have the focal length of mirrors, we can use the lens equation to relate the object and image distances, except that both s and s' are defined as positive to the *left* of the mirror because the action of the mirror is to reflect the light.

Example B.6

For a spherical mirror with a radius of curvature of −16.0 cm, the image is located at $s = 10.0$ cm. What are the image distance and the magnification?

The focal length is

$$f = -\frac{1}{2}r = -\frac{-16}{2} = +8 \text{ cm}$$

Again, a positive focal length means that the focal point is to the left of the mirror, with the concave surface ($r < 0$).

$$\frac{1}{s'} = \frac{1}{f} - +\frac{1}{s} = \frac{1}{8} - \frac{1}{10} = \frac{1}{40} \rightarrow s' = 40 \text{ cm (the image is real)}$$

$$m = -\frac{s'}{s} = -4 \text{ (the magnification is four and the image is inverted)}$$

B.1 OPTICAL INSTRUMENTS

Many optical instruments are used in thermal and flow measurements, ranging from imaging devices (cameras), microscopes, telescopes (as used in lidars) to various beam-conditioning and signal-collection devices. Figure B.6 shows two different designs for a telescope.

As shown in Figure B.6, for telescopes and also for microscopes, one common component of optical instruments is the *eyepiece*. An eyepiece is a lens combination that is used to view the action of the main optical components. The reason an eyepiece is used is that the human eye is also a refractive optical device, and the light

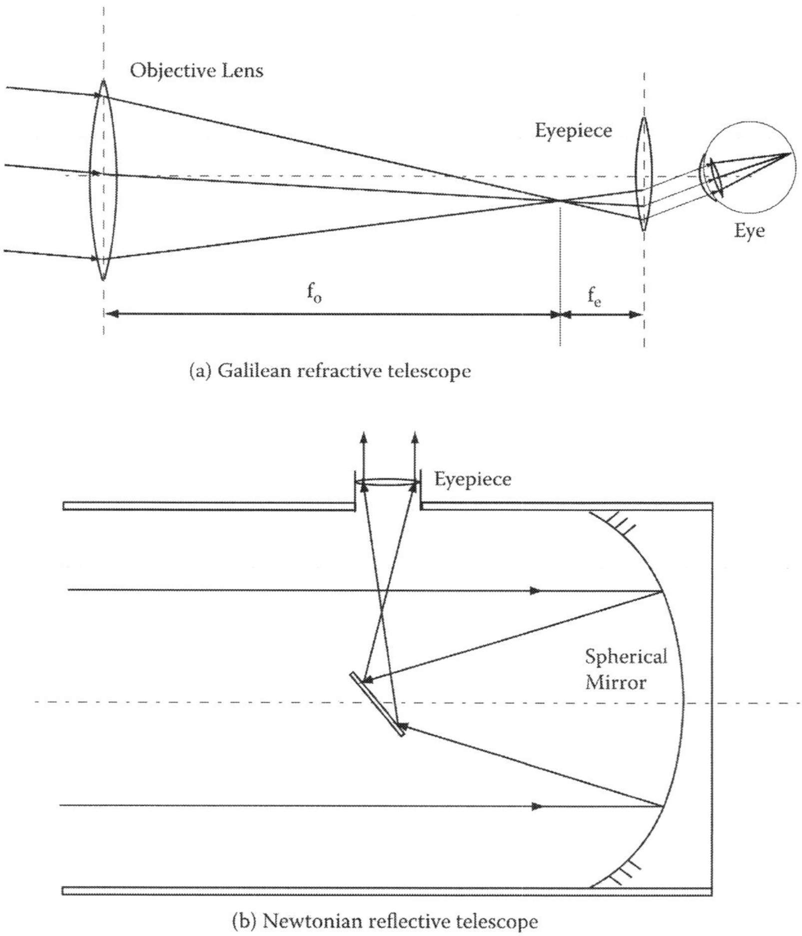

FIGURE B.6 Optical instruments: (a) Galilean telescope and (b) Newtonian telescope.

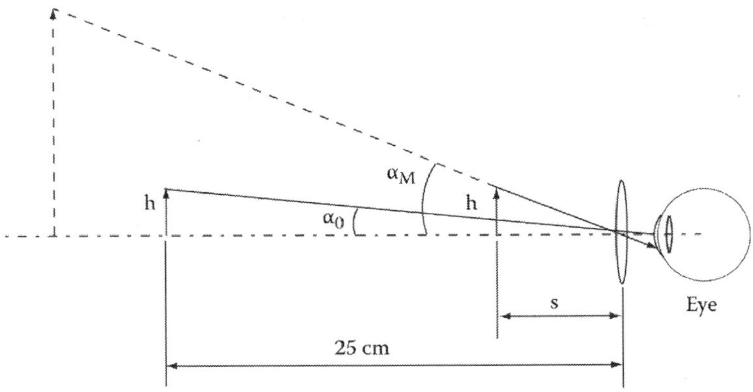

FIGURE B.7 Definition of magnification by an eyepiece.

output should be collimated (parallel) so that the human eye can form an image at an arbitrary distance from the eyepiece. The definition of the magnification due to an eyepiece is illustrated in Figure B.7. First, we start with an unaided viewing of an object with a dimension h, at a distance of 25 cm (this is the nearest position of distinct vision for the human eye). Next, we move this object to a distance s just in front of an eyepiece or a convex lens (or a lens combination) with a focal length of f. If s is smaller than f, then, as shown by the lens Equation B.9, the image distance s' will be negative, meaning that a virtual image will form to the left of the lens, as shown in Figure B.6.

$$\frac{1}{s'} = \frac{1}{f} - \frac{1}{s} = \frac{s-f}{fs} < 0 \quad \text{if} \quad s < f \tag{B.16}$$

Then, the magnification due to the eyepiece is defined as the ratio of the angles α_M and α_o in Figure B.7. Using the small-angle approximation, we then obtain

$$\text{Magnification} = M = \frac{\alpha_M}{\alpha_o} = \frac{h/s}{h/25} = \frac{25}{s} \tag{B.17}$$

Furthermore, if the object is placed at $s = f$, then the outgoing light will be parallel for optimal viewing, and this defines the eyepiece magnification.

$$M = \frac{25}{f} \quad f \text{ in [cm]} \tag{B.18}$$

The microscope magnification is defined in a similar manner, in which the effective focal length of all the lens combinations in the microscope is calculated and substituted into Equation B.18.

$$M_{\text{microscope}} = \frac{25}{f_{\text{eff}}} \quad f_{\text{eff}} \text{ in [cm]} \tag{B.19}$$

For telescopes, the magnification is defined as

$$M_{\text{telescope}} = -\frac{f_{\text{obj}}}{f_{\text{eyepiece}}} \tag{B.20}$$

For microscopes, there is also an objective and an eyepiece lens combination, separated by a distance d. The effective focal length of such lens combinations is

$$\frac{1}{f_{\text{eff}}} = \frac{1}{f_{\text{obj}}} + \frac{1}{f_{\text{eyepiece}}} - \frac{d}{f_{\text{obj}}\, f_{\text{eyepiece}}} \tag{B.21}$$

f_{obj} = focal length of the objective lens combination
f_{eyepiece} = focal length of the eyepiece lens combination

Figure B.8 shows some examples of optical setups, such as beam expansion (Figures B.8a and B.8b) and a signal-collection optical setup (Figure B.8c). In optical diagnostics, it sometimes becomes necessary to increase the probe laser beam diameter,

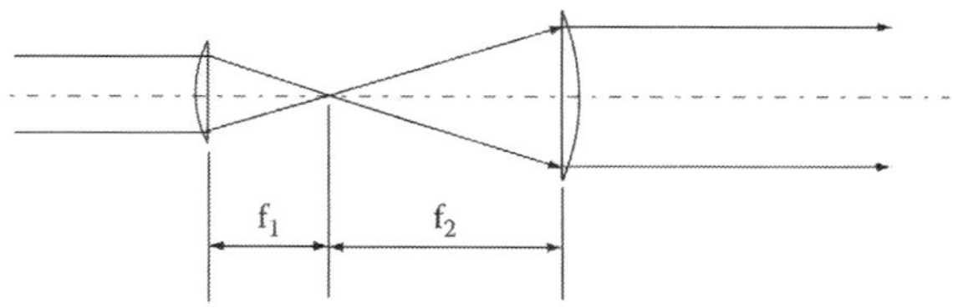

(a) Beam expander using two convex lenses

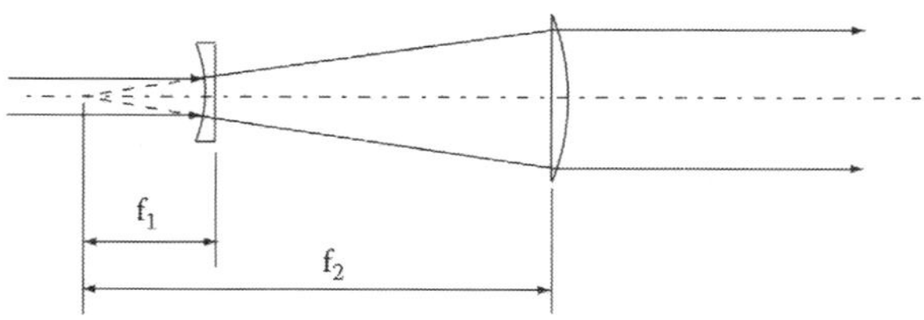

(b) Beam expander using a concave and a convex lens

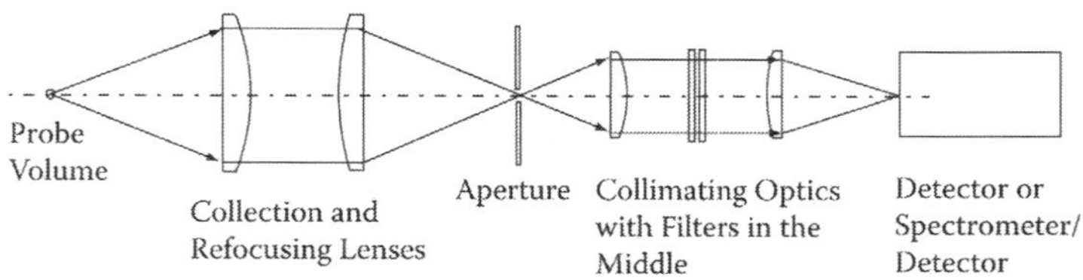

(c) Signal collection optics

FIGURE B.8 Optical instruments: (a) beam expander, (b) beam expander, and (c) signal-collection optics.

in which case beam expanders are used, as shown in Figures B.8a and B.8b. For high-power lasers, the arrangement in Figure B.8b is preferable, as the setup in Figure B.8a focuses the laser to a small spot that may lead to undesirable laser sparks. Figure B.8c shows a complete signal-collection optics. For simplicity, collimating optics may be omitted, in which case the detector or the spectrometer/detector is placed where the aperture is located in Figure B.8c.

B.2 DIFFRACTION

Thus far, we have analyzed various optical systems assuming that the light rays travel in straight paths, i.e., we have been dealing with *geometrical optics*. However, due to the wave nature of light, there are optical phenomena that cannot be explained or analyzed through geometrical optics. *Diffraction* refers to any deviation from geometrical optics that results from the wave nature of light. In more detailed treatments of diffraction phenomena, *Fraunhofer* and *Fresnel* diffractions are considered. The difference is that Fraunhofer diffraction refers to a mathematical description of the far-field wavefronts that can be approximated as planar waves, whereas Fresnel diffraction is a mathematical treatment of near-field wavefronts. Because Fraunhofer diffraction is simpler in mathematical description, we will present some basic diffraction phenomena using the planar wave description of light.

We begin with the *Huygens-Fresnel* principle: Every point on a given wavefront of light can be considered a source of secondary spherical wavelets. Using this principle, we can see what happens to light through a single slit, as shown in Figure B.9. Each point on the slit can be considered a source of a spherical wavefront, and thus the electric field, dE_p, emanating from a location on the slit with a width ds is written as

$$dE_p = \frac{E_s ds}{r} e^{i(kr-\omega t)} \tag{B.22}$$

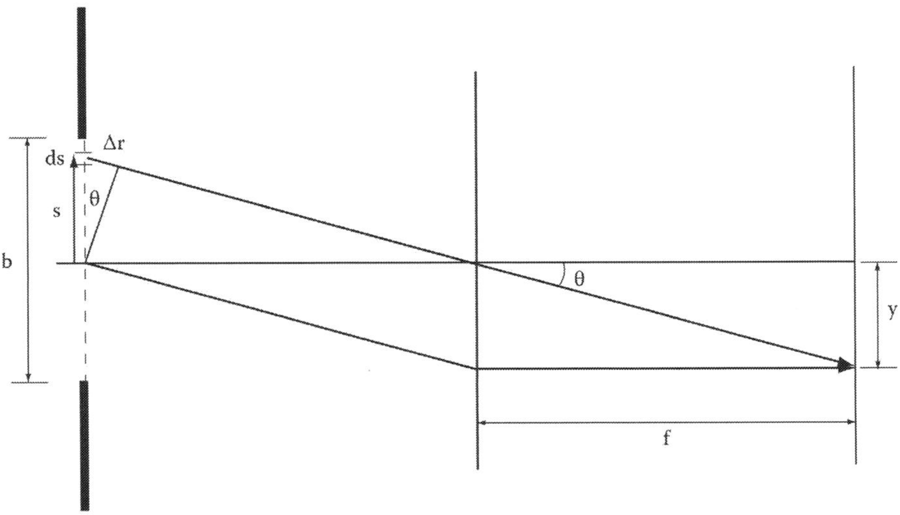

FIGURE B.9 A diffraction through a single slit.

E_s = electric field strength per unit length on the slit
r = distance from the slit to the image plane
k = wavevector = $2\pi/\lambda$
ω = circular frequency = $2\pi f$

The fact that the electric field represents a spherical wave (instead of a planar wave) is written into Equation B.22 by the denominator, r. The light intensity is proportional to the square of the electric field; for spherical waves $I_p \sim E_s^2 / r^2$, so that the intensity for spherical waves decreases as it spreads over an increasing spherical area, $4\pi r^2$.

If we set $r = r_o$ at $s = 0$, then $r = r_o + \Delta r$.

$$dE_p = \frac{E_s ds}{r_o + \Delta r} e^{i[k(r_o+\Delta r)-\omega t]} \approx \frac{E_s ds}{r_o} e^{i[k(r_o+\Delta r)-\omega t]} \tag{B.23}$$

because $\Delta r \ll r_o$.

Now, $\Delta r = s\sin\theta$, so

$$dE_p = \frac{E_s ds}{r_o + \Delta r} e^{i(kr_o-\omega t)} e^{iks\sin\theta} \tag{B.24}$$

Integrating across the slit width where s goes from $-b/2$ to $b/2$,

$$E_p = \int_{-b/2}^{b/2} dE_p = \frac{E_s}{r_o} e^{i(kr_o-\omega t)} \int_{-b/2}^{b/2} e^{iks\sin\theta} ds = \frac{E_s}{r_o} e^{i(kr_o-\omega t)} \frac{e^{\frac{ikb\sin\theta}{2}} - e^{\frac{ikb\sin\theta}{2}}}{ik\sin\theta} \tag{B.25}$$

If we substitute $\beta = kb\sin\theta/2$, and use Euler's notation for the sine function,

$$E_p = \frac{E_s b}{r_o} e^{i(kr_o-\omega t)} \frac{\sin\beta}{\beta} \tag{B.26}$$

Now, the intensity of the light, I_p, is proportional to the square of the electric field strength at the image plane (Equation B.26), and we are interested in its angular dependence at a fixed distance on the image plane. Therefore, the intensity distribution is

$$I_p \sim E_p^2 \sim \left(\frac{\sin\beta}{\beta}\right)^2 \tag{B.27}$$

This means that the intensity will be zero when $\sin\beta = 0$, or

$$\beta = \frac{1}{2} kb\sin\theta = m\pi \quad \text{where } m = \pm 1, \pm 2, \ldots \tag{B.28}$$

Equation B.28 simplifies to the condition that $m\lambda = b\sin\theta$, and if the distance along the vertical image plane is approximated as $y = f\sin\theta$, then the minimum intensity occurs at

$$y_{\min} = \frac{mf\lambda}{b}, \quad m = \pm 1, \pm 2, \ldots \tag{B.29}$$

Therefore, a light-dark pattern is produced by $\left(\frac{\sin\beta}{\beta}\right)^2$ dependence of the light intensity due to the diffraction effect.

A similar treatment of the diffraction pattern of light passing through a circular aperture (as opposed to the two-dimensional rectangular aperture discussed previously) results in the following light intensity distribution:

$$I_p = I_o\left(\frac{2J_1(\gamma)}{\gamma}\right)^2 \tag{B.30}$$

I_o = light intensity at the circular aperture
J_1 = the Bessel function of the first kind
$\gamma = kD\,\sin\theta/2$
D = aperture diameter

The light intensity given by Equation B.30 is plotted in Figure B.10. The angular radius of the primary peak in Figure B.9 is given by $1.22\lambda/D$. Because light passing through an aperture of diameter D has the diffraction-limited angular resolution given by $1.22\lambda/D$, this serves as the minimum resolution achievable with optical devices such as astronomical telescopes. This minimum resolution is referred to as the *Rayleigh criterion*.

$$\text{resolution} = (\Delta\theta)_{\min} = \frac{1.22\lambda}{D} \tag{B.31}$$

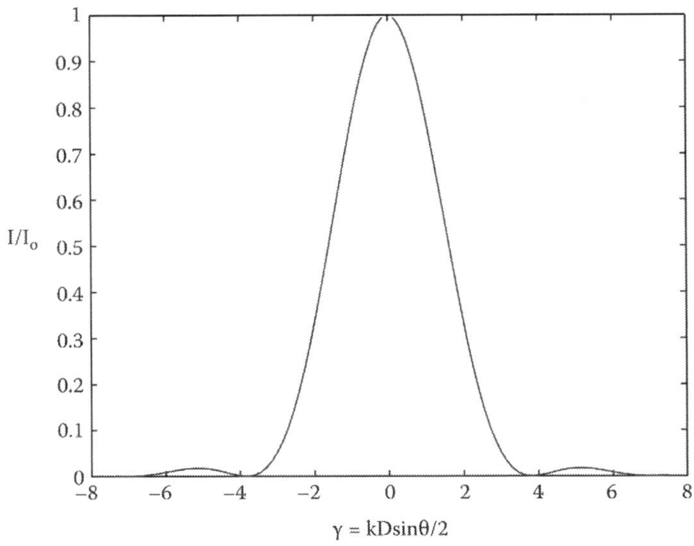

FIGURE B.10 The diffraction pattern through a circular aperture.

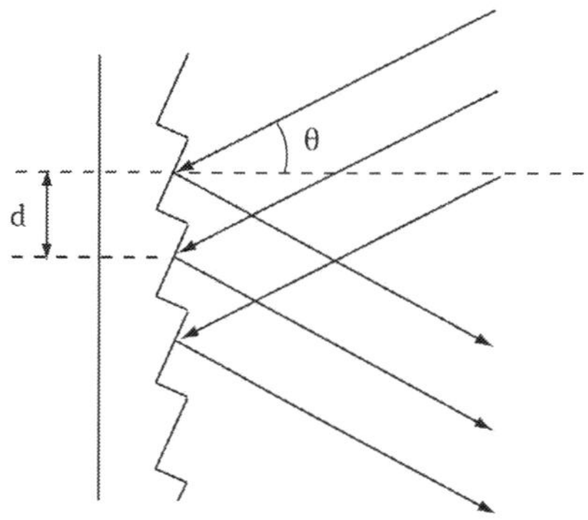

FIGURE B.11 A diffraction grating.

B.3 DIFFRACTION GRATING

A device using diffraction to separate the wavelength components of light is a *diffraction grating*, schematically shown in Figure B.11. The slanted sides of the so-called reflection echelon direct the light at an angle, with a phase difference that is a function of the angle and the distance between the slanted sides. The path difference for the beams is

$$\text{path difference} = d \sin \theta \tag{B.32}$$

Similar in principle to the diffraction described for the single slit, the path difference causes intensity maxima and minima. The maxima occur when $d\sin\theta = m\lambda$ ($m = 1, 2, 3, \ldots$), or

$$\lambda = \frac{d \sin \theta}{m} \tag{B.33}$$

For a fixed m, the wavelength component that is maximized depends on d and the angle, θ, i.e., the wavelength component is separated as a function of the angle, θ. Two widely used devices that use diffraction gratings are *monochromators* and *spectrometers*. Monochromators have a rotating diffraction grating so that, at a given angle of the grating, a fixed wavelength component passes through the device. In spectrometers, the output wavelength of the light depends on the location of the exit slit, due again to the angle-dependent wavelength component reflected from the diffraction gratings.

B.4 INTERFERENCE

Interference, again, is a phenomenon that occurs due to the wave nature of light and is used in various optical diagnostic techniques. For example, in Chapter 4, we saw how the fringe pattern is formed when two monochromatic waves intersect at an angle. We started by writing the electric fields of the two planar monochromatic waves as

$$E_1 = E_o \cos\left(\omega t - k\left(x \cos\frac{\theta}{2} + y \sin\frac{\theta}{2}\right)\right)$$
$$E_2 = E_o \cos\left(\omega t - k\left(x \cos\frac{\theta}{2} - y \sin\frac{\theta}{2}\right)\right), \tag{B.34}$$

As we will see in Appendix C, the foregoing describes a traveling wave with an angular frequency, $\omega = 2\pi f$, and a wavenumber, $k = 2\pi/\lambda$, except that they are oriented at $\pm\theta/2$. Thus, a constant amplitude is found along a path that keeps the argument of the cosine functions constant.

For E_1, the argument is constant along a path such that

$$2\pi ft - \frac{2\pi}{\lambda}\left(x \cos\frac{\theta}{2} + y \sin\frac{\theta}{2}\right) = \text{const}. \tag{B.35}$$

Or, differentiating with respect to time, we get

$$\cos\frac{\theta}{2}\frac{\partial x}{\partial t} + \sin\frac{\theta}{2}\frac{\partial y}{\partial t} = f\lambda = c$$
$$c = \text{speed of light} = f\lambda \tag{B.36}$$

When $\theta/2 = 0$, $\cos\theta/2 = 1$, and $\sin\theta/2 = 0$, Equation B.36 simply states that the wave propagates in the x direction with a speed of c. Similarly, when $\theta/2 = 90°$, $\cos\theta/2 = 0$, and $\sin\theta/2 = 1$, the propagation is in the y direction. In between, the direction of propagation is set by $\theta/2$, as in Equation B.34.

The light intensity, I_L, is proportional to the square of the electric field strength averaged over time:

$$I_L \sim \langle E^2 \rangle \tag{B.37}$$

Using cos(A + B) + cos(A − B) = 2cosAcosB, and dropping the time-dependent term due to time averaging, the spatial variation of the light intensity is then given by

$$I_L \propto E_O^2 \cos^2\left(-kx \cos\frac{\theta}{2}\right)\cos^2\left(ky \sin\frac{\theta}{2}\right) \tag{B.38}$$

We are interested in the interference pattern in the y (vertical) direction, so taking the y-component and using $\cos^2 A = 1 + \cos(2A)$, we get the light intensity variation:

$$I_L \propto E_O^2\left[1 + \cos\left(2ky \sin\frac{\theta}{2}\right)\right] = E_O^2\left[1 + \cos\left(\frac{4\pi}{\lambda_L} y \sin\frac{\theta}{2}\right)\right] \tag{B.39}$$

The first term inside the square brackets represents the constant background intensity, referred to as the pedestal component, and the second term is the spatial variation with respect to the pedestal component in the vertical direction. Equation B.39 represents the mathematical description of the interference pattern when two planar monochromatic waves intersect.

BIBLIOGRAPHY

Jenkins, F.A. and White, H.E., *Fundamentals of Optics*, 4th ed., McGraw-Hill, New York, 1976.

Pedrotti, F.L., Pedrotti, L.S., and Pedrotti, L.M., *Introduction to Optics*, 3rd ed., Pearson-Prentice Hall, Upper Saddle River, NJ, 2007.

Appendix C
Electromagnetics and Electromagnetic Radiation

C.1 MAXWELL'S EQUATIONS

Electromagnetic waves, including light waves, are a manifestation of propagating electric and magnetic fields. The wave equation that describes the propagation of electromagnetic waves is derived from electromagnetic equations, or the so-called Maxwell's equations. The following are the underlying governing equations (and laws) of electricity and magnetism, unified by Maxwell's equations.

Gauss's law states that the net electric flux through any surface is equal to the total charge enclosed by that surface.

$$\nabla \bullet \vec{E} = \frac{\rho_v}{\varepsilon} \quad \text{or} \quad \nabla \bullet \vec{D} = \rho_v \tag{C.1}$$

$\vec{E} = \frac{\vec{F}}{q}$ = electric field intensity $\left[\mathrm{N/C(coulombs)}\right]$ or $\left[\mathrm{V(volts)/m}\right]$

$\vec{D} = \varepsilon\vec{E}$ electric flux density [C (coulombs)/m^2]

ε = electric permittivity [F (farads)/m]

ε_o = electric permittivity in a vacuum = 8.854×10^{-12} [F (farads)/m]

ρ_v = charge density [C/m^3]

Gauss's law is a generalization of Coulomb's force, F, that exists between electric charges q_1 and q_2, which are separated by a distance r.

$$F_r = \frac{q_1 q_2}{4\pi\varepsilon}\frac{1}{r^2} \tag{C.2}$$

The subscript "r" for the force means that it is the radial component of the force vector.

For $q_1 = Q$ and $q_2 = q$, the electric field is then

$$E_r = \frac{F_r}{q} = \frac{Q}{4\pi\varepsilon}\frac{1}{r^2} \quad \text{or} \quad 4\pi r^2 E_r = \frac{Q}{\varepsilon} = \frac{\rho_v V}{\varepsilon} \tag{C.3}$$

where we are writing $Q = \rho_v V$ for a spherical volume V.

Using the divergence theorem (also known as Gauss's theorem), $\int_{\text{Volume}} \nabla \bullet \vec{E} dV = \int_{\text{Surface}} \vec{E} \bullet d\vec{S}$, and Equation C.3, we get

$$\int_{\text{Volume}} \nabla \bullet \vec{E} dV = \int_{\text{Surface}} \vec{E} \bullet d\vec{S} = 4\pi r^2 E_r = \frac{\rho_v V}{\varepsilon} = \int_{\text{Volume}} \frac{\rho_v}{\varepsilon} dV \tag{C.4}$$

Equating the integrands of the first and last terms of the preceding equality, we obtain Gauss's law of Equation C.l.

Gauss's law for magnetic fields states that the net magnetic flux through any surface is zero.

$$\nabla \bullet \vec{B} = 0 \tag{C.5}$$

$$\vec{B} = \text{magnetic flux densitysss } \left[H \text{ (henry)} - A \text{ (amps)} / \text{m}^2\right]$$

Magnetic field lines form closed loops (as seen in elementary science experiments), and for a control volume, the net magnetic flux is zero.

Faraday's law describes the electric field that is induced by an oscillating magnetic field.

$$\nabla \times \vec{E} = -\frac{\partial \vec{B}}{\partial t} \tag{C.6}$$

$$\vec{E} = \text{electric field intensity} \left[\text{V(volts)} / \text{m}\right]$$

The electric field can also be written as $\vec{E} = -\nabla V$, where V=electric potential [V].

Ampere's law prescribes the magnetic field that is induced by an oscillating electric field.

$$\nabla \times \vec{H} = \vec{J} + \frac{\partial \vec{D}}{\partial t} \tag{C.7}$$

$$\vec{H} = \text{magnetic field intensity} \left[\text{A/m}\right]$$

$$\vec{J} = \text{current density} \left[\text{A/m}^2\right]$$

The Lorentz force equation gives the force acting on a particle of charge q due to electric and magnetic field intensities.

$$\vec{F} = q\left(\vec{E} + \vec{v} \times \vec{B}\right) \tag{C.8}$$

$$\vec{F} = \text{force acting upon the particle}$$

$$q = \text{charge of the particle } [C]$$

$$\vec{v} = \text{velocity of the particle}$$

There are also so-called constitutive relations that relate the electric and magnetic properties.

Constitutive relations:

$$\vec{D} = \varepsilon\vec{E} \quad \text{(C.9)}$$

ε = electric permittivity [F (farads)/m]
ε_o = electric permittivity in a vacuum = 8.854×10^{-12} [F (farads)/m]

$$\vec{B} = \mu\vec{H} \quad \text{(C.10)}$$

μ = magnetic permeability
μ_o = magnetic permittivity of a vacuum = $4\pi \times 10^{-7}$ [H/m]

$$\vec{J} = \sigma\vec{E} \quad \text{(C.11)}$$

$$\vec{J} = \text{current density}\left[\text{A/m}^2\right]$$

σ = conductivity [1/Ωm]

We can reduce the vector form of Ohm's law in Equation C.11 to a more familiar form $V = IR$. For a current in the *x*-direction with an applied voltage, V_{ab}, over a distance L, the current density is simply

$$(\text{Current})/(\text{Area}) = \text{I}/\text{Area} = \sigma V_{ab}/L \Rightarrow V_{ab} = I\frac{L}{\sigma(\text{Area})} = IR \quad \text{(C.12)}$$

where $L/(\sigma \times \text{Area})$ is the definition of resistance, R.

Finally, for Maxwell's equations, we have the current continuity equation, analogous to the mass continuity equation in fluid mechanics.

The current continuity equation is

$$\nabla \bullet \vec{J} = -\frac{\partial \rho_v}{\partial t} \quad \text{(C.13)}$$

C.2 ELECTROMAGNETIC WAVE EQUATION

Electromagnetics is relevant for experimentalists, as it appears intrinsically or explicitly in many laboratory instruments. However, in this appendix, we focus on its implications for electromagnetic wave propagation. To do so, let us start by using two of the foregoing equations. From Equations C.6 and C.7 and the constitutive relations C.9, C.10, and C.11, we have

$$\nabla \times \vec{E} = -\frac{\partial \vec{B}}{\partial t} = -\mu\frac{\partial \vec{H}}{\partial t} \quad \text{(C.14a)}$$

$$\nabla \times \vec{H} = \vec{J} + \frac{\partial \vec{D}}{\partial t} = \sigma\vec{E} + \varepsilon\frac{\partial \vec{E}}{\partial t} \quad \text{(C.14b)}$$

Taking the curl ($\nabla\times$) of Equation C.14a, we get

$$\nabla \times \left(\nabla \times \vec{E}\right) = \nabla \times \left(-\mu\frac{\partial \vec{H}}{\partial t}\right) = -\mu\frac{\partial}{\partial t}\left(\nabla \times \vec{H}\right) \quad \text{(C.15)}$$

For the last part of the foregoing equality, we substitute Equation C.14b.

$$\begin{aligned}\nabla\times\left(\nabla\times\vec{E}\right) &= -\mu\frac{\partial}{\partial t}\left(\nabla\times\vec{H}\right) = -\mu\frac{\partial}{\partial t}\left(\sigma\vec{E}+\varepsilon\frac{\partial\vec{E}}{\partial t}\right)\\ &= -\mu\sigma\frac{\partial\vec{E}}{\partial t} - \mu\varepsilon\frac{\partial^2\vec{E}}{\partial t^2}\end{aligned} \tag{C.16}$$

Now, $\nabla\times\left(\nabla\times\vec{E}\right) = \nabla\bullet\vec{E} - \nabla^2\vec{E}$, according to a well-known vector identity. We consider a passive medium, i.e., there is no electric charge in the medium, and therefore the charge density is zero. Using Gauss's law (Equation C.1), zero charge density means $\nabla\bullet\vec{E} = 0$. So, we have, from Equation C.16,

$$\nabla^2\vec{E} = \mu\sigma\frac{\partial\vec{E}}{\partial t} + \mu\varepsilon\frac{\partial^2\vec{E}}{\partial t^2} \quad \text{or} \quad \frac{\partial^2\vec{E}}{\partial t^2} = \frac{1}{\mu\varepsilon}\nabla^2\vec{E} + \frac{\sigma}{\varepsilon}\frac{\partial\vec{E}}{\partial t} \tag{C.17}$$

This becomes the classical wave equation if we consider a medium with zero conductivity ($\sigma = 0$) and a wave speed $c = \dfrac{1}{\sqrt{\mu\varepsilon}}$.

$$\frac{\partial^2\vec{E}}{\partial t^2} = c^2\nabla^2\vec{E} \tag{C.18a}$$

However, for the general case, we retain the first-order time-dependent term that describes the attenuation due to nonzero conductivity.

$$\frac{\partial^2\vec{E}}{\partial t^2} = c^2\nabla^2\vec{E} + \frac{\sigma}{\varepsilon}\frac{\partial\vec{E}}{\partial t} \tag{C.18b}$$

Equations C.18a and b describe the propagation of electromagnetic waves that consist of oscillating electric and magnetic fields. The energy content of electromagnetic waves comes from the strength of both the electric and magnetic fields. Let us consider, for simplicity, a one-dimensional wave equation for the electric field oriented in the y-direction (i.e., $\vec{E} \to E_y$). An electromagnetic wave with such an electric field is said to be *vertically polarized*. For such a polarized wave, the magnetic field is orthogonal to the electric field.

$$\frac{\partial^2 E_y}{\partial t^2} = c^2\frac{\partial^2 E_y}{\partial x^2} \quad (\sigma = 0) \tag{C.19a}$$

$$\frac{\partial^2 E_y}{\partial t^2} = c^2\frac{\partial^2 E_y}{\partial x^2} + \frac{\sigma}{\varepsilon}\frac{\partial E_y}{\partial t} \quad (\sigma \neq 0) \tag{C.19b}$$

It can be shown through substitution that a solution for Equation C.19a is

$$E_y = E_{yo}\mathrm{Re}\left\{\exp\left[i\omega\left(-\frac{nx}{c_o}+t\right)\right]\right\} = E_{yo}\cos\left[\omega\left(-\frac{nx}{c_o}+t\right)\right] \tag{C.20a}$$

E_{yo} = initial amplitude of the electric field
ω = circular frequency = $2\pi f$
n = refractive index of the medium (so that $n/c_o = c$)
c_o = speed of light in a vacuum

Equation C.20a is identical in content to slightly different forms used in Appendix B, and represents an electromagnetic wave traveling with a wave speed (c), frequency (f), and wavelength (λ). The wavelength, λ, is incorporated in the $-\omega nx/c_o$ term, which is equal to $-2\pi fx/c = -2\pi fx/f\lambda = -2\pi x/\lambda$ (since $c = f\lambda$). Table C.1 shows the wavelength, frequency, and other characteristics associated with different electromagnetic waves.

For Equation C.19b, the solution now includes a damping term. Thus, the solution represents a traveling wave whose amplitude decays exponentially with increasing x, depending on k (the absorption index of the medium).

$$E_y = E_{yo}\text{Re}\left\{\exp\left[i\omega\left(-\frac{nx}{c_o}+t\right)\right]\exp\left(-\frac{\omega kx}{c_o}\right)\right\}$$

$$= E_{yo}\text{Re}\,\exp\left\{i\omega\left[-(n-ik)\frac{x}{c_o}+t\right]\right\}E \tag{C.20b}$$

$$= E_{yo}\text{Re}\,\exp\left\{i\omega\left(-\frac{\tilde{n}x}{c_o}+t\right)\right\}$$

TABLE C.1
Electromagnetic Wave Spectrum

Type	Wavelength, λ	Frequency, F [HZ]	Wavenumber, $1/\lambda$ [CM^{-1}]	Associated Phenomena
Cosmic rays	$<10^{-5}$ nm	$>10^{22}$	$>10^{12}$	
Gamma rays	$10^{-5}\sim10^{-1}$ nm	$10^{18}\sim10^{22}$	$10^{8}\sim10^{12}$	Radioactive decay
X rays	$10^{-1}\sim10$ nm	$10^{16}\sim10^{18}$	$10^{6}\sim10^{8}$	Electron collision with molecules
Ultraviolet	$10\sim400$ nm	$10^{14}\sim10^{16}$	$10^{4}\sim10^{6}$	Electronic transitions in gas molecules
Visible light	**400 ~ 760 nm**	**$\sim10^{14}$**	**$\sim10^{4}$**	
Infrared	760 nm ~ 100 μm	$10^{12}\sim10^{14}$	$10^{2}\sim10^{4}$	Rotational transitions in gas molecules
Thermal radiation	100 nm ~ 100 μm	$10^{12}\sim10^{15}$	$10^{2}\sim10^{5}$	
Microwave (radar)	100 μm ~ 100 mm	$10^{8}\sim10^{12}$	$10^{-2}\sim10^{2}$	
Shortwave radio waves	100 mm ~ 100 m	$10^{6}\sim10^{8}$	$10^{-4}\sim10^{-2}$	
Longwave radio waves	100 ~ 1000 m	$10^{4}\sim10^{6}$	$10^{-6}\sim10^{-4}$	

k = absorption index
$\tilde{n} = n - ik$ = complex refractive index

By combining the exponential terms in Equation C.20b, we obtain a form completely analogous to Equation C.20a. The complex refractive index in Equation C.20b now includes the absorption index.

Some interesting relationships between the optical and electromagnetic properties of the media can be obtained by substituting Equation C.20b into Equation C.19b. The derivative terms are

$$\frac{\partial E_y}{\partial t} = i\omega E_y \tag{C.21a}$$

$$\frac{\partial^2 E_y}{\partial t^2} = i^2\omega^2 E_y = -\omega^2 E_y \tag{C.21b}$$

$$\frac{\partial^2 E_y}{\partial x^2} = \frac{-\omega^2 (n - ik)^2}{c_o^2} E_y \tag{C.21c}$$

So, using these derivative terms in Equation C.19b (after converting $c^2 = \mu\varepsilon$), we get an equality

$$-\mu\varepsilon\omega^2 = \frac{-\omega^2 (n - ik)^2}{c_o^2} - i\mu\sigma\omega \tag{C.22}$$

Separating the real and imaginary parts of Equation C.22, we get (after using $\omega = 2\pi c_o / \lambda_o$)

$$\begin{aligned} n^2 - k^2 &= \mu\varepsilon c_o^2 \\ nk &= \frac{\mu\sigma\lambda_o c_o}{4\pi} \end{aligned} \tag{C.23}$$

We can solve for n and k in Equation C.23.

$$\begin{aligned} n^2 &= \frac{\mu\varepsilon c_o^2}{2}\left\{1 + \left[1 + \left(\frac{\sigma\lambda_o}{2\pi\varepsilon c_o}\right)^2\right]^{1/2}\right\} \\ k^2 &= \frac{\mu\varepsilon c_o^2}{2}\left\{-1 + \left[1 + \left(\frac{\sigma\lambda_o}{2\pi\varepsilon c_o}\right)^2\right]^{1/2}\right\} \end{aligned} \tag{C.24}$$

ε = electrical permittivity of the medium
μ = magnetic permeability of the medium
σ = conductivity of the medium

Equation C.24 shows that the optical properties, n and k, are directly related to the material's electromagnetic properties. For media with zero conductivitwy, such as a vacuum, it can be easily verified that $k = 0$ and $n = c_o/c$, so that the complex refractive index contains only the real part, n, and the solution in Equation C.20b reduces to Equation C.20a.

As noted earlier, the energy content of electromagnetic waves comes from both the strength of the electric and magnetic fields. For vertically polarized plane electromagnetic waves, the magnetic field is orthogonal to the electric field, and its strength is given by

$$H_z = \frac{n}{\mu c_o} E_y \tag{C.25}$$

The total energy flux carried by an electromagnetic wave is given by the Poynting vector.

$$\text{Poynting vector } = \vec{S} \equiv \vec{E} \times \vec{H} \tag{C.26}$$

With Equations C.20b and C.25, we can see that the amplitude of the Poynting vector or the energy flux of the electromagnetic wave is

$$\left|\vec{S}\right| = \frac{n}{\mu c_o}\left|E_y\right|^2 = \frac{n}{\mu c_o} E_{yo}^2 \left[\text{Re}\left\{\exp\left[i\omega\left(-\frac{nx}{c_o}+t\right)\right]\right\}\right]^2 \exp\left(-\frac{2\omega kx}{c_o}\right) \tag{C.27}$$

The square of the sinusoidal term (the exponential function with an imaginary argument) averages to 1/2 over its wavelength, so

$$|\vec{S}|_{\text{avg}} = \frac{n}{2\mu c_o} E_{yo}^2 \exp\left(-\frac{2\omega kx}{c_o}\right) \tag{C.28a}$$

Or if we use $\omega = 2\pi c_o/\lambda_o$ in the preceding equation,

$$|\vec{S}|_{\text{avg}} = \frac{n}{2\mu c_o} E_{yo}^2 \exp\left(-\frac{4\pi kx}{\lambda_o}\right) = \frac{n}{2\mu c_o} E_{yo}^2 \exp(-K_{a\lambda}x)) \tag{C.28b}$$

$K_{a\lambda} = 4\pi k/\lambda_o =$ absorption coefficient

At $x = 0$, $|\vec{S}|_{\text{avg},o} = \frac{n}{2\mu c_o} E_{yo}^2$, so that

$$\frac{|\vec{S}|_{\text{avg}}}{|\vec{S}|_{\text{avg},o}} = \exp\left(-\frac{4\pi kx}{\lambda_o}\right) = \exp(-K_{a\lambda}x) \tag{C.29}$$

If we equate the magnitude of the Poynting vector, $|\vec{S}|_{\text{avg}}$, with the spectral intensity of radiation, I_λ, we recover the so-called Beer-Lambert's law (or sometimes called Beer-Lambert-Bourger law) for attenuation of radiation due to absorbing media.

$$\frac{I_\lambda}{I_{\lambda o}} = \exp\left(-\frac{4\pi kx}{\lambda_o}\right) = \exp(-K_{a\lambda}x) \quad \text{Beer-Lambert Law} \tag{C.30}$$

The spectral intensity is the amount of radiative power per unit area per unit solid angle per unit wavelength interval. This definition of the spectral intensity and the angles involved in the solid angle is illustrated in Figure C.1.

$$I_\lambda(\Omega) = \frac{\Delta q}{\Delta A \Delta\Omega \Delta\lambda} = \left[\frac{W}{m^2\ sr\ \mu m}\right] \tag{C.31}$$

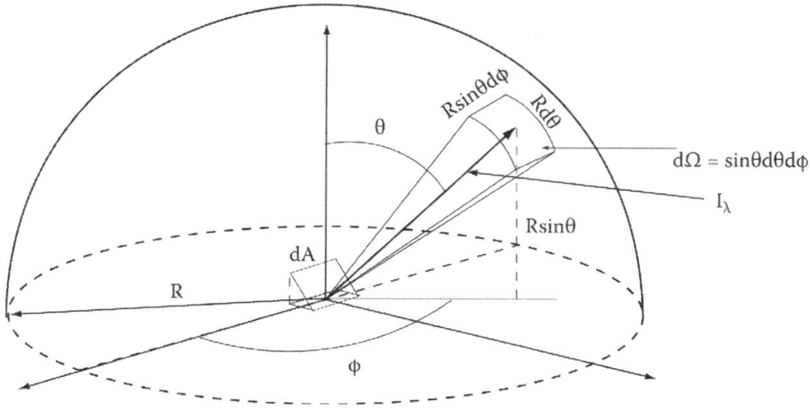

FIGURE C.1 The definition of spectral intensity and solid angle.

$\Delta\Omega$ = solid angle = $\sin\theta d\theta d\phi$ [sr]
ΔA = cross-sectional area perpendicular to the direction of propagation

The Beer-Lambert law states that the radiation exponentially decreases as a function of the distance of propagation and the absorption coefficient, which is usually wavelength-dependent. This relationship is used in absorption measurements, such as absorption spectroscopy, discussed in Chapters 5, 7, and 8.

C.3 RADIATIVE TRANSFER EQUATION

We can expand upon the Beer-Lambert law to include the scattering and emission phenomena. First, we rewrite Equation C.30 in a differential form and use the independent variable s (the path length in the direction of propagation), in place of x.

$$\frac{dI_\lambda}{ds} = -K_{a\lambda} I_\lambda \tag{C.32}$$

It can be verified that Equation C.31 is a solution of Equation C.32. Equation C.32 states that the incremental attenuation occurs due to $K_{a\lambda}$ and the path length, ds. Emission and scattering can also occur in the medium. Emission will add to the intensity, I_λ, whereas scattering can both subtract from and add to I_λ. For example, the light scattered away from the direction of propagation will be subtractive, whereas the light scattered into the direction of propagation will be additive. These two effects are called *out-scattering* and *in-scattering*, respectively. Emission is written as a product of local emissivity and the blackbody radiation, $I_{b\lambda}$ (discussed in the following text). Including these terms in Equation C.32 gives us the grand *radiative transfer equation*, which mathematically derives the change in the spectral intensity, I_λ, in a specific direction, Ω, as a function of the path length, s.

$$\frac{dI_\lambda(\Omega)}{ds} = -(K_{s\lambda} + K_{a\lambda}) I_\lambda(\Omega) + K_{e\lambda} I_{b\lambda} + \frac{K_{s\lambda}}{4\pi} \int_{4\pi} I_\lambda(\Omega') p(\Omega' \to \Omega) d\Omega' \tag{C.33a}$$

It is customary to use Kirchhoff's law, which states that emissivity is equal to absorptivity at the directional and spectral levels. For our purposes, this gives us $K_{e\lambda} = K_{a\lambda}$ for the second term on the right-hand side of Equation C.33a.

$$\frac{dI_\lambda(\Omega)}{ds} = -(K_{s\lambda} + K_{a\lambda})I_\lambda(\Omega) + K_{a\lambda}I_{b\lambda} + \frac{K_{s\lambda}}{4\pi}\int_{4\pi} I_\lambda(\Omega')\,p(\Omega' \to \Omega)\,d\Omega' \qquad \text{(C.33b)}$$

The phase function, $p(\Omega' \to \Omega)$, in the integral term is the ratio of the radiation that comes in from angle Ω' into angle Ω. The integral represents the in-scattering term. It is obtained from the following logic, applied to a volume element of $\Delta V = \Delta s \Delta A$:

Energy is scattered in all directions from radiation from the Ω' direction over a solid angle.

$$\Delta\Omega' = K_{c\lambda}I_\lambda(\Omega')\,\Delta\Omega'\Delta s\Delta A \qquad \text{(C.34)}$$

Energy scattered into the Ω direction from radiation from the Ω' direction from a solid angle $\Delta\Omega'$, if the scattering were isotropic

$$= \frac{\Delta\Omega}{4\pi}K_{s\lambda}I_\lambda(\Omega')\,\Delta\Omega'\Delta s\Delta A \qquad \text{(C.35)}$$

Energy scattered into the Ω direction from radiation from the Ω' direction from a solid angle $\Delta\Omega'$, for nonisotropic scattering

$$= \frac{\Delta\Omega}{4\pi}K_{s\lambda}I_\lambda(\Omega')\,\Delta\Omega'\Delta s\Delta A\; p(\Omega' \to \Omega) \qquad \text{(C.36)}$$

Energy scattered into the Ω direction from radiation from all directions, for nonisotropic scattering

$$= \frac{\Delta\Omega}{4\pi}K_{s\lambda}\int_{4\pi} I_\lambda(\Omega')\,\Delta\Omega'\Delta s\Delta A\; p(\Omega' \to \Omega)\,d\Omega' \qquad \text{(C.37)}$$

The factors ΔA and $\Delta\Omega$ are used to normalize the radiative power to obtain the spectral intensity from this volume element, ΔI_λ, whereas dividing ΔI_λ by Δs gives the gradient term, as in the left-hand side of Equation C.33b. It is understood that the radiation considered in Equations C.34–C.37 is for a specific wavelength interval.

The radiative transfer equation, Equation C.33, is quite general and can be used in a wide range of optical and radiative processes, including optical diagnostics and atmospheric lidars. The biggest challenge lies in the estimation of the phase function and scattering/absorption coefficients of various constituents of the medium.

C.3.1 Blackbody Radiation

The emission term in the radiative transfer equation (Equation C.33) is based on the blackbody radiation multiplied by emissivity. A blackbody is defined as a perfect absorber of radiation at all wavelengths, i.e., its absorptivity is equal to one. For a blackbody in thermal equilibrium in an enclosure (i.e., the blackbody surface temperature is equal to the temperature of the enclosure), it must emit the same amount of radiative energy as it absorbs. Otherwise, there will be an increase in the blackbody temperature, which violates the equilibrium condition. Therefore, emissivity is equal to absorptivity for a given wavelength (and for a given direction), and this is known as Kirchhoff's law. This gives the blackbody emissivity equal to the absorptivity of one. Blackbody radiation, or blackbody spectral intensity, is isotropic and represents the theoretical maximum radiative power from a source at temperature T. The blackbody spectral intensity, $I_{b\lambda}$, is given by Planck's law.

$$I_{b\lambda} = \frac{2C_1}{\lambda_o^5\left(e^{-C_2/\lambda_o T}-1\right)}\left[\frac{\mathrm{W}}{\mathrm{m}^2\ \mathrm{sr}\ \mu\mathrm{m}}\right] \quad \text{Blackbody spectral intensity} \qquad \text{(C.38)}$$

$$C_1 = hc_o^2 = 0.59552137\times10^8\ \mathrm{W\mu m^4}/\left(\mathrm{m^2 sr}\right)$$

$C_2 = hc_o/\mathrm{k} = 14{,}387.752\ \mu\mathrm{m\ K}$
h = Planck's constant = 6.62606876×10^{-34} J s
k = Boltzmann's constant = 1.3806503×10^{-23} J/K
c_o = speed of light in vacuum = 2.99792458×10^{8} m/s

Figure C.2 shows the blackbody spectral intensity at various temperatures. It can be seen that, at low temperatures, the radiation is in the infrared and long-wavelength range. As the temperature increases, not only does the blackbody spectral intensity increase exponentially, but the peak wavelength also decreases. Only near the

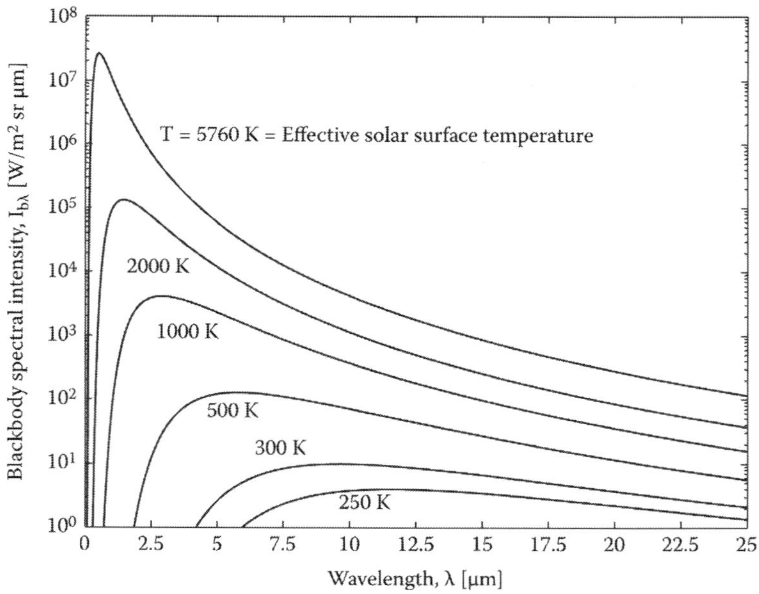

FIGURE C.2 The blackbody spectral intensity at various temperatures.

temperature of 833 K does the blackbody radiation begin to contain visible components, first in the red ($\lambda = 630$–760 nm). This gives rise to the colloquial expression *red hot*. As the temperature increases further, other visible color components are emitted, giving rise to the term *white hot*, as all the visible colors mixed result in white. For solar radiation at the sun's effective radiation temperature of 5780 K, the blackbody radiation peaks in the visible wavelength range, particularly at green (490–540 nm). This explains the biological evolution of human eyes that have peak sensitivity in the green. Because the blackbody spectral intensity gives the radiative power as a function of the wavelength, it can be numerically integrated over a finite or entire wavelength range to find the blackbody radiation over that spectrum. An alternate method is to use the pretabulated integral of the blackbody spectral intensity function, which can be found in radiative heat transfer references (e.g., Siegel and Howell, 2002).

The wavelength at which the blackbody spectral intensity peaks can be found by differentiating Equation C.38 with respect to λT and setting it equal to zero. This gives the trace of λ_{max} for the peak blackbody radiation at temperature T in the following form:

$$\frac{dI_{b\lambda}}{d(\lambda T)} = 0 \Rightarrow \lambda_{max} T = 2897.7686 \ \mu\text{mK}$$

As noted in Equation C.31, the spectral intensity and the blackbody spectral intensity, $I_{b\lambda}$, give the amount of radiative power in a solid-angle interval for some area element that is perpendicular to the direction of radiation. For a horizontal surface of area ΔA_s, $I_{b\lambda}$ can be integrated over the hemisphere by integrating from $\theta = 0$ to $\pi/2$ and $\phi = 0$ to 2π (see Figure C.1). This quantity is called the *spectral hemispherical blackbody flux*, $q_{b\lambda}$ and is defined as

$q_{b\lambda}$ = spectral hemispherical blackbody flux
= (blackbody radiation integrated over the hemisphere)/$\Delta A_s \Delta\lambda$ [W/m^{2}μm]
ΔA_s = elemental area on the emitting surface = $\Delta A/\cos\theta$ (see Figure C.1)

$$q_{b\lambda} = \int_0^{2\pi}\int_0^{\pi/2} I_{b\lambda}\cos\theta \ \sin\theta \ d\theta d\phi = I_{b\lambda}\int_0^{2\pi}\int_0^{\pi/2} \cos\theta \ \sin\theta \ d\theta d\phi = \pi I_{b\lambda} \qquad \text{(C.39)}$$

The fact that $I_{b\lambda}$ is isotropic allows for a straightforward integration of the angle terms in Equation C.39. We can go one more step and integrate $q_{b\lambda}$ over the wavelength to get the *total hemispherical blackbody flux*, e_b, which is then a function only of temperature.

$$q_b = \pi\int_0^{\infty} I_{b\lambda} d\lambda \rightarrow q_b(T) = \sigma T^4 \quad \left[\text{W/m}^2\right] \qquad \text{(C.40)}$$

σ = Stefan-Boltzmann constant = 5.67×10^{-8} [W/m^{2}K^4]

Equation C.40 is the familiar Stefan-Boltzmann law and gives the total blackbody radiative power over all angles and all wavelengths.

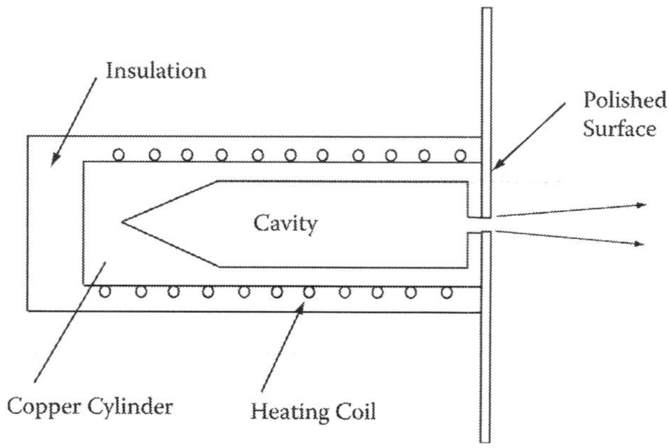

FIGURE C.3 A schematic of a blackbody radiation source.

Blackbody radiation is useful for some optical measurements in calibrating the detector response, as the magnitude of blackbody radiation is precisely given as a function of the wavelength if the source temperature is known. Most real materials deviate from blackbody radiation; however, a radiation source with a large depth-to-diameter ratio emulates blackbody radiation, as the emissivity of such a source approaches one as the depth-to-diameter ratio increases. Figure C.3 shows a schematic of a blackbody radiation source that can be either fabricated in a laboratory or acquired from a commercial vendor.

As noted earlier, real materials do not radiate according to Planck's law (Equation C.38); they radiate less and irregularly as a function of wavelength and direction. The quantity emissivity defines the radiative properties of real surfaces. Emissivity is the ratio of the actual radiative emission to the blackbody radiation at the same temperature, and in the most general form, is a function of wavelength, direction, and temperature.

$$\varepsilon'_\lambda = \text{spectral directional emissivity} = \frac{\Delta q_{\lambda,e}}{\Delta q_{b\lambda}} \tag{C.41}$$

$\Delta q_{\lambda e}$ = spectral emissive flux for solid angle $\Delta\Omega = I_{\lambda,e}\cos\theta\Delta\Omega$
$\Delta q_{b\lambda}$ = spectral blackbody flux for solid angle $\Delta\Omega = I_{b\lambda}\cos\theta\Delta\Omega$

Similar to what we did with the spectral blackbody intensity, emissivity can be integrated over the wavelength and directions to give the total hemispherical emissivity, ε.

$$\varepsilon = \text{total hemispherical emissivity} = \frac{q_e}{q_b} \tag{C.42}$$

$$q_e = \text{Total hemispherical emissive flux} = \int_0^\infty \int_{2\pi} I_{\lambda,e} \cos\theta \, d\Omega \, d\lambda$$

TABLE C.2
The Total Hemispherical (H) and Total Normal (N) Emissivity Data for Some Common Materials

Material	Total Hemispherical (H) or Total Normal (N)	Temperature [K]	Emissivity
Aluminum, highly polished	H	300	0.04
Aluminum, anodized	H	300	0.82
Copper, highly polished	H	300	0.03
Copper, oxidized	H	800	0.58
Stainless steel, polished	N	300	0.17
Stainless steel, highly oxidized	N	800	0.67
Tungsten, polished	H	1000	0.10
Tungsten, polished	H	2500	0.29
Asphalt pavement	H	300	0.55
Brick, red	H	300	0.93 ~ 0.96
Wood	H	300	0.82 ~ 0.92
Cloth	H	300	0.75 ~ 0.90
Concrete	H	300	0.88 ~ 0.93
Glass, window	H	300	0.90 ~ 0.95
Ice	H	273	0.95 ~ 0.98
Paint, black	H	300	0.98
Paint, acrylic white	H	300	0.98
Paint, zinc oxide	H	300	0.92
Paint, acrylic white	H	300	0.98
Paint, zinc oxide	H	300	0.92
Paper, white	H	300	0.92 ~ 0.97
Furnace liner, alumina brick	N	1000	0.33
Furnace liner, kaolin insulating brick	N	1200	0.57
Sand	H	300	0.90
Skin	H	300	0.95
Snow	H	273	0.82 ~ 0.90

q_b = total hemispherical blackbody flux $= \sigma T^4$

It is either the total hemispherical emissivity or the total normal emissivity (the emission in the perpendicular direction to the surface over all wavelengths) that is tabulated in most standard references. The data for spectral emissivity typically require extensive laboratory measurements, whereas directional emissivity is obtained experimentally or estimated using analytical methods. Table C.2 gives the total hemispherical and total normal emissivity data for common materials.

The same concepts can be used to define absorptivity, which is the ratio of absorbed radiative flux to incident radiative flux.

$$\alpha'_\lambda = \text{Spectral directional absorptivity} = \frac{\Delta q_{\lambda,a}}{\Delta q_{\lambda,i}} \tag{C.43}$$

$\Delta q_{\lambda,a}$ = amount of radiative flux absorbed for a given wavelength for incident solid angle $\Delta\Omega_i$
$\Delta\Omega_i$ = incident solid angle of radiation
$\Delta q_{\lambda,i}$ = the incident radiative flux at a given wavelength and for solid angle $\Delta\Omega_i$
θ_i = incident angle of radiation

$$\alpha = \text{Total hemispherical absorptivity} = \frac{q_a}{q_i} \tag{C.44}$$

q_a = total hemispherical flux of radiative power absorbed $= \int_0^\infty \int_{2\pi} I_{\lambda,a} \cos\theta_i \; d\Omega_i d\lambda$

q_i = total hemispherical incident $= \int_0^\infty \int_{2\pi} I_{\lambda,i} \cos\theta_i \; d\Omega_i d\lambda$

Using Kirchhoff's law described earlier in this appendix, absorptivity is equal to emissivity at the spectral, directional level. However, they are not necessarily equal if we integrate either over the wavelength or the solid angles, unless the surface is diffuse (the radiation properties are independent of direction) and gray (the radiation properties are independent of wavelength).

$$\alpha'_\lambda(\lambda,\theta,\phi,T) = \varepsilon'_\lambda(\lambda,\theta,\phi,T) \text{ Kirchoff's law} \tag{C.45}$$

We can further extend these concepts to the definitions of reflectivity and transmissivity. However, because reflective properties depend on the incident angle as well as the reflected angles, we are dealing with two sets of angles, (θ_i,ϕ_i) and (θ_r,ϕ_r). For transmissivity, we deal with incident and transmitted angles at the most fundamental level. We give the fundamental definitions of reflectivity and transmissivity.

ρ''_λ = spectral bidirectional reflectivity = (reflected radiative flux at an angle (θ_r,ϕ_r) for a given wavelength)/(incident radiative flux at an angle (θ_r,ϕ_r) at the same wavelength)

$$\equiv \frac{\Delta q_{\lambda,r}(\theta_r,\phi_r,\lambda)}{\Delta q_{\lambda i}} = \frac{I_{\lambda,r}(\theta_r,\phi_r,\lambda)\ \cos\theta_r \Delta\Omega_r}{I_{\lambda,i}(\theta_i,\phi_i,\lambda)\ \cos\theta_i \Delta\Omega_i} \tag{C.46}$$

τ''_λ = spectral bidirectional transmissivity = (transmitted radiative flux at an angle (θ_t,ϕ_t) for a given wavelength)/(incident radiative flux at an angle (θ_t,ϕ_t) at the same wavelength)

$$\equiv \frac{\Delta q_{\lambda,t}(\theta_t,\phi_t,\lambda)}{\Delta q_{\lambda i}} = \frac{I_{\lambda,t}(\theta_t,\phi_t,\lambda)\ \cos\theta_t \Delta\Omega_t}{I_{\lambda,i}(\theta_i,\phi_i,\lambda)\ \cos\theta_i \Delta\Omega_i} \tag{C.47}$$

From the preceding fundamental bidirectional reflectivity and transmissivity, we can obtain directional reflectivity and transmissivity by integrating the radiative fluxes in the numerators over the hemisphere.

ρ'_λ = spectral directional reflectivity = (total reflected radiative flux over the hemisphere for a given wavelength)/(incident radiative flux at an angle (θ_i,ϕ_i) at the same wavelength)

$$= \frac{\int_{2\pi} I_{\lambda,r}(\theta_r, \phi_r, \lambda) \cos\theta_r d\Omega_r}{I_{\lambda,i}(\theta_i, \phi_i, \lambda) \cos\theta_i \Delta\Omega_i} \tag{C.48}$$

τ'_λ = spectral directional transmissivity = (total transmitted radiative flux over the hemisphere for a given wavelength)/(incident radiative flux at an angle (θ_i, ϕ_i) at the same wavelength)

$$= \frac{\int_{2\pi} I_{\lambda,t}(\theta_t, \phi_t, \lambda) \cos\theta_t d\Omega_t}{I_{\lambda,i}(\theta_i, \phi_i, \lambda) \cos\theta_i \Delta\Omega_i} \tag{C.49}$$

Example C.1

For normally incident light, we can use the foregoing definitions and Beer-Lambert's law (Equation C.30) to relate the absorption coefficient, $K_{a\lambda}$, with the optical properties. For normal incidence and transmittance, Equation C.49 simplifies to

$$\tau'_\lambda = \frac{I_{\lambda,t}}{I_{\lambda,i}} = \exp(-K_{a\lambda} s)$$

Beer-Lambert's law (Equation C.30) was used for the latter part of the preceding equality, where s is the path length.

For a medium with zero reflectivity, we can conclude from the energy balance that the absorbed energy must be the incident energy minus the transmitted energy. Thus, the absorptivity is

$$\alpha'_\lambda = 1 - \tau'_\lambda = 1 - \exp(-K_{a\lambda} s)$$

Using Kirchoff's law (at the spectral, directional level, emissivity equals absorptivity),

$$\varepsilon'_\lambda = \alpha'_\lambda = 1 - \exp(-K_{a\lambda} s)$$

BIBLIOGRAPHY

Incropera, F. P. and DeWitt, D. P., *Fundamentals of Heat and Mass Transfer*, 5th ed., John Wiley, New York, 2002.

Lonngren, K. E. and Savov, S. V., *Fundamentals of Electromagnetics with MATLAB*, SciTech, Raleigh, NC, 2005.

Siegel, R. and Howell, J., *Thermal Radiation Heat Transfer*, 4th ed., Taylor and Francis, Boca Raton, FL, 2002.

Appendix D
Quantum Mechanics and Atomic/Molecular Structure

Atomic and molecular structures play a role in light emission, scattering, and absorption processes, and are thus important for understanding diagnostic methods such as absorption spectroscopy, fluorescence, and Rayleigh and Raman scattering. In this Appendix, we will briefly examine the main concepts of quantum mechanics and atomic/molecular structures.

Quantum mechanics describes the motion of particles at the atomic and molecular scales and is based on the following fundamental postulates:

Postulate 1: The dynamic state of n particles at the atomic and molecular scales can be described by a wavefunction, $\psi(q_1, q_2, q_3, \ldots, q_{3n}, t)$, as a function of the coordinates of the particles, q_i, and time, t, where $i = 1,2,3, \ldots, 3n$. The dynamic state of the n particles involves all the physical variables, such as position, momentum, kinetic energy, potential energy, etc.

Postulate 2: The probability of finding the system in some volume, $dV = dq_1dq_2dq_3$, is given by $\psi\psi^* \, dV$.

If the volume is infinite, then the probability of finding the system in that volume is 1.

$$\int_{-\infty}^{\infty} \psi\psi^* dV = 1 \tag{D.1}$$

Postulate 3: The physical properties of the system are associated with a linear operator (such as a partial derivative, a constant multiplicative factor, etc.). The linear operators in Table D.1 give the corresponding physical properties.

Postulate 4: The average value of a physical property associated with an operator, $\hat{g}$, is given by

$$\bar{g} = \int_{-\infty}^{\infty} \psi \hat{g} \psi^* dV \tag{D.2}$$

where ψ^* is the complex conjugate of ψ. It is assumed that the wavefunction ψ is properly normalized (i.e., as in Equation D.1, $\int_{-\infty}^{\infty} \psi\psi^* dV = 1$).

Postulate 5: The wave function is determined from the Schrödinger equation.

$$\hat{H}\psi = i\hbar \frac{\partial \psi}{\partial t} \quad \text{The Schrödinger equation} \tag{D.3}$$

TABLE D.1
Quantum Mechanical Operators and Associated Physical Properties

Property	Notation for the Operator	Operator
q, Position	$\hat{q}$	q
p, Momentum (mv)	$\hat{p}$	$-i\hbar\frac{\partial}{\partial q} = -i\hbar\nabla$
t, Time	$\hat{t}$	t
E, energy	$\hat{E}$	$-i\hbar\frac{\partial}{\partial t}$
K, kinetic energy ($mv^2/2$)	$\hat{K}$	$-\frac{\hbar^2}{2m}\nabla^2$

Note:
m = particle mass.
$\hbar$ = reduced Planck's constant = $h/2\pi$ = 1.054571×10^{-34} J s = 4.135667×10^{-15} eV s.
h = Planck's constant = 6.626093×10^{-34} J s = 6.582119×10^{-16} eV s.

$\hat{H}$ = Hamiltonian operator ($H = K + P$, where K is the kinetic energy and P is the potential energy).

Example D.1

Kinetic energy can be expressed in terms of the momentum, *p*.

$$K = \frac{1}{2}mv^2 = \frac{(mv)^2}{2m} = \frac{p^2}{2m}$$

The preceding relationship is used to write the operator for kinetic energy in terms of the momentum operator, $\hat{p}$, from Table D.1.

$$\hat{K} = \frac{(\hat{p})^2}{2m} = \frac{(-i\hbar\nabla)^2}{2m} = \frac{\left[-i\hbar\left(\frac{\partial}{\partial x}\vec{i} + \frac{\partial}{\partial y}\vec{j} + \frac{\partial}{\partial z}\vec{k}\right)\right]^2}{2m} = -\frac{\hbar^2}{2m}\nabla^2 = -\frac{\hbar^2\left(\frac{\partial^2}{\partial x^2} + \frac{\partial^2}{\partial y^2} + \frac{\partial^2}{\partial z^2}\right)}{2m}$$

Let us consider a solution of the following form for the Schrödinger equation.

$$\Psi(q,t) = \psi(q)e^{-Et/h} \tag{D.4}$$

Substituting this form into Equation D.3 gives

$$\hat{H}(q)\psi(q)e^{-Et/\hbar} = i\hbar\frac{\partial}{\partial t}\left[\psi(q)e^{-Et/h}\right] \tag{D.5a}$$

$$e^{-Et/\hbar}\hat{H}(q)\psi(q) = i\hbar\psi(q)\left(-i\frac{E}{\hbar}\right)e^{-Et/\hbar} \tag{D.5b}$$

$$\hat{H}(q)\psi(q) = E\psi(q) \tag{D.5c}$$

Equation D.5c is the time-independent form of the Schrödinger equation. Let us examine a simple example to see how the Schrödinger equation can be used to determine the dynamics of a particle.

We can consider a free particle of mass m in a one-dimensional box bounded by $x=0$ and $x=a$. The fact that the particle is free means that no force acts on the particle and, therefore, no potential energy. That is, $P=0$ in the Hamiltonian function ($H=K+P \rightarrow H=K$). From Table D.1, the kinetic energy operator in one dimension is $-\frac{\hbar^2}{2m}\frac{\partial^2}{\partial x^2}$. Thus, we have a simple time-independent Schrödinger equation.

$$-\frac{\hbar^2}{2m}\frac{\partial^2}{\partial x^2}\psi(x) = E\psi(x) \tag{D.6}$$

A solution to this equation is

$$\psi_{\pm} = Ce^{\pm ikx} \tag{D.7}$$

where C is an arbitrary constant and $k = \frac{(2mE)^{1/2}}{\hbar}$. Because the Schrödinger equation is linear, a linear combination of the foregoing solution is the general solution.

$$\psi = Ae^{+ikx} + Be^{-ikx} \tag{D.8}$$

But $e^{\pm ikx} = \sin kx \pm \cos kx$, so that

$$\psi = C \sin kx + D \cos kx \tag{D.9}$$

The boundary condition $\Psi(x)=0$ at $x=0^-$ gives $D=0$. So, the final solution is

$$\psi_n(x) = \left(\frac{2}{a}\right)^{1/2} \sin\left(\frac{n\pi x}{a}\right), \quad n = 1,2,3,\ldots \tag{D.10}$$

The factor $(2/a)^{1/2}$ arises from the normalization condition $\int_0^a |\psi_n(x)|^2\, dx = 1$.

The eigenvalue, E, is obtained by substituting the solution Equation D.10 back into Equation D.6.

$$E_n = \frac{n^2\hbar^2}{8ma^2} \quad n = 1,2,3,\ldots \tag{D.11}$$

The energy levels (Equation D.11) in the preceding text show one of the fundamental characteristics of quantum mechanical particles: the energy levels are quantized. In Newtonian mechanics, the energy levels are apparently continuous; however, at atomic and molecular scales, the energy levels are distinct and discrete, with no energy values allowed between these levels. As we shall see later, this leads to some unique and distinctive spectroscopic properties during energy interactions with quantum mechanical particles.

Now that the wavefunction is found for this particle configuration, all of the physical properties can be computed. For example, the mean momentum and the mean position can be found by using the respective operators in Equation D.2. Using the momentum operator, $\hat{p} = -i\hbar\frac{\partial}{\partial x}$, the mean momentum of the particle is

$$\bar{p} = \int_{-\infty}^{\infty} \psi \hat{p} \psi^* \, dx = -\frac{2i\hbar}{a} \int_{0}^{a} \sin \frac{n\pi x}{a} \frac{\partial}{\partial x} \sin \frac{n\pi x}{a} dx$$

$$= -\frac{2n\pi i\hbar}{a^2} \int_{0}^{a} \sin \frac{n\pi x}{a} \cos \frac{n\pi x}{a} dx \tag{D.12}$$

$$= \frac{i\hbar}{a} \left[\cos^2 \left(\frac{n\pi x}{a} \right) \right]_0^a = 0$$

Because the particle is confined to $x=0$ to a and moves back and forth in both + and − directions, the mean momentum is zero.

The position operator is simply x, and the mean position of the particle is

$$\bar{x} = \int_{-\infty}^{\infty} \psi x \psi^* \, dx = \frac{2}{a} \int_{0}^{a} \sin \frac{n\pi x}{a} x \sin \frac{n\pi x}{a} dx \tag{D.13}$$

$$= \frac{a}{2}$$

The mean position is at the middle of the "box."

D.1 EXAMPLE: A SIMPLE HARMONIC OSCILLATOR

Let us look at the motion of a particle subject to a spring force $F = -kx$. If we define the potential energy, P, as $F = -dP/dx$, then the potential energy function and operator is

$$P = \frac{1}{2} kx^2$$

Substituting this and the kinetic energy operator in Equation D.5c, we get

$$\left(-\frac{\hbar^2}{2m} \frac{\partial^2}{\partial x^2} + \frac{1}{2} kx^2 \right) \psi(x) = E \psi(x)$$

This has a rather lengthy solution procedure but leads to a solution involving a Hermite polynomial, $H_n(x)$.

$$\psi_n(\xi) = \left(\frac{\beta}{\pi 2^n n!} \right)^{1/4} H_n(\xi) e^{-\xi^2/2}, \quad n = 0,1,2\ldots \tag{D.14}$$

$\xi = \sqrt{\beta}$, $\beta = m\beta_o/h$

v_o = frequency of the oscillation

This particle motion also gives quantized or discrete energy levels, depending on the frequency of oscillation, v_o.

$$E_n = \left(n + \frac{1}{2} \right) h v_o \tag{D.15}$$

The preceding linear oscillator can be used as a basic model of diatomic molecule vibration, which also has discrete energy levels depending on the oscillation frequency.

After looking at relatively simple examples, we take a "quantum" leap and apply the Schrödinger equation to the motion of an electron in a hydrogen atom. The electron has kinetic energy due to its orbital motion around the nucleus and is bound to that orbit by the Coulombic force between the positively charged nucleus and the negatively charged electron. The Coulombic force is given by Equation C.2, where the charges, q_1 and q_2, are equal to a single electronic charge, e, and the corresponding potential energy is also given by Equation C.2.

$$\hat{P} = -\frac{e^2}{4\pi\varepsilon_o}\frac{1}{r} \quad \text{Potential energy} \tag{D.16}$$

$$\hat{K} = -\frac{\hbar^2}{8\pi^2 m}\nabla^2 \quad \text{Kinetic energy} \tag{D.17}$$

Using spherical coordinates, the Schrödinger equation becomes

$$-\frac{\hbar^2}{8\pi^2 mr^2}\left[\frac{\partial}{\partial r}\left(r^2\frac{\partial\psi}{\partial r}\right)+\frac{1}{\sin^2\theta}\frac{\partial^2\psi}{\partial\phi^2}+\frac{1}{\sin\theta}\frac{\partial}{\partial\theta}\left(\sin\theta\frac{\partial\psi}{\partial\theta}\right)\right]-\frac{e^2}{4\pi\varepsilon_o r}\psi = E\psi \tag{D.18}$$

Despite its apparent complexity, Equation D.18 has an analytical solution through separation of variables and use of special functions. The solution then has separate dependence on the radius and angular variables.

$$\psi(r,\theta,\phi) = R_n(r)Y_n^l(\theta,\phi), \quad n = 1,2,3\ldots \quad \text{and} \quad l = 0,1,2,\ldots \tag{D.19}$$

Now, the r dependence is given by

$$R_n(r) = P(r)\exp(-kr) \tag{D.20}$$

$$k = \frac{n\pi me^2}{h^2\varepsilon_o}$$

$P(r)$ is a special function called the *finite polynomial*, and n and l are so-called *quantum numbers*. The principal quantum number, n, goes from 1, 2, 3, onward and specifies the discrete radial distances of the electronic shell from the center of the atom. The azimuthal or angular momentum quantum number, l, starts from 0 up to $n-1$. For $n=1$, l can only be 0, whereas for $n=2$, we can have $l=0$ and 1, and so on. The shape of the spherical harmonics function, $Y_n^l(\theta,\phi)$, depends on the azimuthal quantum number, as shown in the following text.

$$Y_n^l(\theta,\phi) = \text{const. as a function of } \theta \text{ and } \phi, \text{ for } l = 0. \text{ Spherical symmetric, } s\text{-orbitals} \tag{D.21a}$$

$$Y_n^l(\theta,\phi) = \begin{matrix} \sin\theta\cos\phi = p_x \\ \sin\theta\sin\phi = p_y, \\ \cos\theta = p_z \end{matrix} \qquad \text{for } l = 1 \text{, two lobes, } p \text{ orbtials} \tag{D.21b}$$

$$Y_n^l(\theta,\phi) = \begin{matrix} 3\cos^2\theta - 1 = d_{z^2} \\ \sin\theta\cos\theta\cos\phi = d_{zx} \\ \sin\theta\cos\theta\sin\phi = d_{yz}, \\ \sin^2\theta\cos 2\phi = d_{x^2-y^2} \\ \sin^2\theta\cos 2\phi = d_{xy} \end{matrix} \qquad \text{for } l = 2 \text{, four lobes, } d \text{ orbtials} \tag{D.21c}$$

Figure D.1 shows the shape of the spherical harmonics for $l=0$, 1, and 2 that are associated with *s*, *p*, and *d* atomic orbitals, respectively. Again, for $n=1$, we can only have Z=0, or the *s*-orbital, giving the first row showing only that orbital in Figure D.1. For $n=2$, we can have *s*- and *p*-orbitals. Although there are three different types of *p*-orbitals, p_x, p_y, and p_z, we only show one representative *p*-orbital in the second row of Figure D.1. Also, to illustrate the difference between the *s*-orbitals for $n=0$ and 1, Figure D.1 shows the two overlapping s-orbitals for $n=2$. Now, for $n=3$, we have the *s*-, *p*-, and the *d*-orbitals, with three possible *s*-orbitals and two representative *p*-orbitals shown in the third row of Figure D.1. In summary, the principal quantum number, *n*, gives the successively higher levels of electron energy at increasing radial distances, whereas the azimuthal quantum number, *l*, dictates the shape of the electron orbit around the nucleus. The energy level is the eigenvalue of Equation D.18.

$$E_n = -\frac{me^4}{8h^2\varepsilon_o^2}\frac{1}{n^2} = -\frac{13.6}{n^2}[\text{eV}], \quad n = 1,2,3,\ldots \tag{D.22}$$

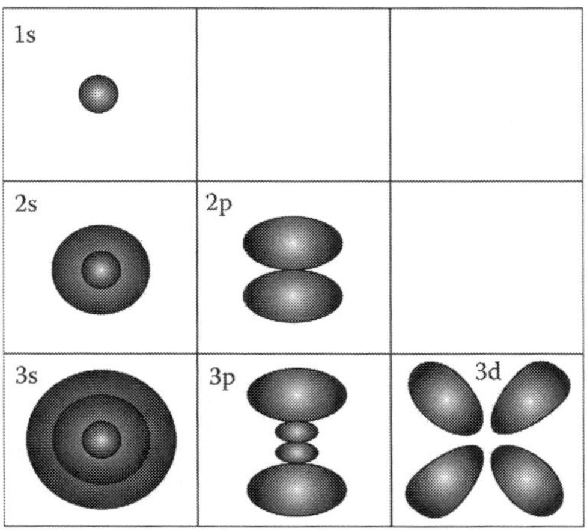

FIGURE D.1 Atomic orbitals.

The electrons occupy certain orbitals, or probabilistic trajectories around the nuclei, that differ in their mean distance from the center of mass of the nuclei and in their shape. The principal quantum number, n, designates different mean distances from the center of mass, starting from 1, 2, 3, and so on. The different shapes are designated by the letters, $s, p, d, f, \ldots$, for atomic orbitals. A more detailed exposition of the quantum mechanical calculations for orbitals and their visualizations can be found on the referenced websites at the end of this Appendix.

For atomic orbitals, the sequence of the energy levels from the lowest level is

$$1s, 2s, 2p_x, 2p_y, 2p_z, 3s, 3p_x, 3p_y, 3p_z, 3d_{xy}, 3d_{yz}, 3d_{xz}, 3d_{zz}, 4s, \ldots$$

For each of these orbitals, at most two electrons with different spin states can occupy a given orbital. The fundamental atomic structure is such that electrons always fill the lowest energy level first before they can occupy higher orbitals. If there are vacancies at an equal energy level, electrons fill the orbitals singly as much as possible before two electrons occupy the same orbital.

Example D.2

From the periodic table, the hydrogen atom has a single electron, helium has two, and so on. For the hydrogen atom, the electronic configuration is $1s^1$, where the superscript denotes the number of electrons in that state. For the first few atoms in the periodic table, the electronic configurations are written as follows:

Element	Symbol	Total Number of Electrons	Electronic Configuration
Hydrogen	H	1	$1s^1$
Helium	He	2	$1s^2$
Lithium	Li	3	$1s^2 2s^1$
Carbon	C	6	$1s^2 2s^2 2p_x^1 2p_y^1$
Nitrogen	N	7	$1s^2 2s^2 2p_x^1 2p_y^1 2p_z^1$
Oxygen	O	8	$1s^2 2s^2 2p_x^2 2p_y^1 2p_z^1$

We can see that the energy states are filled singly at first and then filled with two electrons when more electrons come into the picture.

Now, molecular orbitals can be considered as a superposition of atomic orbitals. We will illustrate the molecular orbitals using a simple homonuclear diatomic molecule, such as O_2 or N_2. In Figure D.1, we saw the shapes of some atomic orbitals, which are designated by the letters, *s, p, d, f,* and so on. For molecular orbitals, the Greek letters σ, π, δ, and ϕ are correspondingly used. The superposition of the orbitals can be achieved in either bonding or antibonding modes. For two atoms with s-orbitals, the bonding and antibonding molecular orbitals are shown in Figure D.2a. The bonding orbitals involve electrons spending most of the time around both nuclei, as shown in Figure D.2a, whereas antibonding orbitals have the electrons spending most of their time near the respective atomic nuclei. Both the bonding and antibonding molecular orbitals have the designation σ because, when

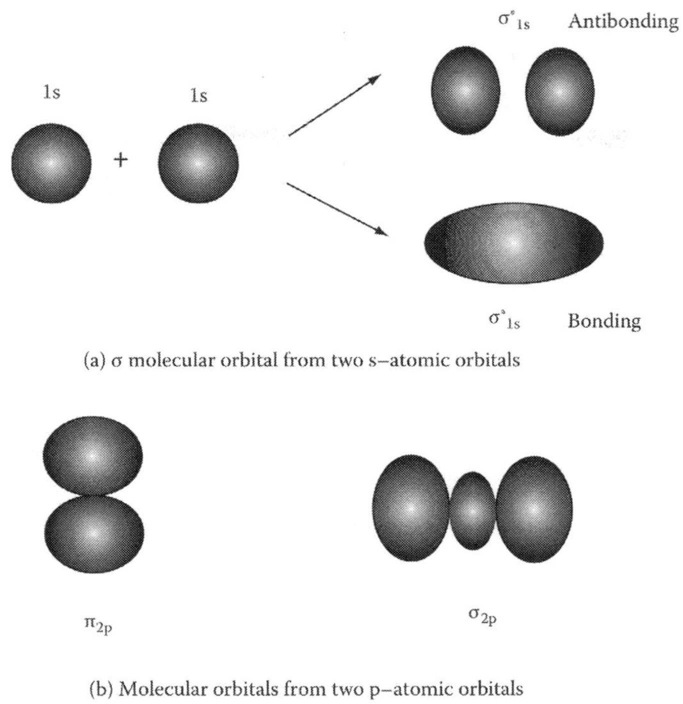

FIGURE D.2 The *s*- and *p*-molecular orbitals.

viewed from the axis connecting the nuclei, the orbitals are symmetric about that axis. Thus, the orbitals shown in Figure D.2a have the designation σ_{1s} for bonding and σ_{1s}* for antibonding. Antibonding does not mean the nuclei will separate, but only that the configuration is slightly less stable and therefore has higher energy than the bonding orbital. For two atoms having p-orbitals, there are more permutations in the possible superposition, and two of them are shown in Figure D.2b.

D.2 VIBRATIONAL AND ROTATIONAL ENERGY OF SIMPLE MOLECULES

In addition to the electronic energy prescribed by the principal and azimuthal quantum numbers, molecules have vibrational and rotational energy due to the appreciable masses of the constituent atomic nuclei. As we saw in a previous example, the vibrational energy of a simple diatomic molecule (with nuclear masses of m_1 and m_2) can be modeled as a harmonic oscillator (Figure D.3a), where the potential energy is given by

$$V = \frac{1}{2}kx^2 = \frac{1}{2}k\left(r - r_e\right)^2 \tag{D.23}$$

r = internuclear distance

r_e = equilibrium internuclear distance

k = force constant = $4\pi 2\mu_{osc}{}^2$

ν_{osc} = vibrational frequency
m = reduced mass = $m_1 m_2/(m_1 + m_2)$

Using the foregoing potential energy in the Schrödinger equation gives the following vibrational energy values:

$$E_v = \text{vibrational energy} = h\nu_{osc}\left(v + \frac{1}{2}\right) \tag{D.24}$$

v = vibrational quantum number

In many spectroscopic applications, we would like to convert the energy levels in Equation D.24 into frequency or wavenumber.

$$G(v) = E_v/hc = \omega\left(v + \frac{1}{2}\right) \quad \left[\text{cm}^{-1}\right]$$
$$\omega = \frac{\nu_{osc}}{c} \quad \left[\text{cm}^{-1}\right] \tag{D.25}$$

In spectroscopy, $G(v)$ is given in units of [cm^{-1}] and is referred to as the *term values* or *wavenumbers*.

For real molecules, the actual vibrational energy may deviate from the harmonic oscillator model, and some correction terms are added to Equation D.25.

$$G(v) = \omega_e\left(v + \frac{1}{2}\right) - \omega_e x_e\left(v + \frac{1}{2}\right)^2 + \omega_e y_e\left(v + \frac{1}{2}\right)^3 + \cdots \tag{D.26}$$

Example D.3

For OH molecules, we have the following parameters for the vibrational energy wavenumbers: $\omega_e = 3739.94$, $\omega_e x_e = 86.35$, and $\omega_e y_e = 0.9046\ \text{cm}^{-1}$. For a transition from $v' = 2$ to $v'' = 1$, the difference in the wavenumbers is

$$G'(2) - G''(1) = \omega_e(v' - v'') - \omega_e x_e\left[\left(v' + \frac{1}{2}\right)^2 - \left(v'' + \frac{1}{2}\right)^2\right]$$
$$+ \omega_e y_e\left[\left(v' + \frac{1}{2}\right)^3 - \left(v'' + \frac{1}{2}\right)^3\right]$$
$$= 3739.94(2-1) - 86.35\left(2.5^2 - 1.5^2\right) + 0.9046\left(2.5^3 - 1.5^3\right)$$
$$= 3405.6\ \text{cm}^{-1}$$

This converts to a wavelength of 2.9363 μm in the infrared region, where most vibrational transitions occur.

For the rotational energy, we can again draw an analogy to the classical mechanical model of a dumbbell rotator, as shown in Figure D.3b. The energy due to the rotation of a rigid rotator is

$$E_r = \frac{1}{2} I \omega_r^2 \tag{D.27}$$

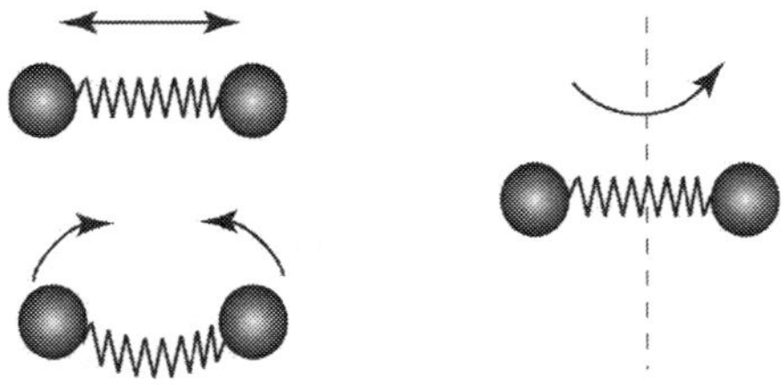

FIGURE D.3 The rigid rotator model of molecular rotation.

I = moment of inertia about the axis of rotation = μr^2
μ = reduced mass = $m_1 m_2/(m_1 + m_2)$
$r = r_1 + r_2$
ω_r = angular velocity of rotation

The angular momentum is $P_r = I\omega_r$. So, the energy of rotation can also be written in terms of the angular momentum.

$$E_r = \frac{1}{2} I\omega_r^2 = \frac{I^2\omega_r^2}{2I} = \frac{P_r^2}{2I} \tag{D.28}$$

The Schrödinger equation then gives the rotational energy levels.

$$E_r = \frac{h^2}{8\pi^2 \mu r^2} J(J+1), \quad J = 0,1,2,\ldots \tag{D.29}$$

J = the rotational quantum number

Similar to vibrational energy, the energy values are divided by hc to give the term values.

$$\begin{aligned} F(J) &= E_l/hc = BJ(J+1)\left[\text{cm}^{-1}\right] \\ B &= \frac{h}{8\pi^2 cI} = \frac{h}{8\pi^2 c\mu r^2} \end{aligned} \tag{D.30}$$

Because real molecules are nonrigid rotators, there are fluctuations in the internuclear distance, r, and corrections are added to Equation D.30 to reflect this effect.

$$\begin{aligned} F(J) &= BJ(J+1) - DJ^2(J+1)^2 + \cdots \\ D &= 4B^3/\omega_r^2 \end{aligned} \tag{D.31}$$

D.3 ENERGY TRANSITIONS AND SPECTROSCOPY

Thus far, we have seen the discrete energy levels due to the electronic, vibrational, and rotational motion of molecules. Molecules can exchange some of this energy through molecular collisions or the exchange of photons or light energy. The energy of the light quanta, or photons, is given by

$$E = hv \tag{D.32}$$

h = Planck's constant = 6.626093×10^{-34} J s = 6.582119×10^{-16} eV s
v = frequency of the photon

This means that a photon of energy hv can be absorbed by the molecule, resulting in a change from one molecular energy state to another, because the total energy of the photon-molecule interaction must be conserved. Conversely, the molecule can undergo an energy shift from a higher energy state to a lower energy state and emit a photon of frequency v corresponding to the energy difference of this downward transition. This is the basis for spectroscopy, wherein the observed frequency of either absorbed or emitted light corresponds to the energy transitions within the molecule and therefore reveals specific information concerning its energy states. Thus, for an energy transition from state 1 to state 2, the following photon energy is observed:

$$\Delta E = E_2 - E_1 = hv \tag{D.33}$$

E_1 = energy of state 1
E_2 = energy of state 2

It can be seen in Equation D.33 that the larger the energy difference, the higher the photon frequency will be, or the wavelength will be smaller for large energy transitions. For molecular energy transitions, the electronic transitions involve the largest energy difference, followed by the vibrational and rotational transitions. The typical ranges of wavelength and wavenumber for various molecular energy transitions are shown in Table D.2.

Not all the energy states can be accessed during energy transitions. The following selection rules specify the range of quantum numbers that can be involved in electronic, vibrational, and rotational energy transitions.

$$\text{Electronic:} \quad \Delta n = 0, \pm 1 \tag{D.34a}$$

$$\text{Vibrational:} \quad \Delta \text{v} = \pm 1 \tag{D.34b}$$

$$\text{Rotational:} \quad \Delta J = \pm 1, \pm 2 \tag{D.34c}$$

Figure D.4 illustrates an energy transition involving electronic, vibrational, and rotational states. The total energy difference is the sum of all the energy differences.

$$\Delta E = \Delta E_e + \Delta E_v + \Delta E_r = hv \tag{D.35}$$

TABLE D.2
Typical Wavelengths and Wavenumbers for Various Molecular Energy Transitions

Transition	Wavelength [nm]	Wavenumber [cm^{-1}]
Electronic, outer shells	$10^2 \sim 10^3$ (uv ~ vis.)	$10^4 \sim 10^5$
Electronic, inner shells	$10^2 \sim 10^3$ (x-rays)	$10^5 \sim 10^6$
Vibrational	$10^3 \sim 10^4$ (Infrared)	$10^3 \sim 10^4$
Rotational	$10^4 \sim 10^5$ (Infrared-microwave)	$10^2 \sim 10^3$

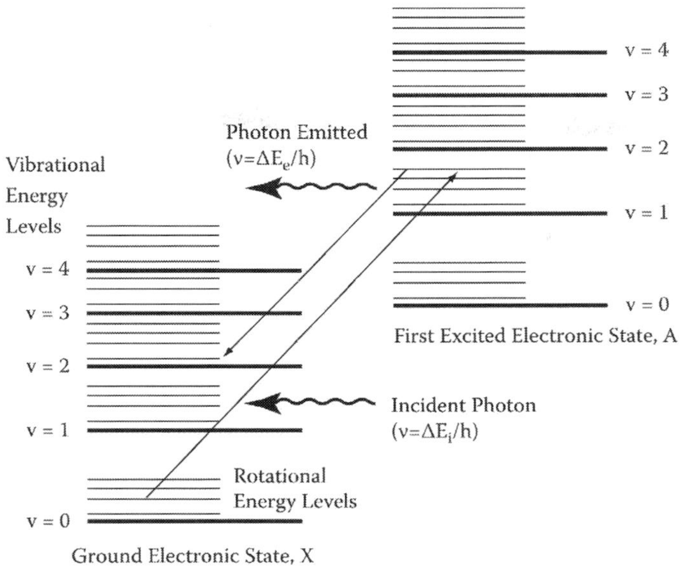

FIGURE D.4 Molecular energy transitions.

BIBLIOGRAPHY

Eckbreth, A. C., *Laser Diagnostics for Combustion Temperature and Species*, 2nd ed., Gordon and Breach, New York, 1996.

Herzberg, G., *The Spectra and Structures of Simple Free Radicals: An Introduction to Molecular Spectroscopy*, Dover, New York, 1971.

https://www.orbitals.com/orb/index.html.

https://www.physics.csbsju.edu/QM/Index.html.

Pauling, L. and Wilson, E. B., *Introduction to Quantum Mechanics, with Applications to Chemistry*, McGraw-Hill, New York, 1935.

Index

K

L

M

N

O

P

Q

R

S

W

Made in the USA
Las Vegas, NV
19 September 2026

e9404190-aa39-44e2-bbde-69343d85265aR01